Empathic Service Design

Empathic Service Design

Challenges in design, analysis and services

Edited by Melanie Sarantou, Satu Miettinen and Titta Jylkäs

BLOOMSBURY VISUAL ARTS
LONDON • NEW YORK • OXFORD • NEW DELHI • SYDNEY

BLOOMSBURY VISUAL ARTS
Bloomsbury Publishing Plc, 50 Bedford Square, London, WC1B 3DP, UK
Bloomsbury Publishing Inc, 1385 Broadway, New York, NY 10018, USA
Bloomsbury Publishing Ireland, 29 Earlsfort Terrace, Dublin 2, D02 AY28, Ireland

BLOOMSBURY, BLOOMSBURY VISUAL ARTS and the Diana logo
are trademarks of Bloomsbury Publishing Plc

First published in Great Britain 2025

Cover design by Louise Dugdale
Cover image © Eugene Mymrin/Getty Images

A catalogue record for this book is available from the British Library.

A catalog record for this book is available from the Library of Congress.

ISBN: HB: 978-1-3504-7639-4
ePDF: 978-1-3504-7640-0
eBook: 978-1-3504-7641-7

Typeset by Integra Software Services Pvt. Ltd.
Printed and bound in Great Britain

For product safety related questions contact productsafety@bloomsbury.com.

To find out more about our authors and books visit www.bloomsbury.com
and sign up for our newsletters.

For Kuan Hui Chen

CONTENTS

Part Three Approaches to Data Analysis for Developing Service Propositions

Part Four Design for Empathic Digital Approaches in Service Improvements

FIGURES

TABLES

CONTRIBUTORS

Bo Gao is currently working as the Senior Research Scientist on the Joint NTU-UBC Research Centre of Excellence in Active Living for the Elderly (LILY), Nanyang Technological University, Singapore. Her research interests include empathetic AI for gerontology, human-centred AI for healthcare, human–computer interaction, service design, etc.

Di Wang is currently working as the Senior Research Fellow in the Joint NTU-UBC Research Centre of Excellence in Active Living for the Elderly (LILY), Nanyang Technological University, Singapore. His research include affective computing, computational neuroscience, knowledge graph, autonomous agents, AI for healthcare and more.

Robin Chan Chung Leung is currently working as a Research Engineer in the Joint NTU-UBC Research Centre of Excellence in Active Living for the Elderly (LILY), Nanyang Technological University, Singapore. His research interests include video game design, UI/UX design, game AI and exergames for the elderly.

Zhiwei Zeng is currently a Presidential Postdoc Fellow at Nanyang Technological University, Singapore. Her current research interests include explainable artificial intelligence (XAI), human–agent interaction (HAI) and the application of XAI and HAI in the assessment and intervention of neurodegenerative diseases and other application scenarios.

Benny Toh Hsiang Tan currently works as a Research Scientist at the Joint NTU-UBC Research Centre of Excellence in Active Living for the Elderly (LILY), Nanyang Technological University, Singapore. His current research interests include UI/UX design, digital healthcare tools, working alliance in digital applications and AI for healthcare.

Yang Qiu is currently working as a Senior Research Engineer in the Joint NTU-UBC Research Centre of Excellence in Active Living for the Elderly (LILY), Nanyang Technological University, Singapore. His research interests include human–computer interaction, experience design, serious games, gamification, playfulness and their applications for healthcare.

Huiguo Zhang is Senior Research Engineer at the Joint NTU-UBC Research Centre of Excellence in Active Living for the Elderly (LILY), Nanyang Technological University, Singapore, and specializes in AIoT for Aging in Place. His key research interests include Tiny Machine Learning (TinyML), AI-powered hardware solutions, and AIoT for healthcare, etc.

Xinjia Yu is a Research Scientist in the Joint NTU-UBC Research Centre of Excellence in Active Living for the Elderly (LILY), Nanyang Technological University, Singapore. She holds doctoral degrees in both International Relations and Human–Computer Interactions. Her research interests include persuasive communicational agent and AI for healthcare and more.

Jun Ji is currently working as the Senior Research Engineer in the Joint NTU-UBC Research Centre of Excellence in Active Living for the Elderly (LILY), Nanyang Technological University, Singapore. His research interests include game software design, intelligent game, gamification for elderly and more.

Bo Huang is currently working as a Research Engineer in the Joint NTU-UBC Research Centre of Excellence in Active Living for the Elderly (LILY), Nanyang Technological University, Singapore. His research interests include web technologies, data analytics and AI for eldercare, along with other areas.

Sing Lee is a senior consultant family physician, researcher, educator and administrator who actively promotes primary care research culture, capacity and capability in Singapore. His research interests are multimorbidity, chronic disease management, digital solutions for primary care and health services research.

Hui Hwang Teong is an internist with special interest in hypertension with the Tan Tock Seng Hospital, Singapore. She has led several studies in hypertension and likes to dabble with digital technology to aid in hypertension management.

Ray Tian Rui Lai is a consultant in the Department of Endocrinology, Tan Tock Seng Hospital, Singapore. His subspecialty clinical and research interests are in diabetes mellitus, diabetes technology and education.

Wai Han Hoi is an endocrinologist and heads the Department of Endocrinology in Woodlands Health Hospital Singapore. Her clinical interests include diabetes mellitus and metabolic diseases as well as patient and medical education.

Chunyan Miao is Associate Vice President (Capability Building), President's Chair Professor and Chair of the School of Computer Science and Engineering at Nanyang Technological University, Singapore. She is a Fellow of IEEE,

Singapore Computer Society (SCS), the Academy of Engineering Singapore (SAEng) and the Institution of Engineers Singapore (IES).

Tokushu Inamura is a designer/engineer born in Aotearoa New Zealand with roots in the Amami Islands and has been Assistant Professor at Kyushu University since 2016. Holding a double Master's degree in Innovation Design Engineering from the Royal College of Art and Imperial College London, he specializes in post-human-centred design.

Kuan Hui Chen is a researcher and service design enthusiast at the University of Lapland, who sadly passed away while conducting a project. We learned a lot from you and were joyful to work with you. Your devotion for digitalization of service design will remain with us forever. This chapter is yours. Rest in peace, our friend.

Kiwoong Nam is a postdoctoral researcher and lecturer at the University of Lapland. He is manager of a Business Finland-funded project called Empathy Business. Dr Nam has worked in a global manufacturing company and service companies. Dr Nam has also worked in two EU-funded academic projects. His current research interests include service design for social innovation, empathic design and design value.

Mariluz Soto is a researcher and the director of the Experience and Service Design Centre in the Faculty of Design at Universidad del Desarrollo, Chile. She has experience as a practitioner in communication and communities. Currently in academia, her research areas are emotions, service prototyping, service design and empathy and nature.

Virve Inget is a PhD candidate currently conducting research at the Department of Industrial Design at the University of Lapland. Virve researches wearable devices, especially smart jewellery. She is a professional goldsmith and holds an MSc degree in Information Processing Science.

Mari Suoheimo is an associate professor of service design at the Oslo School of Architecture and Design, Norway. Her main research focus is on how to approach wicked problems in service design. Suoheimo has several publications in highly ranked journals and is also part of the editorial board of *The Design Journal*.

Jonna Häkkilä is professor of Industrial Design at the University of Lapland, Finland (2014–), and docent in Human–Computer Interaction (HCI) at the University of Oulu, Finland. She leads the user experience research group LUX at University of Lapland and conducts research in the cross-section of design and technology.

Manuela Quaresma is an associate professor of Design at PUC-Rio University. Her expertise includes human factors, ergonomics and human-centred design. She focuses on user research, UX and service design, excelling in complex problem analysis and creating engaging user interactions. She leads the research group in ergodesign, UX and service design at PUC-Rio University.

Thomas Abrell is a service design practitioner and scholar, working for 10 years in product–service systems innovation, currently in customer experience management at Volkswagen Commercial Vehicles. Earlier, he worked in various innovation roles at Volkswagen Group and as Innovation Manager at Airbus. He holds a PhD in Business Innovation (University of St.Gallen).

Masanori Fujita is a professor at the College of International Management at Ritsumeikan Asia Pacific University with a PhD in Engineering. Prof. Fujita is interested in business management, management of technology, innovation, start-up businesses, scientometrics and social network analysis.

Jenny Bergström is a Professor of Design Thinking at Media University of Applied Sciences in Germany and a doctoral student at the University of Lapland in Finland. Her research focuses on the role of materials in the creative process.

Päivi Hanni-Vaara is a Senior Lecturer in Tourism and Hospitality Degree at the Lapland University of Applied Sciences. She is a doctoral candidate at the University of Lapland, Faculty of Art and Design. Her research interests combine elements of experience and service design in digital tourism and hospitality.

Minni Haanpää is a Senior Lecturer in Tourism Research at the University of Lapland, Faculty of Social Sciences. Her research interests include tourism and event experiences, affect, value cocreation and ethnographic methodologies, especially videography and autoethnography.

Heidi Pietarinen is a Professor of Textile Art and Design in the Faculty of Art and Design, University of Lapland, Finland. She works with textile art, bioart and arts-based research. Her research includes multidisciplinary working groups such as Future Bio Arctic Design, High Altitude Bioprospecting and BioARTech Laboratory.

Satu Miettinen is a Professor of Service Design and Dean at the Faculty of Art and Design, the University of Lapland in Finland. She is an active artist and designer in the area of socially engaged art and ecofeminist photography. She also works at BioARTech Laboratory.

Albert Tibaijuka is a senior lecturer at St. Augustine University of Tanzania and a coordinating editor of *African Communication Research Journal*. He earned his PhD in African Studies from the University of Basel, Switzerland, in 2016. His academic experience centres in researching the fields of CSR, corporate relations and mass communication.

Ludovick Myumbo is a senior lecturer at St. Augustine University of Tanzania (SAUT), serving the Faculty of Social Sciences and Communication. He obtained his PhD in Social Sciences from the University of Tampere, Finland, in 2018. His research focuses on gender and social inclusion, sustainable livelihoods and ethics of vaccines.

Luca Simeone is an Associate Professor at Aalborg University in Denmark. His main interest is in the managerial, strategic and organizational aspects of design. Luca has carried out research, teaching and consulting activities in various universities (Harvard, Stanford, MIT, Milan Polytechnic, Malmö University and University of the Arts London), (co)authoring five books and publications for outlets including *Long Range Planning*, *CoDesign*, *Journal of Business Research*, *Journal of Knowledge Management*, *Technovation* and *R&D Management*.

Amélia Steunou is a design researcher specializing in intercultural projects. Her research focuses on crafting culturally sensitive experiences relevant to specific cultural groups, supporting a more inclusive future for service design. She has collaborated on several cross-cultural projects across countries such as France, China, Denmark and Mexico and holds a background in intercultural digital studies and service systems design.

Sheng-Hung Lee is a designer and PhD researcher at the Massachusetts Institute of Technology (MIT) AgeLab and Ideation Lab and Board Director at the Industrial Designers Society of America (IDSA). He is inspired by multiple domains of knowledge and perspectives while working at IDEO. Lee serves as Adjunct Associate Professor at Shih Chien University, Taiwan.

Melanie Sarantou (PhD) is a Professor of Social Design at Kyushu University in Japan. Her interests span the role of improvisation in arts and design practices when the unsuspected emerges. She seeks to understand how haptic and tactile qualities of arts and narrative practices can sustain marginalized communities. Her recent research explores how the arts can mitigate societal challenges.

Amna Qureshi, PhD in Arts from the University of Lapland, Finland, focuses on visual design and design thinking methodologies. She collaborates with marginalized and Indigenous groups using arts-based methods to empower their narratives. Her interdisciplinary approach links art with societal contexts, advancing awareness of visual culture and social issues.

Sherrie Jones, an Indigenous Aboriginal artist working at Arts Ceduna in South Australia, is recognized for her diverse talents. Beyond practising painting, pottery, basket weaving and felt making, she teaches these arts, reflecting her deep cultural connection. Her artworks intricately depict landscapes and ceremonies, emphasizing water's vital role in sustaining life and nurturing her Indigenous cultural heritage.

Anu Kylmänen (FM) is a university teacher of clothing design. She has worked at the University of Lapland since 2013. In her dissertation, Kylmänen studies the effect of clothes on human emotions. In her teaching methods, she has developed, among other things, empathetic approaches, such as visualized empathy.

Johanna Hautamäki (MA) is an RDI expert, service designer and teacher in Finland at Centria University of Applied Sciences Entrepreneurship and Business team, which is part of the Future Working Life research group. Work includes multidisciplinary and anticipatory human-centred development with companies on services, strategies, continuous learning, business operations and organizational culture.

Nicholas Stevens is a designer, researcher and educator currently undertaking his PhD at the University of Oslo. He is interested in exploring the ways the design process can contribute to more sustainable lifestyles. He is particularly drawn to digital products and technologies, their production and influence on the environment and on democracy.

Yangyang Zhao is an Associate Professor at the Department of Informatics, University of Oslo. She is passionate about fuzzy-front end design, human–AI coevolutions, responsible digital transformation, creative (team)work and entrepreneurship.

Alma Leora Culén is a professor at the University of Oslo. Her research is in sustainable interaction design, and her recent work includes reflections on transitions toward more sustainable futures and ways to engage with designing for transitions through design research, practice and education.

Ella Björn is a doctoral researcher at the University of Lapland in the Faculty of Art and Design. She is currently working as a project planner for empathy business, a project funded by Business Finland. Her research interests include service design, empathy and participatory design.

Moe Shimomura (PhD) is an Assistant Professor and the Deputy Director of the International Office at the Faculty of Design, Kyushu University, Japan. Her research interests include service design within public sectors, design education and intercultural communication. She is currently involved in the practice and research of community design for care.

Leon Loh is an Assistant Professor in the Faculty of Design at Kyushu University, Japan. His primary research interest is design education. His research focuses on product design education at the university level and design education in elementary, junior high and high school to promote critical thinking and creativity.

Yanfang Zhang is an associate professor at the Faculty of Design at Kyushu University, Japan, specializing in universal design and societal design and design for SDGs. Yanfang wants to explore combining the practice and education of universal and humanized design with research to contribute more deeply to society.

Noriko Takano is a researcher in the Faculty of Medical Science at Kyushu University, Japan. She uses the mouse genome to conduct research on the molecular mechanisms of DNA damage and its repair systems. She also dedicates her efforts to research on STEAM education and genetic education.

Tori Simpson is an Interdisciplinary Designer. She holds an MASc in Global Innovation Design from Royal College of Art and Imperial College London, where she led projects in Japan, Singapore, the Philippines and the UK. Her social innovation work adopts various contexts, including disaster-preparedness, medical care, ecological conservation and education.

Harry Hosker is a designer for ecological conservation. He has an MASc in Innovation Design Engineering from Royal College of Art and Imperial College London, where he has engaged stakeholders across academia and industry. His work aims to engage emerging technologies with impactful conservation initiatives worldwide.

Yasuyuki Hirai has been a Professor in Inclusive Design at the Faculty of Design at Kyushu University in Fukuoka, Japan since 2000. He has over 20 years of experience as a designer, working for KOKUYO, Japan's largest office furniture manufacturer and IDEO in the USA. His expertise is in social innovation, inclusive design and design thinking and office design for innovative approaches to realize human and knowledge-centric solutions. His design works have received numerous design awards, including the Red Dot Award in Germany, the Good Design Award and the Kids Design Award in Japan.

Emma Kirjavainen is a PhD student at the University of Lapland, Finland, where she conducts research in the Lapland User Experience design research group (LUX). Her research interests include user-centric design of social robots and smart environments.

Ashley Colley is a University researcher and docent in creative technologies at the University of Lapland, Finland. He specializes in interactive and creative technologies and human–computer interaction.

Titta Jylkäs is an accomplished service designer and academic who focuses on digital services and the application of AI in service systems. She is an independent design consultant in Finland and an associate professor of design and technology at the Sino-German Faculty of Branding at Zhejiang Wanli University in China. Previously, she worked in large corporations as an in-house designer, a consultant at a design agency, a service designer in the public sector, and a senior researcher and lecturer in academia.

PEER REVIEWERS

Akane Matsumae
Satoru Tokuhisa
Ashley Colley
Jonna Häkkilä
Amna Qureshi
Kiwoong Nam
Maria Grinuik
Leon Loh
Moe Shimomura
Yanfang Zhang
Yangyang Zhao
Bo Gao
Florian Sametinger
Heidi Pietarinen
Tokushu Inamura
Mari Suoheimo
Mariluz Soto
Sheng-Hung Lee
Tarja Pääkkönen
Johanna Hautamäki
Luca Simeone
Albert Tibaijuka
Ludovick Myumbo
Vikki Eriksson
Maria Huhmarniemi
Mari Mäkiranta
Melanie Sarantou
Satu Miettinen
Titta Jylkäs

1

Introduction

Melanie Sarantou, Satu Miettinen and Titta Jylkäs

Over the past two decades, the central role of empathy in design has been discussed (Coleman et al., 2003). Thus, readers may wonder why another book on empathy is needed. The need for empathy is often mentioned in everyday design practice. As designers, we say that we need to use empathy, and its power has been evidenced in recent decades (Kouprie & Visser, 2009; Sarantou & Miettinen, 2022; Tunstall, 2014). However, applying empathy systematically during the design process needs to be considered. It needs to be backed up by research work. In addition, there needs to be adequate experience in applying design tools to benefit from empathy and create innovation that will respond to the need or problem that has arisen. This book presents several aspects of design work that utilize empathy, considering the limitations and critical aspects that one needs to address in empathic design.

It is widely accepted that a lack of empathy in design will cause products to fail (Dam & Siang, 2020). However, empathy is equally important in the design of services because, if delivered effectively, when empathy is embedded in the action, trustworthy services that earn loyalty are delivered (Bates & Petouhoff, 2022). Another important aspect of empathy is the role it plays in design processes. Empathic designers follow their intuition and are in touch with their emotions during design processes, bringing them closer to users and guiding interpretive approaches alongside objective ones (McDonagh & Thomas, 2011). Because of the deep embeddedness of artificial intelligence (AI) in the future aspects of service design, especially in terms of human decision-making, empathy plays a central role in design research (García-Corretjer et al., 2023). Therefore, the motivation for adding a book on empathy to the plethora of those already in existence is straightforward: the everyday needs of users for empathic experiences and services drive innovation for future services; thus, continuing the development of practical knowledge about empathic service design for a fast-changing world is needed. Responding to the everyday problems of users is still creating the best innovation that can change the individual service experience for the better and drive societal change. The research questions guiding the book's content are 'What challenges prevent empathic service design?', 'How is digitalization challenging empathy?', 'How can new theories of empathy enable transformation in service contexts?', 'How can new approaches to analysis guide empathic service delivery?', 'What frameworks can be used to evaluate the effectiveness of AI tools in promoting empathy within service design practices?', and 'How can service designers effectively integrate empathy into the design of AI-driven digital services?'

Business, social and technological innovations contribute to the well-being of society at large, and services keep societies functioning (Ratten, 2013). People require education, well-being, healthcare, security, transportation, housing, reliable food sources and energy, all of which are delivered through service ecosystems. In the current times of digitalization and AI, we need interpreters of the everyday needs of human beings in service delivery (Makhubu, 2015), which are more often realized through more-than-human–human collaboration. More-than-human theories are defined as 'posthumanist theories [that] highlight understanding of knowledge and being as constantly becoming and intrinsically entangled' (Ylirisku et al., 2024, p. 2). A forthcoming chapter by Simpson et al. in this publication illustrates an analytical approach comprising arts-based methods and collaborative analysis that designers can use to sensitize themselves and participants in preparing for designing with more-than-humans. Hence, a new approach to empathy is needed, and for this purpose, we must operate on several levels with a more holistic, cultural and philosophical understanding of empathy.

To implement service ecosystems that are human-centred and address the service issues arising from a more global and digital world (Bates & Petouhoff, 2022), empathy must be operationalized and actioned through design research. There is an eminent need for tools and methodologies to be utilized for developing services driven by human necessities and that contribute not only to well-being but also to sustainable development goals and global equity. Empathy for the environment is essential and thus eco-centred, and inclusive design, for example, should be considered in parallel with service design. Soto's chapter (this publication) explores '*empathy in service prototyping from the perspective of nature-centric thinking*', providing an integrated perspective on how to modify the service prototype process by elucidating the relationships between cocreation, empathy and nature.

Bates and Petouhoff (2022) posit that most companies fail to critically view their businesses from both employees' and customers' perspectives. However, to operationalize and action empathy, both of these perspectives should be central to each organization. More complex approaches to data collection and analytical approaches that can build connections between organizations, employees and users are needed to harness these perspectives. Changes in how design research is executed are needed more than ever. Research should not only be based on more thorough analysis but also have more collaborative, multisensory and personalized approaches.

New mechanisms, such as digital interfaces, and methods for empathy building, such as artistic practices, can be employed to stimulate empathic dialogues between users and organizations. Sarantou and Miettinen (2023) have established a model for mediating empathy in design and business settings through participatory design, arts-based methods, narratives/storytelling and technological/digital approaches. This publication will go further by exploring and reporting – through ample visual representations of the design research processes under discussion – how arts-based and digital tools can enhance empathy.

Notably, although empathy needs to be embedded in services, limitations to people's ability to have empathy exist (Bloom, 2017; Waytz, 2016). This aspect requires acknowledgement from organizations so that they can extend care to their employees, and should be considered in design research (Akimenko et al., 2022). Although limitations to empathy do not exist according to Cameron et al. (2017), people's goals, values, choices and beliefs affect their ability to empathize (Betzler & Keller, 2021). Western dualism upholds the bias dividing emotions and rationality (Cameron et al., 2017; Heylighen & Dong, 2019), but empathy requires a more holistic understanding as a complex phenomenon, which is a choice that people make (Betzler & Keller, 2021). Bloom (2017) argues for rational compassion, which is a skill that service designers can foster, adopt and proactively use (see also Kukar, 2018). This is an area in desperate need of design research.

Empathy is an integral part of service prototyping and prototyping in general. The role of empathy should be a focal interest in prototyping processes because it enables the discovery of user needs while acknowledging the challenges users may face in their experiences with services or products. Prototyping is a process in which users and stakeholders are engaged in service development. Even the development process or the research process itself proposes several roles that demand empathy, as Moe Shimomura, Leon Loh, Yanfang Zhang and Noriko Takano (Chapter 9) explain. Service design can play a critical role in engaging stakeholders in the improvement of, for example, education through early prototyping of learning materials (Shimomura et al., this publication).

Service design processes use workshops as a platform for codesigning solutions and learning from one another (Kuure, 2020). This book highlights the methods and tools that can be used for future service prototyping to better consider the feelings and experiences from which users draw and those that influence their behaviours. The visual approach of the publication illustrates how the captured interactive qualitative data of users' service journeys can be integrated and adopted by educational institutions and organizations as a means for new knowledge creation and learning models. In service design, cocreation (Vink & Oertzen, 2018) and codesign (Trischler et al., 2019) with users are the primary service development strategies. The codesign approach used in service design fosters a strong peer-to-peer learning process, and prototyping in simulation environments (e.g. the service innovation laboratory at the University of Lapland, also referred to as SINCO; Kuure et al., 2014) enables a technology-aided and experiential learning process.

The Structure of the Book

This book has four parts in which themes, including the challenges faced in the applications of empathy and service design, the implications for creating empathic services within sensitive contexts, how more empathic futures may be created with novel approaches to data analysis and the role of design tools and empathic digital services, will be explored. Part One focuses on the challenges experienced in service design, for example, addressing and emphasizing empathy fatigue more profoundly. The challenges address how empathy is often a self-centred approach that is applied incorrectly or superficially by designers without considering the needs of the participants and communities with which they work. Part Two elaborates on the challenges with empathy in various marginalized contexts. How empathy can be deeply embedded in user experiences is often overlooked in systemic service flows, where people are often seen as enablers of services in scripted ways that lack considerations of empathy. The last two parts of the book present various case studies from different global locations, illustrating how empathy is used as a tool or an approach, providing actionable insights for practitioners and researchers.

Parts One & Two **Challenges with empathy**	**Parts Three & Four** **Propositions and improvements**
1. Superficial approaches to empathic design	3. Approaches to data analysis
2. Sensitive contexts	4. Digital approaches

FIGURE 1.1 *Presentation of the book parts. Parts One and Two address challenges with empathy in service design, whereas Parts Three and Four present propositions and improvements for empathic service design.*

The book will critically re-examine the role of empathy. First, critiques of empathy in design processes need to be offered the opportunity to highlight the problems of the concept in design. Parts One and Two address challenges with empathy by questioning how involving the people who use the services in design processes can transform service delivery. Therefore, the emphasis of this book will be on the sources of concern around empathy and the risks that users, communities, organizations and the environment face when empathy is not deeply integrated into services using technology. Those challenges related to empathy as a superficial approach used by designers, driven by their own needs and understanding of the skill rather than a compassionate understanding of those they are designing with, will be addressed. Although all chapters discuss specific processes or activity-oriented themes, the structure provided by the book parts will assist readers in understanding the challenges with empathy and the implications of service delivery within sensitive contexts.

Parts Three and Four of the book emphasize the plurality of design tools and approaches that can be used to analyse data for improved problem identification and enhancement of digital services. Part Four presents several company case studies; it focuses on how empathy can be practised in concrete ways and how challenges with empathy can be addressed through technology-oriented development. The four book parts are illustrated in Figure 1.1.

Challenges in the Applications of Empathy and Service Design

When designing with empathy, it is important to consider both the challenges and limitations of empathy. Empathy is not a trivial tool nor a magic wand that will bring change on its own. One needs to understand what kind of empathy is needed and how to apply this when also considering the project size. When one is scaling up work, it may become difficult to keep tabs on empathy. Thomas Abrell discusses the challenges of empathy in Chapter 2, 'User Empathy in the Early Phase of Product–Service Systems Development'. He highlights the complexity of applying empathy when projects have multiple stakeholders and extend into the future. Designers may work closely with users in the early phase development of the project, but how to hold on to this when the project is progressing and how to communicate with various departments and decision-makers should be explored. Abrell is proposing that research rigour in service blueprinting and prototyping should have convincing evidence and tested solutions. Another solution may be strategic prototyping, which focuses on working with stakeholders, focusing on communication and decision-making.

Empathy connects and crosscuts with a variety of industries, such as original equipment manufacturers or financial service providers, as illustrated in the chapters of Abrell and Lee. To develop these services, designers engage in an empathy-driven innovation process in which prototyping is a tool that uses proactive empathy. Hautamäki (Chapter 3) presents a case study in which an overall service prototyping infrastructure is set to promote innovation through empathy. Inget et al. (Chapter 6) take the innovation aspect further through empathy and service prototyping; they conceptualize, prototype and design aesthetically pleasing wearables, helping them navigate pandemic-related challenges while also designing for practical use contexts.

Hautamäki presents another set of challenges in Chapter 3, where she discusses how service prototyping has developed at Centria UAS. She describes empathy experiments conducted in the simulation and prototyping environment of the SILMU. These experiments can be real-life situations where empathic design methods are applied as a way to foster active learning and multidisciplinary ways of working while increasing a shared understanding of a service experience, a fact-based user journey and even future user needs. Empathy experiments emphasize the principles of a safe space. Hautamäki highlights the difficulty of distinguishing between the presuppositions and facts as well as not only responding to the user needs with empathy but also creating actual design solutions. Hautamäki and her team are using empathy experiments to push participants beyond these challenges.

It is good to remember that, when we apply empathy with service and experience design, extra care is needed due to the drastic paradigm shift from designing products to designing experiences. This is what Sheng-Hung Lee reminds us of in Chapter 4, 'Design for Longevity'. Lee goes back to Suri's (2001) reminder of bias that empathy may depend on cultural differences, varying knowledge levels, education backgrounds, expertise and life and work experiences, or they could mislead or reach incorrect conclusions. This is a good list for all of us to keep in mind. Lee also proposes the use of tangible objects for empathy-building in service design, which allows more space, exploration and comprehension to think about the service at hand.

In Chapter 5, Zhang and Fujita present longitudinal research addressing the relationships among a design process, social implementation and the development of trust in citizen-participatory workshops. They argue that the use of empathy may foster the social implementation or innovation of workshops. Their longitudinal work highlights their empathic and communication skills as well as their ability to build trusting relationships. It is also good to consider the timeline when collaborating with communities. Trusting relationships take time to develop and understand the community's needs and motivations as well. Of course, it is the best-case scenario to achieve social implementation during and after the design process.

In Chapter 6, Inget and co-authors discuss how to work with low-fi prototypes for smart jewellery. Their work reminds us of the use of empathy in hands-on work, an experience shared by Lee in Chapter 4. They also present an important challenge in empathy design. One needs to consider different cultural contexts when designing workshops that include the use of empathetic methods. In their example, place-specific pictures can provoke very different experiences, here depending on the context. These are important aspects of situated design and have an impact on how empathy is experienced.

Implications for Creating Empathic Services within Sensitive Contexts

Sensitive contexts, such as indigenous research and sensitive cultural settings, pose distinctive challenges to designers and researchers. Designers and researchers need to acknowledge the position of power they have over research communities and research environments in all phases of their research. For example, in the data collection and analytical phases, researchers usually extract knowledge from research communities, after which the data are interpreted by outsiders (Seppälä et al., 2021). Conversely, in collaborative approaches, in which the participants contribute to data generation and subjective interpretations forthcoming from the data (Gough & Madill, 2012), not only can rich findings be established (Sarantou et al., 2019), but dialogues within the research community can also be continued, and potential biases in the data sources can be uncovered through respectful exchanges (Sarantou et al., 2022) and the acknowledgement of emotions and lived experiences between researchers and communities (Soto et al., 2021).

Openness to feedback and a willingness to adjust findings can further elicit empathy between stakeholders in research (Belman & Flanagan, 2010; García-Corretjer et al., 2023). This has affects how research reports are created because research communities can have immediate input into readability and visualization processes, thereby making the research more useful through effective dissemination (Heer et al., 2005). These considerations have a direct bearing on the ethical underpinnings of the research process.

In Chapter 7, 'The Role of Cultural Sensitivity in Fostering Empathy in Service Design', Steunou and Simeone explore a service design process conducted in China and Denmark, demonstrating how four cultural sensitivity-related processes – recognition, learning, respect and adaptation – are crucial for building actors' trust and empathy. Their chapter details how cultural factors are crucial in producing user-friendly experiences, permitting user appropriation, improving the appeal of the service and, in general, developing empathy. Qureshi and Jones explain in Chapter 8, 'The Role of Empathy in Reliving Indigenous Cultural Heritage: Insights from Arts Ceduna, South Australia', how to understand the effects of empathy on the reinterpretation of Indigenous cultural heritage better; they do this by using participatory observation and qualitative interviews with an Indigenous artist community in Ceduna, South Australia. The results show that empathy supports social justice, strengthens Indigenous people and aids in the preservation of culture, making it easier to incorporate traditional and modern aspects.

In the two chapters focusing on different cultural settings in educational practices, the need for more empathic approaches is explored. Chapter 9, 'Creating an Empathy Network: Insights into a Service Design Intervention in a High School in Japan' by Shimomura et al., explores, here through a case study carried out in a Japanese high school, fresh interpretations and understandings of empathic service design. The design researchers worked in conjunction with educators to sensitively cocreate instructional materials and programmes by utilizing action research and taking into account the input from surveys and in-class observations to better develop educational outcomes that would serve both students and educators. In another complex educational setting, Chapter 10, 'Empathy and Distance Learning' by Kylmänen, explains how the Carefree Room Project drew from stories to foster environments as a way to enhance and support students' well-being in unique circumstances (such as distance learning). The project used art and science to open new channels for encouraging candid communication and a more cooperative and empathic learning environment.

Nature-centric thinking and more-than-human empathy are thoughtfully explored in Chapters 11 and 12 as sensitive contexts for creating more empathic services. Soto explains in Chapter 11, 'Service Prototyping with Empathy and Nature-Centric Thinking', how this type of prototyping engages bodies and interactions with memories in natural settings by using collages to convey their sentiments. With participants, Soto explores how their connections with the environment and themselves can be gently nudged through nature simulations in service prototype workshops, connecting conscious thoughts, emotions and bodies in service design settings. Bergström et al. explore in Chapter 12, 'More-Than-Human Empathy in Arts-Based Material Research Processes', the functioning of creative processes in fostering more empathic bonds between people and the environment. The authors provide two case studies using a theoretical framework that includes creative processes, a material-based approach and empathic exploration. Renewable gluten plastic was used in the project 'The Secret Life of a Material', whereas immersive and interactive experiences with Lake Tjörnin in Reykjavik were explored in the project 'The Biggest Bread Soup in the World'. According to the authors, the bird life around this lake, which 'behaves like a living thing', reminds us how vulnerable our ecosystems are. Considering the impact of design on non-humans provides new perspectives on the need for transformative service design.

Empathic Futures with Novel Approaches to Data Analysis

Designing futures with novel approaches to data analysis involves using innovative and creative methods to leverage user data for informed decision-making. These data can be based on user behaviour or feedback, sensory or visual data or social media. Digitally and AI-enhanced approaches to data analysis can include advanced machine learning, algorithms and predictive modelling to obtain accurate proofs of concepts or prototypes for service development (Alexopoulos et al., 2019). The aim of data analysis in these contexts is to enable highly personalized experiences for users (Lehrer et al., 2018). Empathy can support participant output through quality insights, building on research understanding. As a result, algorithms and machine learning can provide users with product suggestions, which are used as evidence during social media use when users receive advertisements for products and services based on internet search histories. Novel approaches to data analysis can shape a wide range of future scenarios in service design (Karpen et al., 2017; Ojasalo et al., 2015), including personalized experiences to address global challenges.

As technology and methodologies continue to evolve, data analysis plays a pivotal role in driving innovation and decision-making in diverse forms of service delivery. For example, data analysis is essential for transport services to monitor environmental changes and predict natural disasters. Complex societal issues can be addressed through digital and AI-enhanced networks or system analysis (Lee et al., 2020). AI can handle data processing and analysis at scale, and humans can offer unique interpretive and domain-specific expertise for insightful decision-making. Complex challenges can be addressed, and data can be responsibly used via interdisciplinary and collaborative approaches to data analysis (Slota et al., 2023).

Although the human role is central to data analysis, human–machine interaction and collaboration are gaining a stronghold in research. In design research, especially when systems approaches are embraced, complex mixed methods datasets are often used to understand problems. AI and machine learning can assist designers in generating different datasets that can be pieced together in analysis. Empathy and data analysis may seem like unrelated concepts, but they can complement each other in several meaningful ways, particularly when human-centric and creative arts-based approaches are favoured (Akimenko et al., 2023). Researchers can use empathy in human-centred approaches to understand user needs through specific data collection choices that can be more context-related and empathetic to specific problems (Roelofsen, 2018; Ventura & Bichard, 2017). Deeper insights can be garnered through, for example, raising the appropriate dialogue and eliciting storytelling instead of adopting intimidating interview or checkbox approaches to data collection (Akimenko et al., 2023; Sarantou et al., 2019).

Although AI and machine learning assist designers and researchers with complex analysis, the human role in data analysis steers the adoption of empathic and collaborative approaches towards work and research, brings about ethical decision-making and plays a critical role in interpretation (Barnes & Du Preez, 2015). In Chapter 14, 'Effective Analytical Approaches as Precursors for Proof-of-Concept Generation', Björn et al. discuss how affinity analysis methods can generate a proof of concept to investigate more empathic services and tools for building workable and practical service models that provide users with more empathic outcomes. With empathy incorporated into analytical practices, the human aspects of data generation and interpretation can be valued, while awareness of more-than-human roles and inputs in research processes can be raised (Homewood et al., 2021), leading to more responsible, accountable and ethical outcomes.

Chapter 13, 'Empathic Approaches to Qualitative Data Analysis: Calling for a Deeper Understanding of Research Problems' by Myumbo and Tibaijuka, aims to uncover the underlying possibilities for a deeper comprehension of research problems – particularly the capacity to identify participants' inner emotional states; their study examines researchers' experiences with data processing and discursive methods to better articulate participants' life realities. These approaches can illustrate more sensitivity to cultural, social and ethical considerations when gathering data, and trust can also be established. By emphasizing qualitative approaches to analysis, such considerations, which are integral to understanding how services are experienced by various users with different emotional and lived experiences, can be integrated into empathic service design.

Chapter 15, 'Extending Visual Analysis Techniques for Empathy Building and Cross-Cultural Design' by Simpson et al., explores the role of empathy building within workshop practices and how empathy among facilitators and towards participants can guide postworkshop data analysis. The chapter envisions more cohesive futures in which better relationships between humans and non-humans can inspire transformative service design.

Design Tools and Empathic Digital Services

When designing future digital services, the role of designers is to ensure that human emotion and empathy are considered. This book looks at the ways to utilize service design and prototyping as strategies to ensure that empathy plays a strong role in digital service design. The role of service designers in technology (especially AI-driven design processes) is to address users' needs, which is often considered part of the technology-heavy process (Jylkäs et al., 2019). When addressing and minimizing the risk involved in how users will adopt designs for future services, designers in these spaces adopt tools that enable them to sketch future scenarios, address human experiences and facilitate meaningful design results.

Designers are discussing the impact of digital technologies and AI in the design process and design outcomes. Regarding technology tools, designers address not only their design practices but also the evolving relationship between empathy and AI. Although AI systems can be programmed to recognize and respond to human emotions in various ways, they do not possess true empathy, as humans do. In Chapter 16, 'Empathic Development for Service Prototyping: Challenges and Opportunities that Emerged During Design Thinking Workshops with Analogue and Digital Tools', Chen and Nam discuss the role of technologies such as VR and AI in design processes and the gap they pose in facilitating emotional interactions. This publication presents examples through case studies on how to utilize different technological tools to foster empathy in a creative process. Regardless of the lack of emotional recognition in AI, various AI solutions can contribute to personalizing user experiences, training and simulation. However, these issues raise ethical questions regarding privacy, consent and the potential for the misuse of emotional data (Hand, 2018; Latif et al., 2022).

AI can play a role in augmenting and enhancing empathy through emotion recognition via facial expressions, tone of voice and other physiological cues. Because of AI's lack of cognition and emotions, its ability to exhibit empathy is limited to data inputs and programmed functions. AI systems can mimic empathy, especially through smartphones, through a combination of natural language processing, machine learning and emotional modelling (D'Alfonso, 2020). In Chapter 20, 'The Role of Emojis in Fostering Empathy in Chatbot Interactions', Kirjavainen et al. also address the role of emotions in design tools by providing an example of integrating emojis in chatbot conversation and exploring where the line of appropriate and inappropriate usage of emojis occurs.

Service design offers organizations various approaches for how to remain abreast of rapid developments in digital technologies and consumer cultures by responding to users' needs; this can be done by using ethnographic tools, such as observations, interviews – or 'service safaris' – and service visualizations (Harwood et al., 2020), where the service is tested and evaluated from the user perspective to collect user insights and understand their mental models (Vink et al., 2019). While the use of AI has increased, questions related to the ethical and sustainable use of technology have also arisen. In Chapter 17, 'Empathy and Sustainability at the Core: Leveraging AI Tools in Service Design', Stevens et al. investigate the lessons learned from using generative AI (GAI), such as ChatGPT, in design processes while also providing a framework in which GAI enriches the understanding of empathy in complex design processes in the context of sustainability.

AI and different digital approaches can provide opportunities for diversity, equity and inclusion (DEI) in service design. It is important to emphasize a diversity-in-use perspective in design to ensure that users or workshop participants can use design in ways that address the full intersectional scope of who they are while ensuring that they have the possibility of delivering equal outcomes to others (Gomez et al., 2021). This book presents case studies that address the DEI perspective in design and acknowledges that these contexts hold the privilege of being located in the Global North. In Chapter 19, 'Demonstrating a Two-Method Experiment: The Use of a Digital Tool and Empathy Map in Designing Empathic Phygital Experiences in a Peripheral Destination', Hanni-Vaara et al. explore digitally mediated service experiences in a physical environment. They present a two-method data collection model for experimenting with empathetic phygital (physical + digital) services. In Chapter 18, on the other hand, 'Empathetic AI: Integrating Artificial Intelligence and Service Design for 3H Care in Singapore', Gao et al. address the role of empathic AI in healthcare services. They present a four-layered A3C system framework that includes user needs, data, intelligent agents and 'HAI touchpoints' (human-centred AI touchpoints). The framework enables dynamic adjustments to user states and incorporates AI within the information architecture.

Acknowledgements

The book is part of the Business Finland-funded RRF project *'Empathy Business. How to Digitalise Service Prototyping and Business through Creativity'* with grant agreement No. 7425/31/2022 from 01.11.2022 to 31.12.2024.

References

Akimenko, D., Sarantou, M., & Miettinen, S. (2022). Collaborative art and storytelling as an empowering tool for social design. In M. Sarantou & S. Miettinen (Eds.), *Empathy and business transformation*. Routledge. https://doi.org/10.4324/9781003227557

Alexopoulos, C., Lachana, Z., Androutsopoulou, A., Diamantopoulou, V., Charalabidis, Y., & Loutsaris, M. A. (2019, April). How machine learning is changing e-government. In *ICEGOV '19: Proceedings of the 12th international conference on theory and practice of electronic governance* (pp. 354–363). Association for Computing Machinery.

Barnes, V., & Du Preez, V. (2015). *Mapping empathy and ethics in the design process*. Design Education Forum of Southern Africa. http://hdl.handle.net/11189/5927

Bates, T., & Petouhoff, N. (2022). *Empathy in action: How to deliver customer experiences at scale*. https://empathyinactionbook.com/
Belman, J., & Flanagan, M. (2010). Designing games to foster empathy. *International Journal of Cognitive Technology*, 15(1), 11.
Betzler, M., & Keller, S. (2021). Shared belief and the limits of empathy. *Pacific Philosophical Quarterly*, 102(2), 267–291.
Bloom, P. (2017). *Against empathy: The case for rational compassion*. Random House.
Cameron, D. C., Inzlicht, M., & Cunningham, W. A. (2017, March 2). *Does empathy have limits?* The Conversation. https://theconversation.com/does-empathy-have-limits-72637
Coleman, R., Lebbon, C., Myerson, J. (2003). Design and empathy. In J. Clarkson, S. Keates, R. Coleman, & C. Lebbon (Eds.), *Inclusive design*. Springer. https://doi.org/10.1007/978-1-4471-0001-0_29
D'Alfonso, S. (2020). AI in mental health. *Current Opinion in Psychology*, 36, 112–117.
Dam, F. R., & Siang, T. Y. (2020). *What is empathy and why is it so important in design thinking?* Interaction Design Foundation. https://www.interaction-design.org/literature/article/design-thinking-getting-started-with-empathy
García-Corretjer, M., Ros, R., Mallol, R., & Miralles, D. (2023). Empathy as an engaging strategy in social robotics: A pilot study. *User Modelling and User-Adapted Interaction*, 33(2), 221–259.
Gomez, K., Gomez, L. M., & Worsley, M. (2021). Interrogating the role of CSCL in diversity, equity, and inclusion. In U. Cress, C. Rosé, A. F. Wise, & J. Oshima (Eds.), *International handbook of computer-supported collaborative learning* (pp. 103–119). Computer-Supported Collaborative Learning Series 19. https://doi.org/10.1007/978-3-030-65291-3_6
Gough, B., & Madill, A. (2012). Subjectivity in psychological science: From problem to prospect. *Psychological Methods*, 17(3), 374–384. https://doi.org/10.1037/a0029313
Hand, D. J. (2018). Aspects of data ethics in a changing world: Where are we now? *Big Data*, 6(3), 176–190.
Harwood, T., Garry, T., & Belk, R. (2020). Design fiction diegetic prototyping: A research framework for visualising service innovations. *Journal of Services Marketing*, 34(1), 59–73. https://doi.org/10.1108/JSM-11-2018-0339
Heer, J., Card, S. K., & Landay, J. A. (2005, April). Prefuse: A toolkit for interactive information visualisation. In *Proceedings of the SIGCHI conference on human factors in computing systems* (pp. 421–430). https://doi.org/10.1145/1054972.1055031
Heylighen, A., & Dong, A. (2019). To empathise or not to empathise? Empathy and its limits in design. *Design Studies*, 65, 107–124.
Homewood, S., Hedemyr, M., Fagerberg Ranten, M., & Kozel, S. (2021, May). Tracing conceptions of the body in HCI: From user to more-than-human. In *Proceedings of the 2021 CHI conference on human factors in computing systems* (pp. 1–12). Association for Computing Machinery.
Jylkäs, T., Augsten, A., & Miettinen, S. (2019, November). From hype to practice: Revealing the effects of AI in service design. *Conference Proceedings of the Academy for Design Innovation Management*, 2(1), 1203–1216.
Karpen, I. O., Gemser, G., & Calabretta, G. (2017). A multilevel consideration of service design conditions: Towards a portfolio of organisational capabilities,

interactive practices and individual abilities. *Journal of Service Theory and Practice*, 27(2), 384–407.
Kouprie, M., & Visser, F. S. (2009). A framework for empathy in design: Stepping into and out of the user's life. *Journal of Engineering Design*, 20(5), 437–448.
Kukar, P. (2018). Against empathy: The case for rational compassion. *Philosophical Psychology*, 31(3), 479–482. https://doi.org/10.1080/09515089.2018.1442923
Kuure, E. (2020). *Service design workshops in design practice* [Doctoral dissertation, University of Lapland]. https://urn.fi/URN:ISBN:978-952-337-235-1
Kuure, E., & Miettinen, S. (2013). Learning through action: Introducing the innovative simulation and learning environment service innovation corner (SINCO). In T. Bastiaens & G. Marks (Eds.), *Proceedings of e-learn 2013 – World conference on e-learning in corporate, government, healthcare, and higher education* (pp. 1536–1545). Association for the Advancement of Computing in Education (AACE). https://www.learntechlib.org/primary/p/115095/
Kuure, E., Miettinen, S., & Alhonsuo, M. (2014). Change through service design – Service prototyping as a tool for learning and transformation. In Y. Lim, K. Niedderer, J. Redström, E. Stolterman, & A. Valtonen (Eds.), *Design's big debates – DRS international conference 2014*. https://dl.designresearchsociety.org/drs-conference-papers/drs2014/researchpapers/18
Latif, S., Ali, H. S., Usama, M., Rana, R., Schuller, B., & Qadir, J. (2022). AI-based emotion recognition: Promise, peril, and prescriptions for prosocial path. Preprint arXiv:2211.07290.
Lee, J. H., Hsu, C., & Silva, L. (2020). What lies beneath: Unravelling the generative mechanisms of smart technology and service design. *Journal of the Association for Information Systems*, 21(6), 1621–1643.
Lehrer, C., Wieneke, A., Vom Brocke, J. A. N., Jung, R., & Seidel, S. (2018). How big data analytics enables service innovation: Materiality, affordance, and the individualisation of service. *Journal of Management Information Systems*, 35(2), 424–460.
Makhubu, R. L. (2015). *A systematic modelling approach to interpreting service delivery. Advances in communications and media research* (Vol. 11). Nova Science Publishers.
McDonagh, D., & Thomas, J. (2011). Design + empathy = intuitive design outcomes. *The Design Journal*, 14(2), 147–150.
Ojasalo, K., Koskelo, M., & Nousiainen, A. K. (2015). Foresight and service design boosting dynamic capabilities in service innovation. In R. Agarwal, W. Selen, G. Roos, & R. Green (Eds.), *The handbook of service innovation* (pp. 193–212). Springer.
Ratten, V. (2013). The development of social e-enterprises, mobile communication and social networks: A social cognitive perspective of technological innovation. *Journal of Electronic Commerce in Organisations (JECO)*, 11(3), 68–77.
Reynolds, D., & Reason, M. (Eds.). (2012). *Kinesthetic empathy in creative and cultural practices*. Intellect Books.
Roelofsen, M. (2018). Performing 'home' in the sharing economies of tourism: The Airbnb experience in Sofia, Bulgaria. *Fennia – International Journal of Geography*, 196(1), 24–42. https://doi.org/10.11143/fennia.66259

Sarantou, M., & Miettinen, S. (Eds.). (2022). *Empathy and business transformation*. Taylor & Francis.

Sarantou, M., Qureshi, A., Jones, S., & Gunter, S. (2023). 'I will now be more aware': The intersection of human–computer interaction and Indigenous artmaking in living cultural heritage. INTERACT 2023 19th IFIP TC13 International Conference, York, UK, 28 August – 1 September 2023, Proceedings, Parts I–IV.

Sarantou, M. A., Sillgren, S. K., & Pokela, L. M. (2019). In her lap: Embodied learning through making. In T. Jokela & G. Coutts (Eds.), *Relate north* (pp. 84–106). ISEAS.

Seppälä, T., Sarantou, M., & Miettinen, S. (Eds.) (2021). *Arts-based methods for decolonising participatory research*. Routledge. https://doi.org/10.4324/9781003053408

Slota, S. C., Fleischmann, K. R., Lee, M. K., Greenberg, S. R., Nigam, I., Zimmerman, T., Rodriguez, S., & Snow, J. (2023). A feeling for the data: How government and nonprofit stakeholders negotiate value conflicts in data science approaches to ending homelessness. *Journal of the Association for Information Science and Technology*, 74(6), 727–741.

Soto, M., Mollenhauer, K., Miettinen, S., & Sarantou, M. (2021). Building a community through service design and responsiveness to emotions. In T. Seppälä, M. Sarantou, & S. Miettinen (Eds.), *Arts-based methods for decolonising participatory research*. Routledge. https://doi.org/10.4324/9781003053408

Suri, J. F. (2001). The next 50 years: Future challenges and opportunities for empathy in our science. *Ergonomics*, 44(14), 1278–1289. https://doi.org/10.1080/00140130110105850

Trischler, J., Dietrich, T., & Rundle-Thiele, S. (2019). Co-design: From expert- to user-driven ideas in public service design. *Public Management Review*, 21(11), 1595–1619. https://doi.org/10.1080/14719037.2019.1619810

Tunstall, E. D. (2014). *Undesigning apathy: Designs for systems of caring*. https://theconversation.com/un-designing-apathy-designs-for-systems-of-caring-22866

Ventura, J., & Bichard, J. A. (2017). Design anthropology or anthropological design? Towards 'social design'. *International Journal of Design Creativity and Innovation*, 5(3–4), 222–234.

Vink, J., Edvardsson, B., Wetter-Edman, K., & Tronvoll, B. (2019). Reshaping mental models – enabling innovation through service design. *Journal of Service Management*, 30(1), 75–104.

Vink, J., & Oertzen, A. S. (2018, July). Integrating empathy and lived experience through cocreation in service design. In *ServDes2018. Service Design Proof of Concept, Proceedings of the ServDes. 2018 Conference* (pp. 18–20). Linköping University Electronic Press.

Waytz, A. (2016). The limits of empathy. *Harvard Business Review*, 94(1), 68–73.

Ylirisku, H., Slotte Dufva, T., & Snellman, M. (2024). Editorial: Arts education in a more-than-human world. *Research in Arts and Education*, 2024(2), 1–12. https://doi.org/10.54916/rae.145827

PART ONE

Challenges with Empathy and Implications for Service Design

2

User Empathy in the Early Phase of Product–Service Systems Development:

Challenges with User Research, Service Blueprinting and Service Prototyping

Thomas Abrell

Abstract

Driven by new digital technologies and changing market dynamics, original equipment manufacturers face challenges and opportunities to develop product–service systems instead of isolated products. However, to offer relevant services, these companies need to understand their users. Empathy plays an important role in this because discovering user needs includes two dimensions of empathy: to feel what a user might feel in a certain situation and to understand how a user might experience the world from their point of view. To achieve this, service design draws heavily on user empathy gained from user research, sequencing services in service blueprints and service prototyping to explore, communicate and refine service concepts.

This article contributes to the body of literature on service design and innovation by conceptually exploring the relationship of these three core activities of service design to user empathy and discussing the benefits and challenges of empathy in the early phase of product–service systems development.

Introduction

The world is changing from isolated products and services to integrated product–service systems, often driven by smart, connected products (Porter & Heppelmann, 2014). This change requires traditional 'product' companies to become increasingly service oriented as they integrate digital technologies into their products that enable entirely new value propositions (Gebauer et al., 2021). Such emerging product–service systems can be product, use, or result oriented (Baines et al., 2007) and are complex to design. From a customer and user perspective, on the other hand, simplicity is needed because products and services are used to solve problems or satisfy needs. Therefore, they need to fit the users' context, and companies aiming to provide meaningful product–service systems need to build up empathy with their users. This is particularly challenging in an environment where a multitude of interconnected products and services need to function together, their complexity is high, and the answers to user needs may be well situated in the future. Design approaches to innovation, such as service design, are often described as highly empathic (Koskinen & Battarbee, 2003). However, this may not be the case by default (Steen, 2008). Instead, conscious effort needs to be undertaken to acquire and maintain empathy when designing new product–service systems.

The present article is a conceptual exploration of empathy in the early phase of product–service systems development. Its aim is to advance the understanding of empathy in relation to core activities and context for developing more empathic products and services. Therefore, the article highlights the benefits and challenges of empathy in three core activities of a service design process in the fuzzy front end of product–service systems development: user research, service prototyping and service blueprinting. To approach this topic, a targeted literature review (Huelin et al., 2015) was conducted, and relevant articles were selected to discuss empathy challenges in the early phase of product–service systems development. In the first part of the present article, the three core activities are introduced along the UK Design Council's (2024) double diamond process for innovation. In the second part, empathy is discussed in the context of the early phase of product–service systems development.

Although studies have highlighted the positive connection between design and empathy, few studies have investigated empathy challenges in design processes. The author of the current article has been working for original equipment manufacturers for the past 10 years in a variety of innovation roles in the early phase of product–service systems development. Thus, building on the targeted literature review, the present article highlights challenges observed in practice based on the experience of the author with empathy in creating new product–service systems. Articulating these challenges from the author's practice is meant to inspire future research.

Core Activities of Service Design in the Early Phase of Product–Service Systems Development

To discuss benefits and challenges for empathy, it is important to lay the foundation by conceptually exploring the core activities of service design in such an early development phase. This section starts by introducing empathy in design as well as the UK Design Council's (2024) double diamond process. Then, during the double diamond process, each core activity is introduced and described.

Empathy in the Double Diamond Process

Empathy and design are deeply intertwined. Koskinen and Battarbee state, 'Clearly, designers think about the user in all phases of the design process. Empathy is ubiquitous' (2003, p. 49); this emphasizes that it is crucial to obtain the users' perspective to make informed development decisions. In this context, being empathic is described as 'being more sensitive to users, be able to understand them, their situation, and feelings' (Kouprie & Sleeswijk Visser, 2009, p. 438). Empathic design consists of activities and a process (Leonard & Rayport, 1997). In the present article, the UK Design Council's (2024) double diamond process, which was originally published in 2003, is used to illustrate the core activities in the early phase of product–service systems development.

According to Koskinen and Battarbee (2003), the methods of empathic design are best suited to this early, conceptual part of development. The double diamond process consists of four phases: 'discover', 'define', 'develop' and 'deliver'. The 'discover' and 'develop' phases diverge, while the 'define' and 'deliver' phases converge (UK Design Council, 2024). The double diamond framework has been applied to service design (Stickdorn, 2010). Figure 2.1 provides a summary of the aims of the core activities mapped along the UK Design Council's (2024) double diamond process.

The following empathy aims can be derived from the discussion in the literature and from a practical perspective: In the 'discover' phase, the aim of empathy is to understand the context of users, including their motives, emotional states, possible use of product–service systems and challenges. In the 'define' phase, design teams face the challenge of converging rich empirical data into a narrower problem statement to be explored. User understanding is narrowed to graphics and representations to communicate to the wider audience and convey empathy to others who were not directly involved in the process. The aim is to turn the empathy gained during the process into innovation opportunities. In the 'development' phase, the challenge is to create suitable solution ideas and address the feelings and needs of users. Often, multiple concepts are further developed. In the 'deliver' phase, the

Double Diamond Process

UK Design Council 2024, www.designcouncil.org.uk

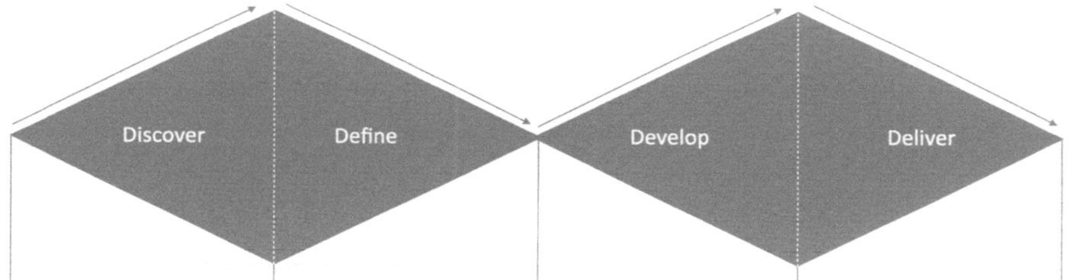

Aims	Discover	Define	Develop	Deliver
Service Prototyping	• explore (Floyd, 1984) • reach user understanding (Soto et al., 2023)	• create representations of ideas (Coughlan et al., 2007) • explore new ideas and solutions (Rahman et al., 2017)	• experiment (Floyd, 1984) • solve problems and communicate (Bogers & Horst, 2014) • co-create solutions with users (Boukhris et al., 2017)	• evolve (Floyd, 1984) • iterate design solutions (Stickdorn & Schneider, 2010) • evaluate early (Beaudouin-Lafon & Mackay, 2007) to create a good solution-need fit with users
Service Blueprinting	• explore aspects of service (Stickdorn & Schneider, 2010) • identify key touchpoints (Radnor et al., 2014) and service failures (Shostack, 1984)	• identify improvement potential (Chuang, 2007) • capture and describe research data insights (Selby, 2020) • provide overview (Potthoff et al., 2018)	• refine service (Radnor et al., 2014) and align touchpoints (Ryu et al., 2019) • design interactions (Shostack, 1984) • Transfer feelings and emotive responses (Selby, 2020)	• provide roadmap for service delivery (Stickdorn & Schneider, 2010) • make processes and interactions visible (Shostack, 1984)
User Research	• generate holistic understanding (Stickdorn & Schneider, 2010) • understand latent user needs, use context (Fraser, 2009) and user emotions (Brown, 2009) • investigate users' opinions and what they do (Daae & Boks, 2015)	• achieve state of empathy to discover opportunities (Brown, 2009) • translate observations into insights (Brown, 2009) • inform and inspire design processes (Van Dijk, 2010)	• co-create solution ideas with users (Wetter Edman, 2011) • identify all possible problems and needs (Leonard & Rayport, 1997)	• translate insights into products and services (Brown, 2009) • receive feedback from users (Brown, 2009)
Empathy	• understand user problems, feelings, emotions	• convey empathy gained during research to others • turn empathy into innovation opportunities	• create personalized services • address feelings and needs of users	• tweak product-service-system to answer needs • build technical/ organisational enablers to act empathetically

FIGURE 2.1 *Aims of core activities and empathy during the double diamond process. Analysis of the literature. Process: UK Design Council (2024). Mapping the aims of the core activities to the process is based on the author's practical perspective. Empathy aims are a summary of the literature and the practical perspective of the author.*

aim of empathy is to tweak the product–service system to answer needs and build technical and organizational enablers to act empathetically.

Service Prototyping

Prototyping in design can be defined as the activity of creating representations of ideas before a final artefact exists (Coughlan et al., 2007). This representation may take into account rough approximations of the features of a product, service or system (Otto & Wood, 2001). How prototypes are designed and used depends on the desired outcomes. Houde and Hill (1997) distinguish between role, implementation, look and feel prototypes. Role prototypes are focused on user utility, look and feel prototypes are focused on concrete experiences with the prototype, while implementation prototypes highlight a potential solution to a technical problem (Houde & Hill, 1997). Because services are intangible in nature, service design needs to rely on different prototyping approaches. Prototyping methods such as storytelling and enacting services are commonly used for prototype services (Stickdorn & Schneider, 2010). Service prototyping can occur at different phases of a design process (Soto et al., 2023). Soto et al. (2023) argue that prototyping enhances the ability to empathize in service design.

In the 'discover' phase, prototypes are used exploratively to clarify requirements (Floyd, 1984). Although prototyping is often attributed to later phases of the design process, it is also an essential tool for reaching user understanding (Soto et al., 2023). In the 'define' phase, prototyping is used to create representations of ideas before the entire product–service system is designed (Coughlan et al., 2007). In this phase, prototyping is also used to explore new ideas and solutions (Rahman et al., 2017). In the 'develop' phase, prototyping includes experimentation to determine the fit of the solution to its context (Floyd, 1984). In this phase, prototyping is used to solve problems and communicate across disciplines (Bogers & Horst, 2014). These solutions are often co-created with users (Boukhris et al., 2017). In the 'deliver' phase, prototypes evolve towards a functioning system (Floyd, 1984); they are used to iterate design solutions (Stickdorn & Schneider, 2010) and allow for early evaluation (Beaudouin-Lafon & Mackay, 2007) to create a good solution-need fit with the user.

Service Blueprinting

Similar to product innovation, service innovation also requires a structured approach. Service blueprinting is a technique introduced by Shostack (1982, 1984) that takes into account the sequence of a service, customer actions and interactions with company employees and products as well as supporting processes. The service blueprint is a two-axis representation of service processes (Fließ & Kleinaltenkamp, 2004) with a time sequence of actions on the horizontal axis and different action areas on the vertical axis. Service blueprinting has been proposed as a tool for service innovation (Bitner et al., 2008; Shostack, 1982, 1984) and for efficiently managing service processes (Fließ & Kleinaltenkamp, 2004). Service blueprinting highlights the role of the customer who uses the service (Shostack, 1982, 1984).

In the 'discover' phase, service blueprinting is used to explore aspects of the service that need to be revised and map how a user perceives a service (Stickdorn & Schneider, 2010). Oftentimes, the service blueprint is in this phase merely a rough draft (Stickdorn & Schneider, 2010). It is used to identify key touchpoints (Radnor et al., 2014) and possible service failures (Shostack, 1984). In the 'define' phase, service blueprinting helps identify the most crucial service failures and touchpoints to further improve (Chuang, 2007). It is also used to present user research data insights to designers and other stakeholders in the innovation process (Selby, 2020). Furthermore, service blueprinting helps present an overview of the entire service (Potthoff et al., 2018). In the 'develop' phase, the tool is used to refine a service, moving beyond mapping existing services to creating future ones (Radnor et al., 2014). Service blueprints enable to map both digital and physical interactions in one canvas, hence helping align touchpoints (Ryu et al., 2020). Therefore, service blueprinting enables the design of interactions between users and employees (Shostack, 1984) and transfers feelings and emotive responses (Selby, 2020). Finally, in the 'deliver' phase, it is used to provide a roadmap for service delivery (Stickdorn & Schneider, 2010) and make processes and interactions visible (Shostack, 1984).

User Research

User research is an umbrella term related to seeking knowledge about or from users (Khodakarami & Chan, 2014). What characterizes user research in design is its future orientation and aim to achieve empathy through the research (Brown, 2009). User research is conducted throughout the entire design process, but it varies from the designer participating in the world of the user to the user participating in design activities (Muller & Kuhn, 1993). While at the beginning broad user and problem understanding was the most important aspect, it later becomes crucial to focus the research on receiving feedback on potential solutions.

In the 'discover' phase, the aim is to generate a holistic understanding of the user and of empathy (Schneider & Stickdorn, 2010). Observation is a central technique used to achieve user empathy because it allows the researcher to experience what the user experiences in their own context (Brown & Katz, 2011; Leonard & Rayport, 1997). The aim is to reach a deep understanding of latent user needs, the use context (Fraser, 2009) and user emotions (Brown, 2009). Daae and Boks (2015) distinguish between methods that investigate users' opinions and methods that investigate what users do. In the 'define' phase, the aim is to reach a state of empathy to discover opportunities, translate observations into insights and connect with people (Brown, 2009). The aim is to use these insights to inform and inspire design processes (Van Dijk, 2010). In the 'develop' phase, the aim is to co-create solution ideas with users (Wetter Edman, 2011). This phase is also used to follow up on open questions that come through a deeper understanding of users and problems to identify all problems and needs (Leonard & Rayport, 1997). In the 'deliver' phase, insights are translated into products and services, and feedback on specific design solutions is collected (Brown, 2009). As touchpoints mature, user research changes. For example, when an application is developed and in use, it is possible to deploy analytics to systematically analyse user behaviour.

Empathy in Product–Service Systems Development

Product–service systems require different products and services to work together and, therefore, need to be designed integrally (Marques et al., 2013). Additionally, the digital capabilities of a product as well as the touchpoints of smart product–service systems need to be taken into consideration (Abrell & Durstewitz, 2024; Zheng et al., 2019). The customers' experience often involves interactions with different partners in the value chain, such as retail partners; interactions with front-line personnel for service delivery; different departments within the organization, such as customer care; digital touchpoints, such as applications and websites; and physical products. Table 2.1 provides a summary of the benefits and challenges of empathy.

There are benefits and challenges to the process and context. Assuming that the mirroring hypothesis of modularity theory (Henderson & Clark, 1990; Sanchez & Mahoney, 1996) expands to innovation of product–service systems, the multitude of different touchpoints involved often leads to disrupted customer experiences because there are multiple teams at work. Although each of the elements may be its own development project, which, in turn, has different subprojects, the present article is concerned with achieving empathy during the entire process. Therefore, achieving empathy in such a context is a communication and handover challenge because different teams may be responsible for different aspects of development. Another challenge is the different time horizons of the different development parts because product development often requires more time upfront than does service development (Abrell & Durstewitz, 2024). The different time horizons of products and services are demanding to look far into the future regarding building empathy, requiring the transfer of today's insights into tomorrow's products and services.

TABLE 2.1 *Summary of the benefits and challenges for empathy. The literature was analysed, and the author's practical perspective was added*

	Benefits of empathy derived from the literature	**Challenges for empathy observed in practice**
	Process and context	
Large-scale development effort	–	Mirroring hypothesis (Henderson & Clark, 1990; Sanchez & Mahoney, 1996): Difficult to keep empathy throughout the entire design process which involves multiple development teams. Practical perspective: Communication challenge of user perspective across teams. Handover challenge between teams responsible for certain aspects of development.
Time horizons	–	Different time horizons of products, services, and software (Abrell & Durstewitz, 2024)
Double diamond process	User research as essential part of the double diamond process (Design Council, 2024), entire processes can be adapted to develop empathy (Barnes & Du Preez, 2015)	Although described as iterative (Stickdorn, 2010), in practice a dedicated user research phase is often carried out early in the project.
	Core activities	
Service prototyping	Prototyping makes future situations tangible and, therefore, supports building empathy (Fulton Suri, 2003; Soto et al., 2023; Wright & McCarthy, 2008)	Service prototyping challenges: inconsistency, authenticity, validity, intangibility and time (Blomkvist & Holmlid, 2010). Practical perspective on related empathy challenges: Only a limited number of people take part in such prototyping efforts. The combination of tangible elements (such as screen flows) with intangible elements (such as users' reactions) is difficult to document and communicate. Product–service systems are typically complex; therefore, it is difficult to prototype an entire scenario and communicate it later.

TABLE 2.1 *continued*

	Benefits of empathy derived from the literature	**Challenges for empathy observed in practice**
Service blueprinting	Making the user experiences tangible and observable by others. Blueprint reported to foster empathy (Baranova et al., 2010)	Practical perspective: Complexity of product–service systems may induce the risk of losing focus on the user perspective. Conscious effort is required to take into account emotional aspects.
User research	Immersing into real-world context through user research (Fulton Suri, 2003). Tools to do so including ethnographic approaches, design probes, contextual interviews, among others (Wright & McCarthy, 2008)	Practical perspective: Possibility of losing empathy during the process of converging rich qualitative data.

The double diamond process itself can be ambiguous in terms of empathy. On the one hand, user research is an essential part of the double diamond process (Design Council, 2024), and the entire process can be adapted to develop empathy (Barnes & Du Preez, 2015). However, despite the highlighted importance that the process is iterative (Stickdorn, 2010), projects in practice often have a dedicated research phase up front, and access to users can be limited during development. If there is a handover between teams, the empathy built up by a research team during problem exploration must be communicated to a wider stakeholder audience.

In the following section, the benefits and challenges of the core activities are introduced. Service prototyping makes future situations and the users' point of view tangible and, therefore, supports empathy building (Fulton Suri, 2003; Soto et al., 2023; Wright & McCarthy, 2008). Blomkvist and Holmlid (2010) state that service prototyping is a collaborative effort and is mainly used for learning or communicating, often involving both. Prototyping occurs in any phase of the design process but has two functions: exploring and evaluating. Exploring is described as generating insights and developing thinking while evaluating is about testing and collecting feedback (Blomkvist & Holmlid, 2010). Tools such as enacting the service experience help the design team and stakeholders comprehend complex systems from the users' point of view.

Blomkvist and Holmlid (2010) point out several challenges of service prototyping: inconsistency, authenticity, validity, intangibility, and time.

Inconsistency refers to the human element in delivering a service, while authenticity refers to the artificial situation created in a prototyping setting. Validity refers to whether a simulation can represent the real situation, while intangibility and time refer to the nature of services as being intangible and time-bound (Blomkvist & Holmlid, 2010). These general challenges of service prototyping also address building empathy. Although designers acknowledge the value of a holistic perspective, prototypes are often built for single interfaces (Blomkvist & Holmlid, 2010). If only a part of the service can be prototyped because of complexity, the involved stakeholders who are only present in the prototyping session may fail to empathize with the user. Moreover, typically only a limited number of people take part in such prototyping efforts. The combination of tangible and intangible elements, such as touchpoints, actors and systems, to name a few, is possible for prototypes but difficult to document and communicate. Although certain artefacts, such as physical or digital products, endure, the users' reactions and the interplay of these artefacts are intangible. Additionally, the complexity of product–service systems makes it difficult to prototype such a system as a whole.

Service blueprinting helps make these intangible user actions tangible and observable to others. Like other approaches, service blueprinting relies on visualization to effectively work together with stakeholders (Bitner et al., 2008). Therefore, the blueprint is reported to foster empathy (Baranova et al., 2010).

In practice, one of the greatest challenges of service blueprinting is the complexity of product–service systems. Because of the several dimensions that need to be taken into account, it is often not feasible to develop service blueprints for all possible service encounters, forcing design teams to set priorities. As a consequence, multiple blueprints are being developed for specific challenges. By their nature, service blueprints involve both the customer perspective and provider perspective. This can become very complex, and design teams risk losing focus on the user perspective while exploring the systems, products, procedures and employee interactions that are required to provide a service. Therefore, conscious effort is required to take into account the emotional aspects.

User research is crucial for building empathy because it is about immersing in the real world and understanding users (Brown, 2009; Fulton Suri, 2003). Therefore, user research helps build user empathy by building a rich understanding of the user, context, and challenges for the user, drawing on tools such as ethnographic approaches, design probes, and contextual interviews (Wright & McCarthy, 2008). The holistic approach of user research leads to rich empirical data that helps the design team build empathy.

However, the holistic approach to user research makes it difficult to convey the nuanced understanding built up by the design team to others. Particularly in the converging phases of the double diamond process, design teams must condense their information. Service designers often use visualizations to communicate their information, which helps them maintain empathy throughout the design process (Segelström, 2009). There is, however, a risk when condensing the insights about the user because they may look like information derived by market research studies but without the rigour that is often applied in this field. By condensing user insights, one risks losing the ability to build empathy for users. Therefore, keeping empathy while making condensed user insights accessible to others is a core empathy challenge of user research.

Conclusion

The three core activities of service design – service prototyping, service blueprinting and user research – can lead to building up and maintaining empathy throughout the design process. This article has provided an overview of the aims of these core activities in the double diamond process, discussed the benefits and challenges for empathy in the context of product–service systems development and added a practical perspective. The three core activities should not be seen in isolation but as mutually reinforcing activities throughout the entire design process to achieve better user empathy and design product–service systems that empathetically answer to the needs of users. For example, user research can be dominant in the first phase to understand the users and their context to build up empathy; the results can be mapped in service blueprints throughout the research; and ideas can be prototyped. These early prototypes can, in turn, be used to further co-create with users to better understand their perspective. By using these complementary core activities, the strengths of each can be combined, such as the analytical nature of a service blueprint combined with a service prototype built upon user research and tested with users. However, each design process is different. By understanding the aims of the core activities as well as the benefits and challenges of empathy, design teams can better decide how to tailor their process.

Limitations

The current article has conceptually explored user research, service prototyping, and service blueprinting from the perspective of empathy and related challenges from the practical perspective of the author in the early phase of product–service systems development. As the current article is conceptual in nature, future empirical studies are needed. Although this development perspective sheds light on how to gain empathy and maintain it throughout the design process, the present article has not addressed the empathy perspective of service design for already existing services that are used in the field. In particular, the continuous improvement of customer relationship management systems and the increasing generation of user data and analysis tools that are more sophisticated – as well as generative artificial intelligence – need to be explored further for their potential to deliver empathic services to users.

Takeaway

The early phase of product–service systems development is a challenging environment for empathy: Several teams work on aspects of a product–service system, results need to be handed over to others, and the needs of users well in the future must be considered. Service prototyping, service blueprinting and user research are core activities in this early phase. Therefore, it is important to understand their aims, benefits, and challenges. These activities should not be seen in isolation but rather as mutually reinforcing throughout the entire design process to achieve better user empathy and design product–service systems that empathetically answer to users' needs.

References

Abrell, T., & Durstewitz, M. (2024). Leveraging user-centricity for OEMs through product, software, and service development. In S. West, J. Meierhofer, & T. Buecheler (Eds.), *Smart Services Summit: Building resilience in a changing world*. Springer.

Baines, T. S., Lightfoot, H., Steve, E., Neely, A., Greenough, R., Peppard, J., Roy, R., Shehab, E., Braganza, A., Tiwari, A., Alcock, J., Angus, J., Bastl, M., Cousens, A., Irving, P., Johnson, M., Kingston, J., Lockett, H., Martinez, V., Michele, P., Tranfield, D., Walton, I., & Wilson, H. (2007). State-of-the-art in product-service systems. *Proceedings of the Institution of Mechanical Engineers, Part B: Journal of Engineering Manufacture*, 221(10), 1543–1552.

Baranova, P., Morrison, S., & Mutton, J. (2010). *Service design in higher and further education*. JISC Briefing Paper. https://vwcrmhandbook.pbworks.com/w/file/fetch/52382067/Service%20Design%20in%20Higher%20and%20Further%20Education.pdf (accessed 8 February 2025).

Barnes, V., & Du Preez, V. (2015, September). Mapping empathy and ethics in the design process. In *Design Education Forum of Southern Africa: 7th International DEFSA Conference Proceedings* (pp. 1–11). Design Education Forum of Southern Africa (DEFSA).

Beaudouin-Lafon, M., & Mackay, W. E. (2007). Prototyping tools and techniques. In A. Sears & J. A. Jacko (Eds.), *The human–computer interaction handbook* (pp. 1043–1066). CRC Press.

Bitner, M. J., Ostrom, A. L., & Morgan, F. N. (2008). Service blueprinting: A practical technique for service innovation. *California Management Review*, 50(3), 66–94.

Blomkvist, J., & Holmlid, S. (2010). Service prototyping according to service design practitioners. In *Conference Proceedings, ServDes. 2010, Exchanging Knowledge* (Vol. 2, pp. 1–11). Linköping University Electronic Press.

Bogers, M., & Horst, W. (2014). Collaborative prototyping: Cross-fertilization of knowledge in prototype-driven problem solving. *Journal of Product Innovation Management*, 31(4), 744–764.

Boukhris, A., Fritzsche, A., & Möslein, K. (2017). Cocreation in the early stage of product-service system development. *Procedia CIRP*, 63, 27–32.

Brown, T. (2009). *Change by design*. HarperCollins.

Brown, T., & Katz, B. (2011). Change by design. *Journal of Product Innovation Management*, 28(3), 381–383.

Chuang, P. T. (2007). Combining service blueprint and FMEA for service design. *The Service Industries Journal*, 27(2), 91–104.

Coughlan, P., Suri, J. F., & Canales, K. (2007). Prototypes as (design) tools for behavioral and organizational change: A design-based approach to help organizations change work behaviors. *The Journal of Applied Behavioral Science*, 43(1), 122–134.

Daae, J., & Boks, C. (2015). A classification of user research methods for design for sustainable behaviour. *Journal of Cleaner Production*, 106, 680–689.

Fließ, S., & Kleinaltenkamp, M. (2004). Blueprinting the service company: Managing service processes efficiently. *Journal of Business Research*, 57(4), 392–404.

Floyd, C. (1984). A systematic look at prototyping. In R. Budde, K. Kuhlenkamp, L. Mathiassen, & H. Züllighoven (Eds.), *Approaches to prototyping* (pp. 1–18). Springer. https://doi.org/10.1007/978-3-642-69796-8_1

Fraser, H. M. (2009). Designing business: New models for success. *Design Management Review*, 20(2), 56–65.

Fulton Suri, J. (2003). Empathic design: Informed and inspired by other people's experience. In I. Koskinen, T. Mattelmäki, & K. Battarbee (Eds.), *Empathic design* (pp. 51–58). IT Press.

Gebauer, H., Paiola, M., Saccani, N., & Rapaccini, M. (2021). Digital servitization: Crossing the perspectives of digitization and servitization. *Industrial Marketing Management*, 93, 382–388.

Henderson, R. M., & Clark, K. B. (1990). Architectural innovation: The reconfiguration of existing product technologies and the failure of established firms. *Administrative Science Quarterly*, 35(1), 9–30.

Houde, S., & Hill, C. (1997). What do prototypes prototype? In M. Helander, T. K. Landauer, & P. Prabhu (Eds.), *Handbook of human–computer interaction* (pp. 367–381). Elsevier.

Huelin, R., Iheanacho, I., Payne, K., & Sandman, K. (2015). What's in a name? Systematic and nonsystematic literature reviews, and why the distinction matters. *The Evidence Forum*, May 2015, 34–37.

Khodakarami, F., & Chan, Y. E. (2014). Exploring the role of customer relationship management (CRM) systems in customer knowledge creation. *Information & Management*, 51(1), 27–42.

Koskinen, I., & Battarbee, K. (2003). Introduction to user experience and empathic design. In I. Koskinen, T. Mattelmäki, & K. Battarbee (Eds.), *Empathic design* (pp. 37–50). IT Press.

Kouprie, M., & Sleeswijk Visser, F. (2009). A framework for empathy in design: Stepping into and out of the user's life. *Journal of Engineering Design*, 20(5), 437–448.

Leonard, D., & Rayport, J. F. (1997). Spark innovation through empathic design. *Harvard Business Review*, 75(6), 102–115.

Marques, P., Cunha, P. F., Valente, F., & Leitão, A. (2013). A methodology for product–service systems development. *Procedia CIRP*, 7, 371–376.

Muller, M. J., & Kuhn, S. (1993). Participatory design. *Communications of the ACM*, 36(6), 24–28.

Otto, K., & Wood, K. (2001). *Product design: Techniques in reverse engineering and new product design*. Prentice-Hall.

Porter, M. E., & Heppelmann, J. E. (2014). How smart, connected products are transforming competition. *Harvard Business Review*, 92(11), 64–88.

Potthoff, T., Siemon, D., Wilms, K., Möser, S., Hellmann, M., Stieglitz, S., & Robra-Bissantz, S. (2018). Collaborative service blueprinting for design thinking: Evaluation of a digital prototype. In *Proceedings of the 51st Hawaii International Conference on System Sciences* (pp. 4584–4593).

Radnor, Z., Osborne, S. P., Kinder, T., & Mutton, J. (2014). Operationalizing coproduction in public services delivery: The contribution of service blueprinting. *Public Management Review*, 16(3), 402–423.

Rahman, A., Razek, A., & van Husen, C. (2017, June). Innovation by service prototyping design dimensions and attributes, key design aspects, and toolbox. In *2017 International Conference on Engineering, Technology and Innovation (ICE/ITMC)* (pp. 571–576). IEEE.

Ryu, D. H., Lim, C., & Kim, K. J. (2020). Development of a service blueprint for the online-to-offline integration in service. *Journal of Retailing and Consumer Services*, 54, 101944.

Sanchez, R., & Mahoney, J. T. (1996). Modularity, flexibility, and knowledge management in product and organization design. *Strategic Management Journal*, 17(S2), 63–76.

Segelström, F. (2009, November). Communicating through visualizations: Service designers on visualizing user research. In *DeThinking Design, ReThinking Services–First Nordic Conference on Service Design and Service Innovation* (pp. 175–185). Linköping University Electronic Press.

Selby, A. (2020). From provocation to persuasion: The changing faces of illustration in a service design experience. *CONFIA*, 8(1), 271–284.

Shostack, G. L. (1982). How to design a service. *European Journal of Marketing*, 16(1), 49–63.

Shostack, G. L. (1984). Designing services that deliver. *Harvard Business Review*, 62(1), 133–139.

Soto, M., Miettinen, S., & Alhonsuo, M. (2023). The prototyping service experience as a tool of empathy and research: From Finland to Chile. In C. Cipolla, C. Montálvao, L. Farias, & M. Quaresma (Eds.), *ServDes 2023: Entanglements and Flows: Service Encounters and Meanings: Conference Proceedings* (pp. 440–457). Linköping University Electronic Press. https://doi.org/10.3384/ecp203023

Steen, M. G. D. (2008). *The fragility of human-centred design*. IOS Press.

Stickdorn, M. (2010). The iterative process of service design thinking. In M. Stickdorn & J. Schneider (Eds.), *This is service design thinking* (pp. 124–135). BIS Publishers.

Stickdorn, M., & Schneider, J. (Eds.). (2010). *This is service design thinking*. BIS Publishers.

UK Design Council. (2024). *Framework for innovation*. https://www.designcouncil.org.uk/our-resources/framework-for-innovation/

Van Dijk, G. (2010). Design ethnography: Taking inspiration from everyday life. In M. Stickdorn & J. Schneider (Eds.), *This is service design thinking* (pp. 108–115). BIS Publishers.

Wetter-Edman, K. (2011). *Service design – a conceptualization of an emerging practice*. https://gupea.ub.gu.se/bitstream/handle/2077/26679/gupea_2077_26679_1.pdf

Wright, P., & McCarthy, J. (2008). Empathy and experience in HCI. In *CHI 2008 Proceedings: CHI 2008* (pp. 637–646). ACM Press.

Zheng, P., Wang, Z., Chen, C. H., & Khoo, L. P. (2019). A survey of smart product-service systems: Key aspects, challenges and future perspectives. *Advanced Engineering Informatics*, 42, 100973.

3

Challenges with Empathy: Evolving Service Prototyping at Centria UAS

Johanna Hautamäki

Abstract

Centria University of Applied Sciences (UAS) is a multidisciplinary organization with close connections to working life. We are building our operations with multidisciplinary human interaction, cocreation and increasing empathy in mind. Our service simulation and prototyping environment, the SILMU Service Innovation Laboratory for Multidisciplinary Cocreation, is a development-through-trial-driven process based on the Centria Education and Research, Development and Innovation (RDI) strategy and on the needs of companies in the region. Our approach is that, because reacting to the current state and working-life culture is not sustainable, we need to anticipate future needs and skills for working life. The objective of this chapter is to identify and reflect on the challenges of empathic service design through a practical case: the development trajectory of SILMU operations at Centria. The development of the SILMU is our contribution to developing practical knowledge on empathic service design. The challenges with empathy are identified and reflected upon in this chapter by describing the practical development work and the methods used in evolving the service prototyping at Centria UAS.

The Development Process Started with a Strategy-Driven Approach

Evolving service prototyping and including service design in Centria's operations started in line with strategic development in Centria's education and RDI operations. The bases for UAS operations in Finland are defined by the law: the two main tasks are education and RDI, with RDI including regional development (Universities of Applied Sciences Act, 2014). As a higher education organization, this includes meeting development challenges and renewing the structures of businesses and working life in general. Therefore, merely reacting to the needs of the current environment and working-life culture is not sustainable or according to the mandate.

> *The future does not exist in the present, but anticipation does. The form the future takes in the present is anticipation.*
>
> (Miller, 2018, p. 27)

The SILMU service simulation environment was built as one of the labs at Centria to serve the needs of both education and RDI. Initially, the development of the SILMU responded to Centria's strategy in 2016 (Centria, 2012) in two ways. Firstly, by being customer- and student-oriented as well as cocreative with stakeholders. Secondly, by anticipating the skills needed in the working life to support regional competitiveness. The UAS must form knowledge of the future needs for skills in working life, which we must be able to meet with education and research as well as with continuous learning. Based on this knowledge, it is possible to act in an anticipatory way. This strategic development also requires cooperation between RDI and education operations and the ability to perform multidisciplinary work. Increasing empathy between actors and stakeholders as part of communication is a way to increase and deepen joint understanding and make personal connections to the user perspective (Kouprie & Sleeswijk Visser, 2009, p. 439). Empathic methods and service prototyping are ways to communicate in multidisciplinary teams and learn more about user needs and everyday life. The purpose is to create empathy not only for understanding user needs, but also for achieving a holistic view; empathy for different stakeholders' perspectives can be viewed and experienced in the same way. Visualizing and storifying different perspectives as well as experiencing them firsthand with empathic bodily methods, increases the understanding between stakeholders and provides a shared language to communicate, hence lowering silos.

Although Centria has no degree programme generally on design, service design is seen as part of the development of RDI operation skills and in studies of different degree programmes, such as engineering, entrepreneurship and social and health care. Service design is both a method for learning and a development approach for use in working life. Service design and empathic methods provide tools for multidisciplinary communication and increasing empathy among stakeholders (Kouprie & Sleeswijk Visser, 2009, p. 439). The SILMU service simulation and prototyping environment is a home base for building this competence as a multidisciplinary development-through-trial-driven process. The environment consists of physical equipment and a set of methods. In this context, creating empathy towards someone's everyday life is seen as a basis for learning, communicating or designing. Empathy is a way to transmit culture, share needs and build trust while facilitating communication, growth and change (Keefe, 1976, p. 10). Empathy methods are ways to facilitate empathy and to learn how to empathize; they provide a way to immerse oneself in a specific situation, cultural approach or experience. We can never fully understand or replicate the feelings or experiences of another person, but with these methods, we can get closer to being able to empathize. Empathy experiments are situations in which one or more empathy methods are applied to action to learn in real-life situations (Figure 3.1).

The goal of developing SILMU 10 years ago was to create a prototyping environment for services for both RDI and education activity needs. We started to build an experiential learning environment in cooperation with the University of Lapland Service Innovation Corner (SINCO) Lab. The SINCO lab is a service simulation environment and service design of university RDI activities. SINCO was a central partner in developing the technical side of SILMU at the beginning and in cooperation while starting at Centria with SILMU. The first cases were based on Centria's strategy for increasing lifelong learning as well as education and RDI integration (Figure 3.2).

The starting point for development work was to map out what types of simulations are used in education at Centria in different degree programmes and what needs there are for the future. The information was gathered

FIGURE 3.1 *Makerspace empathy experiment. A Centria HealthLab RDI expert empathizes with elderly persons' everyday life situations with a set of equipment designed to simulate limitations in sensory and mobility impairments. Image by Johanna Hautamäki, original video by Anttoni Porri.*

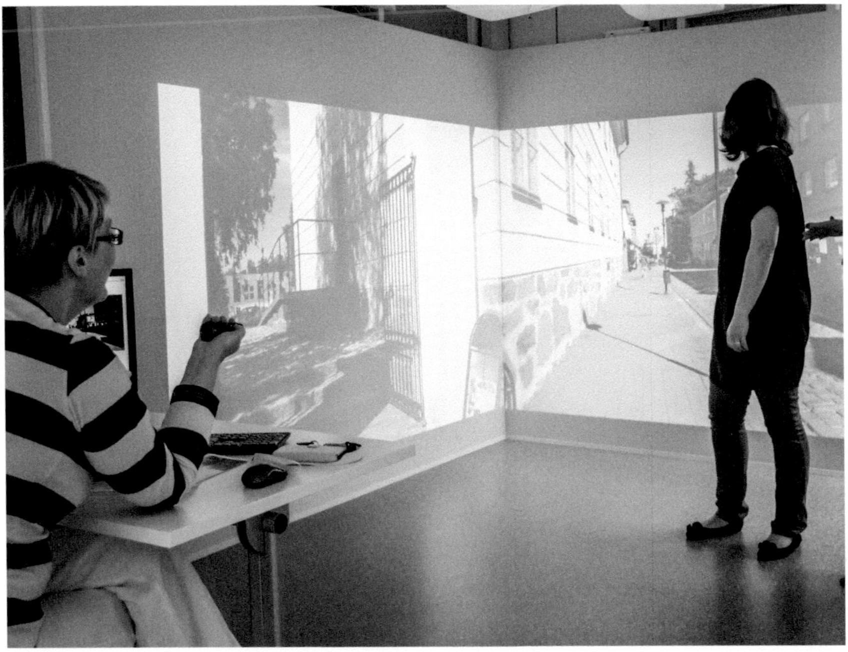

FIGURE 3.2 *The starting point for developing the SILMU in 2015 was an ordinary classroom in which simulation equipment was built. A view of a simulation in that room involving developing the pedestrian zone experience in the City of Kokkola. Image by Johanna Hautamäki.*

using a staff survey. The results showed that simulation was already used in various ways, such as simulating computer programs, equipment operation and control, gamified business operations, patient care situations and virtual models for testing the strength and functionality of machines (Hautamäki et al., 2020, p. 21).

The simulation activities most similar to the needs of the SILMU at Centria were those in the nursing education simulation. The first case involved cooperation between an RDI project and a nursing study course. This made it possible for a development-through-trial with the nursing degree programme simulation-based learning. This work started with digitally implemented visual simulations and storytelling for nursing students to develop a simulation method for SILMU. In some cases, we also added drama methods for interaction points with the patient. In these cases, we adapted elements from nursing simulations and aimed to increase students' empathy for the patient experience throughout the care path. The created service prototypes represented the care path and were used as a method for the students to visualize and reflect on the theoretical data, making a logically chosen path of all possible ways to treat a specific type of patient. The different methods of simulation, visualization and prototyping serve

FIGURE 3.3 *A simulation of an elderly person's home at SILMU. In this setup, nursing students were working with a case on elderly care home visits in the role of nurses, while the teacher was taking the role of the patient. Image by Johanna Hautamäki.*

as ways to foster active learning and multidisciplinary ways of working, increasing a shared understanding among participants. The same approach was subsequently applied to entrepreneurship education with the customer-oriented development of services.

With these development-through-trial cases in nursing and entrepreneurship study courses, we were able to construct cases together with students and teachers, and as a result, the development-through-trial simulation process evolved (Figure 3.3). An important part of this development process was always the surveys or reflective discussions with the participants after the workshops; alongside the observations during the workshops, it provided crucial information for development. Based on this work, we adopted three phases for the simulation in SILMU:

1. Defining the starting point: Script/document of the starting situation for a simulation
2. The simulation: Roles for participants as active participants or story tellers/observers
3. Closing phase: A reflective learning discussion.

Based on the observations and feedback from the participants, visualizing the fact-based care path, including story elements, provides patient care training in a context and more of a real-life situation to empathize with the patient experience. This way of telling the patient's journey as a visual story from the patient's perspective or using drama methods to create the actual care situation allowed the students to empathize with the patient experience on a deeper level than merely doing the procedure as a training exercise. Visualizing the care path also helped students position their tasks on the whole care path and provided them with an overall picture of patient care. Based on the reflective discussions with participating students and teachers, the process of creating and going over simulated patient journeys, including visual and story elements, deepened the learning because students needed to form a fluid patient journey and make consistent choices on the individual care process. The goal is not only to learn the theory but also to connect it to a specific type of patient case and care process. This development-through-trial led us to improve our storytelling and simulation methods. Bringing the deeper background story to the everyday level or the patient's life created more empathy than just simulating the care process. This method of simulation learning included more empathy-creating elements than regularly used simulation methods.

There is also a recognized need for more customer-oriented thinking and empathic methods among IT and business students at Centria. In the development-through-trial process, creating profiles and mapping out customer journeys were used as an introduction to the service design approach and as an exercise on being more user centric or empathic (Figure 3.4). These were both used as quick methods to try out or as a starting point

FIGURE 3.4 *One of the first methods used with students was a profile in different forms. The picture is a screenshot of a more rarely used profile form in a workshop, a video profile presenting an elderly woman. Image and original video by Johanna Hautamäki.*

for a larger project. The larger projects consisted of cocreative workshops or making observations in an authentic environment and creating visual service prototypes in the SILMU based on them.

For example, in larger projects, IT and nursing students participated in piloting multidisciplinary cooperation. The task was to create a service prototype for an app that would gather blood pressure measuring data via Bluetooth from the measuring device. The main part of the cooperation was performed in the form of a workshop in which students in mixed groups created personas and a visual service prototype in the SILMU about the nurses' experiences with using the app. In the workshop, nursing students took the role of users, and IT students took the role of app developers. Both gained experience in multidisciplinary work and empathic service design methods. The nurses learned how to communicate with tech experts in their work, and the IT students learned how to be more user-driven and start requirement engineering in a user-centred way.

Based on these experiences, empathic and prototyping methods such as these activate participation, and the participants become excited and engaged easily in doing this. On the other hand, this type of experience poses the danger of remaining at a superficial level of assumptions and not seeing the difference between presuppositions and facts. Deeper engagement requires more time and effort and it is more difficult to incorporate the deeper engagement into study courses than just individual workshops.

The approach to the simulation method with SILMU evolved based on these development-through-trial cases, and this can be described as imitating and creating everyday life situations for learning and development purposes. To achieve this goal, we used various methods for storytelling and prototyping to create empathy and experiences. The SILMU lab itself consists of both virtual and physical elements, making rapid prototyping and experiences for small groups possible (Hautamäki et al., 2020, pp. 9, 21–22, 54).

Development Based on Working-Life Needs, Regional Development

After starting operations with internal development cases, the next phase focused on cases with companies based on their actual needs or problems. These two phases of development overlapped, and strategy-based development continued while working-life cases started. Based on the Centria's new strategy of 2018–22 (Centria, 2018), the aim was to expand the work for regional development and lifelong learning outside degree programmes and into working-life situations. The long-term goal has been based on lifelong learning and inclusion within companies, supporting people in working life and anticipating which skills will be needed in the future. In this way, it would support the region's small- and medium-sized companies' (SMEs') competitiveness and access to a skilled workforce. Activating the staff in the development work and increasing empathy within the organizations towards the customers and stakeholders is one way to support this skill development and competitiveness.

The perspective of development work in SILMU with SMEs has evolved since 2016 from an individual service perspective to an organizational culture and strategic-level development perspective. Empathic service design has been at its core, especially in the cases presented in this chapter. During 2016–19, in the European Social Fund (ESF)–funded project 'AppSim–Creating competence through applied simulation pedagogy, digitality and service design', SILMU lab and the methods were developed in an intensive period with five working-life partners. In this project, the cases based on SMEs helped develop the methods and cocreation models further in a working-life-driven way with a living lab model. The core of the model is operating in an open innovation ecosystem in real-life situations to work in a development-through-trial and cocreative way (Heikkanen & Österbeg, 2012, pp. 6–11). The development-through-trial process emphasizes interaction and learning from the process (Antikainen et al., 2019, p. 6). The benefit of this type of development work is twofold. It is included in the living lab operational model that all stakeholders will bring and share, and develop knowledge and resources together (Leminen, 2015, p. 147). The working-life partners provide a real-life testbed with their operations and services as well as with knowledge of current and upcoming needs, and while participating in this cooperation, they can increase their skills and develop their operations (Hautamäki et al., 2020, p. 6).

Specifically, one of the SMEs in the Living Lab Cooperation, a private sector elderly care home, had the goal of developing an active employee-oriented innovation organizational culture. The company wanted to develop an active organizational culture in which the staff wanted to openly develop their own work and was encouraged to do so. A case example of the empathy methods used in action is a staff orientation day, where empathy methods were used for learning to use safety equipment and to empathize in a care activity situation from the residents' view. The exercise consisted of two parts, where, in both parts, the staff was putting themselves in the role of the resident. In the first part, they learned how to use safety equipment – a fire mattress – and in the second part, they performed a care activity involving lifting a person with a patient lift from a bed to a wheelchair. One of the goals of these exercises was to build confidence so that, in cases of emergency, staff members could take action. The other goal was, by empathizing with the patients' experience in the care situation, to improve work techniques and develop overall conditions in the nursing home.

At the beginning of the case, one of the main challenges was to find the best way to encourage staff to immerse themselves in empathy experiments and to create a safe space to participate with ease. To lower the threshold, we decided to shoot films showing how empathy experiments work; these films featured the head nurses themselves. The aim was to show the team spirit

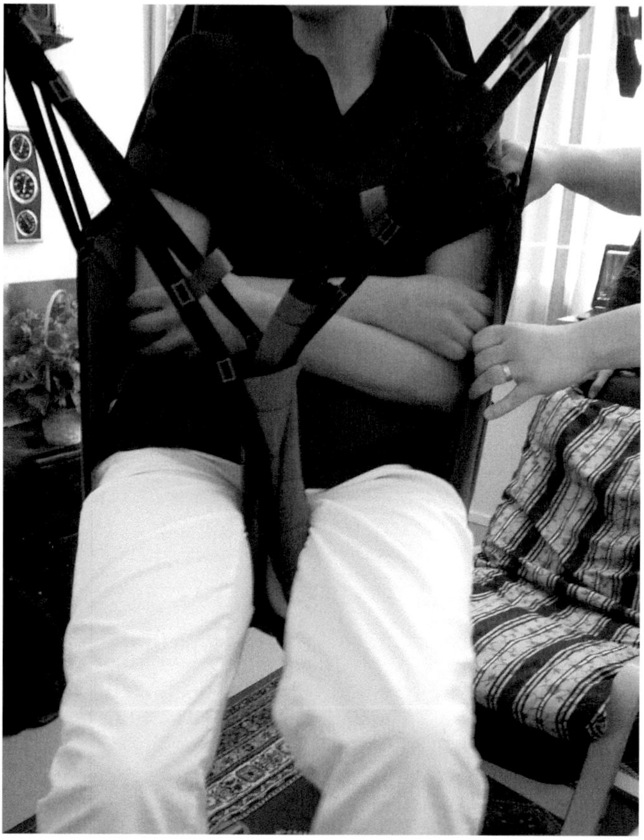

FIGURE 3.5 *One of the elderly care home nurses participated in the empathy experiment, in which she was lifted from a bed to a wheelchair with a patient lift. In this situation, she empathizes with the patient's emotions and physical experience of being lifted with the equipment and, through this, develops working methods. Image by Johanna Hautamäki.*

and create an encouraging atmosphere for the situation when participants watched them at the beginning of the experiment.

This case illustrates how empathy methods are a working tool for developing an organizational culture through experience and increasing the active attitude towards the work. Based on the reflective discussions with feedback from the participants, head nurses and CEO, the inspirational videos worked well to encourage participation and create a safe space for immersion in the situation. Participation helped create ideas on how to improve the work and showed why some procedures need to be handled in a certain way, for example, when making the bed with a safety mattress. Participation in a resident's role provides firsthand insights into how the residents can feel during a care procedure, and based on this, one can improve their own actions accordingly (Figure 3.5). The CEO of the nursing home decided, based on experience and feedback, that the empathy experiments would be taken as part of the orientation of new employees to foster the right kind of organizational culture and to learn in action. In conclusion, the importance of creating safe and encouraging spaces for experiments cannot be emphasized enough. It was a key to success. The experiments ended by having a positive effect on the organizational culture and the common spirit of the workplace.

Makerspace: Make Ideas and Experiences into Reality!

By renewing space and mapping out what could lay ahead for us, SILMU is now in a new starting point. Centria's new strategy 2025–8 (Centria, 2024) – reorganizing spaces by combining SILMU in a multidisciplinary way with other activities at Centria – is bringing a brand new start with new possibilities for growth. The new multidisciplinary working space is called Centria Makerspace. It will combine multiple operations: IT education and RDI; business and entrepreneurship education and RDI; Centria HealthLab research group; future working-life research group; and SME development services. This will create an interesting combination with synergies for new cooperations. With this new operational model, we are still focusing on

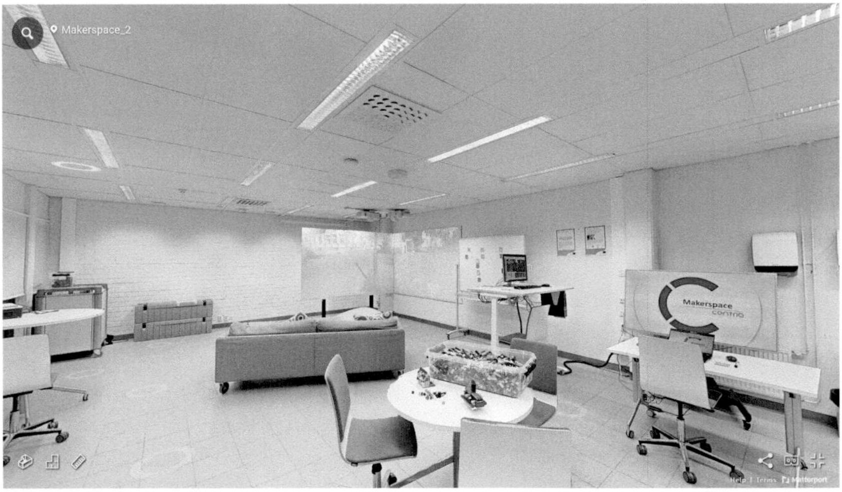

FIGURE 3.6 *The new Makerspace is now in action. Simulation activities and other prototyping methods consist of both physical equipment and sets of methods. A screenshot of a virtual model of the space. Model by Anttoni Porri.*

multidisciplinary work and increasing our understanding of empathy and prototyping. Our future service prototyping and use of empathic methods will be based on the experience we have gained in our previous work. We are continuing the development-through-trial process, together with education and working life, using the living lab model. Networks of cooperation, both nationally and internationally, are bringing interesting new possibilities. The profile of the new Centria Makerspace includes prototyping experiences and services in different ways, both physically and virtually (Figure 3.6). The Makerspace is still going to be a home base for creating knowledge and increasing skills.

Conclusion

In these cases, the development-through-trial process has proven to be a good basis for developing methods, operating models and skill levels in the Makerspace context. Identifying challenges and trying out solutions are part of this process. The first general challenges we aimed to overcome were the silos between the disciplines, degree programmes and education and working life. Empathic methods, as well as increased communication and techniques for communicating findings and ideas, have proven to be effective.

The first identified main challenge with empathy methods is that there is a clear gap in the working process between the first experiences of empathic knowledge or experience and turning it into actual design solutions, a change in a service or in a behaviour. Creating empathy is only the first step, and actions or changes in behaviour need to follow this. In the cases we have worked on with SILMU, we have learned that it is easy to perform empathy experiments and use empathic methods for raising awareness or interest, but the challenge can lie in putting them into action. Creating a deeper understanding by using methods, as well as empathic experience and understanding, takes time, an open attitude and willingness; not taking enough time for the process can be seen as one of the barriers to empathy (Kouprie & Sleeswijk Visser, 2009, p. 447). Cases such as IT and nursing workshops, as well as the introduction of service design methods to students from different degree programmes, have raised this challenge. In the workshops, participants create profiles and customer journeys from their own perspectives and learn to use the method. This is a self-centred approach and only the first step towards being user centred or creating actual data-based knowledge. It is not enough to increase empathy merely lightly with these types of workshops, although they can be the first steps. Achieving deeper empathy and a user-centric approach require more time and effort.

The second identified main challenge within the cases is identifying the presuppositions people using the empathy methods have and separating them from actual knowledge or facts. In the very early stages of our SILMU development process, we encountered the challenge that participants could not always distinguish between assumptions and facts about the users. In these described cases, profiles and journey mapping are used in the first working phases for visualizing and sharing knowledge and creating a shared understanding or new ideas within participants. In these situations, the information that is shared is based on personal experience and practical knowledge. As in the IT and nursing students' workshop, the nursing students provided information about the user profile and the context in which the app would be used based on their own experiences. This is also often the case with SMEs; the first workshops are based on their existing expertise. The next workshops then aim to include either users or research data. Presuppositions can include earlier research, pre-existing theoretical structures, personal knowledge, experience or beliefs (Ashworth & Lucas, 2000, p. 298). There is a need to create self-awareness of presuppositions and positioning among the participants individually and at a more general level as a group or organization. A person's own perspective can distort or mislead the interpretation with empathic perceptions or experiences. There is at least a need for a reflective discussion to follow the empathic methods being used to compare experiences and created impact. We argue that, to rely more on the empathy method results, there needs to be interaction and the pursuit of a less self-centred perspective, with transparency on the presuppositions and source of information. This, combined with multifaceted constructions, such as emotional interest, sensitivity, self-awareness, personal experience, a variety of shared perspectives (Smeenk et al., 2018, p. 18) and transferable information, provides more reliable information and a deeper understanding. We identify a need for a tool or a template to be used for identifying and discussing the presuppositions and perspectives used alongside the empathy methods.

Using empathy methods is an effective way to help participants become closer to their users and gain a new perspective. All efforts in this direction are still in development for participants; identifying the context and extent of this development is the key. In a setting like SILMU, where we are planning the working process, facilitating workshops and using empathic methods, in this sense we are responsible for the quality of the knowledge and skills created. There lies a responsibility to have participants reflect on the experience and gained knowledge and set it in a real-life context. Therefore, we continue to challenge participants to move past the assumptions to the actual data and experiential level.

Takeaway

Creating empathy is the first step, and actions or changes in behaviour need to be followed. Two key challenges identified with the cases in this chapter are taking the time and effort to form a deep enough empathic knowledge and to identify the participants' presuppositions in the situation. At their best, empathic methods are a way to bridge silos between disciplines and increase cultural understanding and communication.

References

Antikainen, R., Kangas, H-L., Alhola, K., Stenvall, J., Leponiemi, U., Pekkola, E., Rannisto, P.-H., & Poskela, J. (2019). *Kokeilukulttuuri Suomessa – nykytilanne ja kehittämistarpeet*. Finnish government report and research publication series. https://julkaisut.valtioneuvosto.fi/bitstream/handle/10024/161281/2-2019-KOKSU_raportti_.pdf

Ashworth, P., & Lucas, U. (2000). Achieving empathy and engagement: A practical approach to the design, conduct and reporting of phenomenographic research. *Studies in Higher Education*, 25(3), 295–308. https://doi.org/10.1080/713696153

Centria University of Applied Sciences. (2012). *Strategy 2016*. Centria University of Applied Sciences.

Centria University of Applied Sciences. (2018). *Centria University of Applied Sciences strategy 2018–2022*. Centria University of Applied Sciences.

Centria University of Applied Sciences. (2024). *Centria University of Applied Sciences strategy 2025–2028*. Centria University of Applied Sciences.

Hautamäki, J., Mäkitalo, T. & Hautala, M. (2020). *Tulevaisuuden työelämää varten*. Centria University of Applied Sciences. http://urn.fi/URN:NBN:fi-fe2020052739352

Heikkanen, S., & Österberg, M. (2012). Mikä on living lab? In S. Heikkanen & M. Österberg (Eds.), *Living Lab ammattikorkeakoulussa* (pp. 6–11). HAAGA-HELIA University of Applied Sciences. https://urn.fi/URN:ISBN:978-952-6619-17-0

Keefe, M. (1976). Empathy: The critical skill. *Social Work*, 21(1), 10–14. https://doi.org/10.1093/sw/21.1.2

Kouprie, M., & Sleeswijk Visser, F. (2009). A framework for empathy in design: Stepping into and out of the user's life. *Journal of Engineering Design*, 20(5), 437–448. https://doi.org/10.1080/09544820902875033

Leminen, S. (2015). *Living labs as open innovation networks – Networks, roles and innovation outcomes*. Aalto University Publications Series. https://urn.fi/URN:ISBN:978-952-60-6375-1

Miller, R. (2018). Futures literacy: Transforming the future. In R. Miller (Eds.), *Transforming the future anticipation in the 21st century* (pp. 1–12). UNESCO. https://unesdoc.unesco.org/ark:/48223/pf0000264644

Smeenk, W., Sturm, J., Terken, J., & Eggen, B. (2018). A systematic validation of the Empathic Handover approach guided by five factors that foster empathy in design. *CoDesign*, 15, 1–21. https://doi.org/10.1080/15710882.2018.1484490

Universities of Applied Sciences Act. 932/2014, § 4. https://www.finlex.fi/fi/laki/alkup/2014/20140932

4

Design for Longevity: An Empathic Service Design Framework

Sheng-Hung Lee

Abstract

To create a longevity service that enhances quality of life, service designers need to understand people and intentionally practice empathy when developing experiences. Thus, this research explored the following question: 'How can we develop an empathic longevity service from prototype to eventual experience?' To address this question, a controlled experiment was designed that incorporated tech- and touch-based service encounters. The qualitative descriptive study consisted of surveys, in-field experiments and semistructured interviews. The quantitative analysis covered six descriptive statistics (DS) criteria and a correlation coefficient (r). The study proposes a conceptual longevity service model that emphasizes empathy-building techniques by using tangible artefacts (Longevity Planning Blocks). The purpose is to enable researchers, designers and educators across industries and academia to envision an empathic longevity service that is respectful, meaningful and enjoyable.

Introduction

Most countries have entered an era of ageing populations, and some have even entered an era of superageing populations (WHO, 2022). Maintaining a high quality of life for individuals who are living beyond 100 years of age requires comprehensive, strategic longevity planning and literacy that covers not only people's financial, physical and mental health but also integral

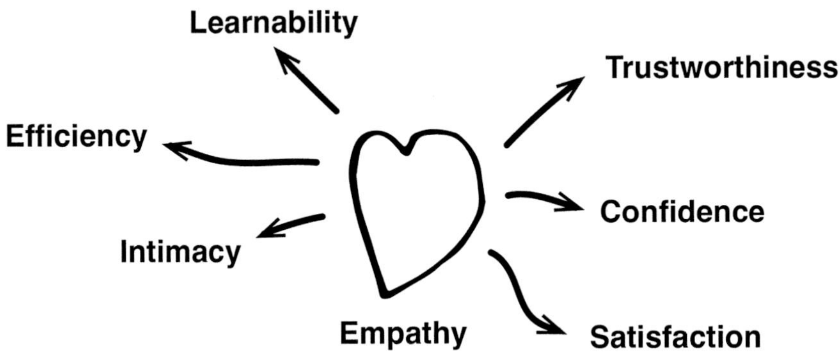

FIGURE 4.1 *In this preliminary study, we measured empathy through six design attributes: learnability, efficiency, intimacy, trustworthiness, confidence and satisfaction.*

measures of well-being: mobility, community, family, risk, investment, education and future outlook.

The Design for Longevity (D4L) study aims to develop a longevity coaching service – or a sustainable, holistic planning model that extends beyond traditional financial planning (Lee, 2023a; Lee, 2023b) – to empower individuals to better prepare for their future longevity (Houston, 2014; Mauterstock Jr., 2020). The study applied and analysed pilot test results obtained from 30–45-minute experiments with twelve participants in Boston. The objective was to address the research question: 'How can we develop an empathic longevity service from prototype to eventual experience?' (Lee, 2023; Lee, Coughlin, Yang, et al., 2023; Mager, 2009). For this experiment, the author proposed two distinct service encounters: tech- and touch-based groups (Dolata et al., 2019; Giebelhausen et al., 2014; Wu et al., 2022). Six design attributes – learnability, efficiency, intimacy, trustworthiness, confidence and satisfaction (Nielsen, 2010) – were identified as figures of merit to gauge the overall empathic quality of the longevity coaching service prototype (Figure 4.1).

The present study employed both qualitative and quantitative methods (Friese, 2014). Two research hypotheses were formulated: (1) One can use six design attributes – learnability, efficiency, intimacy, trustworthiness, confidence and satisfaction – to ascertain whether an experience is empathic; (2) The current study used these design attributes to determine whether our longevity coaching service prototype was empathic.

The researchers built twelve Longevity Planning Blocks (LPBs). To build these design artefacts (Turkle, 2007), the researchers used an 8,000-day framework that divides retirement into four distinct stages (Coughlin, 2019a, 2019b). Each intentionally designed tangible artefact, or LPB, features eight thought-provoking questions (Lee, Coughlin, Balmuth, et al., 2023) to empower participants and facilitators to empathically navigate sensitive and personal longevity-related discussions (Lee, Yang, et al., 2023).

For this study, LPBs were used to facilitate participants' expression of ideas and abstract concepts and, as a service prototyping tool, to explore and quantitatively assess the six design attributes of the empathic quality of longevity services.

Literature Review

Empathy involves putting ourselves in the shoes of others to ensure that we meet people's needs and expectations with respect. Oxford Reference (2024) defines empathy as 'the ability to imagine and understand the thoughts, perspectives and emotions of another person'. Empathy is also recognized as a particular kind of imagination (Suri, 2005). Suri (2001) explained that empathy resembles a creative balance intertwined with subjective imagination stemming from detailed life observation. Because emotion is critical for human connection (Collins, 2019), the ability to understand emotion – that is, empathy – plays a vital role in maintaining the functioning of society.

Scholars such as Batson (2009) and Davis (1996) have described empathy as a dual process of establishing an understanding of the target audience, coupled with a motivation to act on that understanding. In ergonomics and product design, Suri's (2003) concept of empathic design aims to

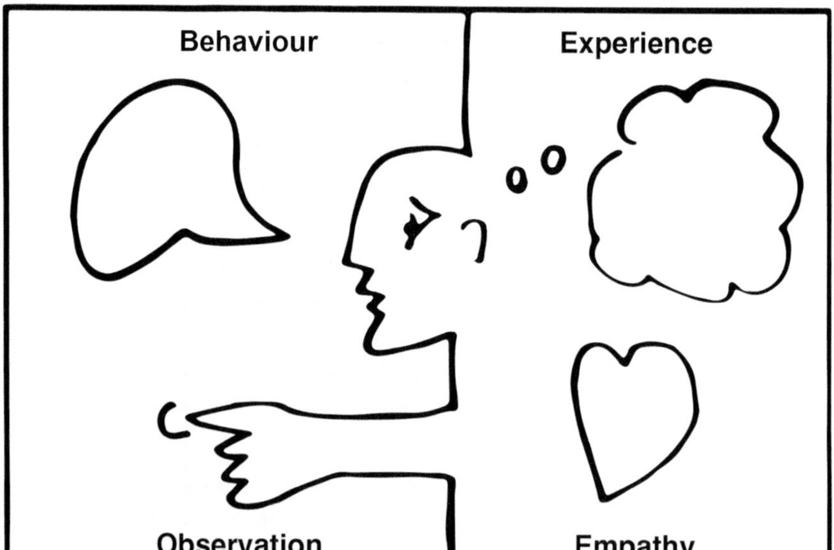

FIGURE 4.2 *The interplay of empathy with behaviour, experience and observation: 'We comprehend people's thoughts and emotions by empathizing with their actions and words' (Suri, 2001, p. 1283).*

utilize our knowledge to inform and inspire the creation of practical and enjoyable things (Figure 4.2). Suri (2001) proposes three primary potentials for empathy in ergonomics: (1) enabling us to gain a deeper understanding of target users in specific contexts; (2) strategically assisting in developing effective connections with key stakeholders, sponsors and colleagues involved in implementation; and (3) enhancing the communication channel for ergonomics-related issues by fostering empathy with users. The process of studying empathy encompasses establishing rapport and deriving meaning from subjective observations, leading to a better grasp of the complexity and sensitivity of human experiences. This approach can empower researchers to gain holistic insights (Suri, 2001) to enrich the comprehensively embodied understanding of people's concerns (Suri, 2003).

The concept of applied empathy has long been utilized as a research tool in the field of ergonomics and product design. However, we consider the significant shift from focusing on product-centred tangible design features to prioritizing service- and experience-driven economics. This transition underscores the need to adapt and expand this potent approach and mindset to meet the demands of a changing design landscape. It is a paradigm shift that takes into account multiple layers of purpose, people, products, processes and platforms. Leveraging empathy as a research tool and a mindset requires not only observation, but also the ability to predict the feelings and thoughts of others, encompassing their mental status, emotions, values, motivations, personal preferences, goals and inner conflicts (Suri, 2003). Academics and industries have shown increased interest in exploring how service prototyping can apply empathy and enhance research based on collaborative practice to improve people's experience (Porcini, 2023; Soto et al., 2023).

Empathy is commonly used both as a research tool and a mindset, typically positioned at the often ambiguous start of the design process (Kouprie & Visser, 2009; Leonard & Rayport, 1997). To better simulate and realize early-stage ideas, designers conduct rapid prototyping to test not only the features and mechanisms of products, but also the flows and qualities of services and experiences (Lee, Patskanick, et al., 2023). Generally, prototypes serve as useful instruments for thinking throughout design phases (Brown, 2019), with the prototyping phase allowing designers to learn and implement transformative changes (Kuure et al., 2014). When designing for empathy in complex systems within organizational or governmental platforms, scholars have underscored the significance of intellectual empathy, the empathizing process and empathic design tools (Sustar & Mattelmäki, 2017). These elements play an integral role in deliberately situating empathy to reframe, understand and solve challenges while balancing business objectives with human values. Furthermore, Tan et al. (2019) demonstrate that empathy is equally important in cocreated service experiences. Vink and Oertzen (2018) explore the theme of lived experience as an approach to cocreation, examining its relationship with the interdependent nature of empathy.

Empathy can be viewed as an individual's intuitive ability to identify and connect with other people's inner states based upon meticulous observation of their outwards verbal and behavioural expressions (Suri, 2003). Suri (2001) has embraced empathy as a partial and legitimate tool. However, not all applications of empathy are flawless. Empathy carries the potential for drawbacks. Researchers' and designers' interpretations of target users' reactions and interactions may be biased regarding cultural differences, varying knowledge levels, education backgrounds, expertise and life and work experiences, or they could mislead or reach incorrect conclusions (Suri, 2001). Excessive empathy for the emotions of target users might cause researchers and designers to lose sight of holistic project objectives and missions. A striking balance between empathy and objective observation is integral, and achieving this equilibrium relies on the experiences of researchers and designers (Suri, 2003). Further studies could delve into the nature of empathic development throughout the design and creative process, investigating ideal approaches for integrating empathy as a tool, mindset and action throughout the design, constraints and alignment of empathically generated design results with user satisfaction (Hess & Fila, 2016).

Methods

The controlled longevity coaching service experiment was designed to incorporate two distinct service encounters: a tech-based service and a touch-based service. The aim was to explore the following research question: 'How can we develop an empathic longevity service, from prototype to eventual experience?' We aimed to address this research question experimentally by utilizing six design attributes under the two research hypotheses. Evidence-driven research approaches were employed, consisting of a qualitative descriptive study, including an initial survey, a tech- or touch-based service experiment and a semistructured interview (Figure 4.3). The qualitative analysis was complemented by numerical analysis, covering six descriptive

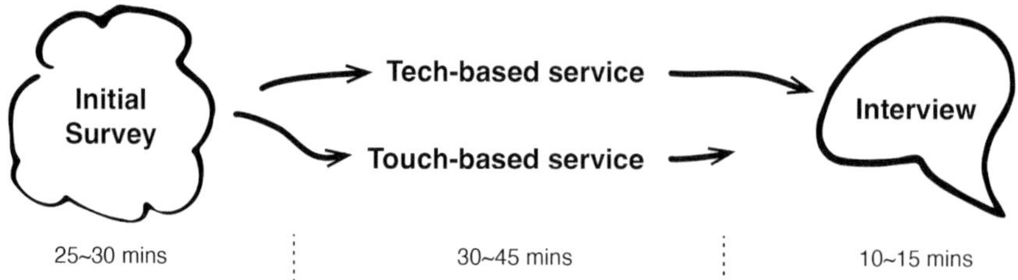

FIGURE 4.3 *Overview of the service prototyping research flow.*

statistics (DS) criteria and correlation coefficients (*r*) to evaluate the research question and hypotheses.

The initial (pre-experiment) survey aimed to determine participants' financial conditions, longevity literacy, experimental expectations, demographics and other concerns. This enabled the researchers to make informed decisions about potential adjustments to the service prototype. Seven participants joined the tech-based group, with the interaction self-navigated without any in-person facilitator. The five-person touch-based group was guided by an in-person facilitator. The groups were provided with the same longevity coaching information, props, experience flow and structure for analysis.

Experiment and Interview

The longevity coaching service experiment was designed for a 30–45-minute immersive experience (Figure 4.4) consisting of five actions: an experiment briefing, an LPB introduction, a four-stage retirement introduction, an exercise and a wrap-up. In this prototype, two service encounters – tech based and touch based – were tested as approaches to gauge participants' perceptions regarding complexity, privacy and personal information and whether to engage with in-person guidance while learning the concept of longevity planning. In the tech-based group, a participant was provided with a controller to self-navigate

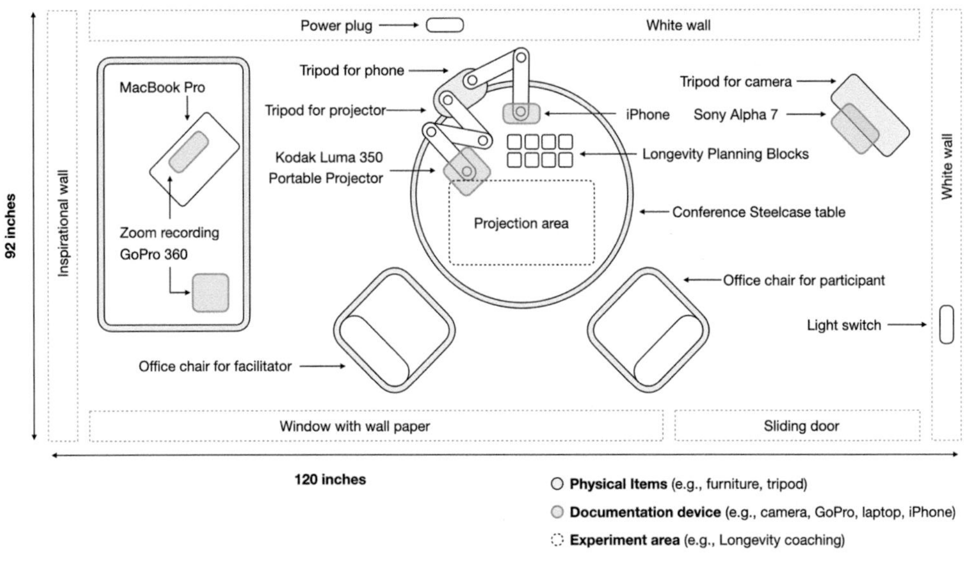

FIGURE 4.4 *Floorplan of the experiment space with documentation devices associated with the zones.*

FIGURE 4.5 *The setup of the tech-based service experiment with twelve LPBs at the MIT AgeLab.*

FIGURE 4.6 *A participant in the tech-based service group controlled the pace of the longevity coaching service.*

the longevity coaching service without any in-person facilitator support (Figure 4.5). In the touch-based group, a participant was accompanied by a facilitator throughout the experiment.

Empathy, trust and reliability were established through either a robust, secure automated longevity system or in-person professionals who had a wealth of knowledge. This controlled experiment was intended to observe and gain insights into how participants interact with the concept of a longevity coaching service, particularly when discussing sensitive topics and sharing intimate data. In addition to the survey and the experiment, the study incorporated 10-minute semistructured interviews that delved into participants' intimate stories, pain points, ideas and reflections that could not be captured through other qualitative or quantitative research methods.

The study offered a range of formats and adaptable touchpoints to gather participants' verbal, behavioural and emotional information and insights. By providing Post-it® notes, we prepared participants for open-ended questions and encouraged them to express themselves through words and drawings as a means of taking notes. The post-experiment interview aimed to create a safe, inviting space for participants to discuss their impressions, suggestions for this service experience and stories immediately after the experiments. In the tech-based group, the participants controlled the pace of content consumption and could always backtrack to review content and videos on the digital canvas (Figure 4.6). The touch-based group also had a facilitator periodically check in with participants to ensure smooth progression through the content on the digital canvas.

The author consulted with three financial advisers from Edward Jones Investments and Raymond James to gain insights into the current landscape of financial planning services, customers' pain points, advisers' limitations and organizational challenges that required transformative solutions. The experiment flow was planned through the lens of human-centred design and empathy. Terminology was employed carefully using 'longevity planning' rather than 'financial advising' because the experiment was intended to help prioritize participants' needs over selling financial products. The ambiance of a dimly lit, closed and quiet environment was intentionally created to establish a safe, confidential physical and mental space where participants felt comfortable sharing their intimate stories.

Descriptive Statistics Criteria and Correlation Coefficient Analysis

The study applied DS to preliminarily quantify the research results from participants (P). The DS describes the characteristics and content of the data samples. In this section, we gathered and analysed ten valid responses from twelve participants, except for data from P3 and P10. The survey time data for P3 and P10 suggest that they likely did not complete the survey. The five

TABLE 4.1 *The five datasets were used to measure the quality of the longevity coaching service*

Dataset	Figure of merit
D1	$\#\,questions = \dfrac{Total\ number\ of\ questions\ asked\ by\ a\ participant}{Longevity\ coaching\ service\ experiment}$
D2	$\#\ touching\ LPBs = \dfrac{Total\ number\ of\ times\ LPBs\ touched\ by\ a\ participant}{Longevity\ coaching\ service\ experiment}$
D3	$\#\ \text{using Post-it}^{\circledR} = \dfrac{Total\ number\ of\ Post\text{-}it\ notes\ used\ by\ a\ participant}{Longevity\ coaching\ service\ experiment}$
D4	$Survey\ time = \dfrac{Total\ time\ cost\ of\ finishing\ filling\ out\ the\ initial\ survey\ by\ a\ participant}{Initial\ survey}$
D5	$Experiment\ time = \dfrac{Total\ time\ cost\ of\ finishing\ a\ longevity\ coaching\ service\ experiment}{Longevity\ coaching\ service\ experiment}$

DS criteria, including the mean (*M*), median (*Mdn*), mode (*Mo*), standard deviation (*SD*), variance (σ^2) and skewness (α^3), were applied to help us understand the data characteristics before conducting the correlation coefficient (*r*) analysis.

We also defined five datasets from D1 to D5 in Table 4.1. D1 is defined as the total number of questions a participant asked during each experiment. D2 represents the total number of times a participant interacts with or touches LPBs. D3 quantifies the total number of Post-it® notes used per experiment. D4 captures the total time it took for a participant to complete the initial survey using Qualtrics survey software. D5 records the total time participants took to complete the longevity coaching service experiment. Because of the project's limited sample size and scope, the goal was to conduct a correlation analysis to assess the relationships between datasets.

Results

The study recruited twelve participants whose average age was 33.1 years old – 45.5 per cent male and 54.5 per cent female – primarily based in Boston. Approximately 30 per cent of participants had savings accounts equivalent to approximately 3–6 months of their essential living expenses, 23.3 per cent adhered to a monthly budget, 20 per cent had a retirement fund or retirement savings account (IRA, 401k or 401b), 16.7 per cent had high-yield savings accounts, and only 10 per cent had a financial adviser or financial planning experience.

FIGURE 4.7 *The participant attached his Post-it® responses to each of the eight questions on the LPBs.*

The utilization of LPBs highlighted the usefulness of tangible artefacts. In the tech-based group, one participant attached his eight Post-it® responses to the eight questions on the LPBs (Figure 4.7). In the touch-based group, the participants regarded LPBs as a medium to facilitate sensitive conversations with longevity coaches. Most participants emphasized the power of tactility. One participant said, '*Every time I flipped the block in my hand, it sparked new ideas and made me feel closer to tackling these complex and personal questions*'. Another participant shared, '*The dimensions of the cubes are just right. I can place them on a table to form more productive discussions with my longevity coach*'.

Table 4.2 shows the six DS criteria employed to present the data characteristics compared with the results of the correlation coefficient analysis. On average, participants asked approximately 39.67 questions per experiment. Each participant used an average of 10.17 Post-it® notes per experiment. The typical time to finish one experiment was close to 33.29 minutes per participant, on average, as regulated by the original time constraint. The average time for completing one initial survey was approximately 15.17 minutes per participant, which aligned with the anticipated time frame. The SD of D4 was relatively small, measuring 3.59 because its *M* (15.17) and *Mdn* (13.1) were very close, resulting in the data points of D4 being more concentrated and narrowly spread in a relatively tight cluster compared with the other datasets. Conversely, D2 exhibited a larger SD of 23.05, indicating the dispersion of data points across D2.

TABLE 4.2 *Experimental results analysed from five descriptive statistics*

	D1 **# questions**	**D2** **# touching LPBs**	**D3** **# using notes**	**D4** **Survey time**	**D5** **Experiment time**
M	39.67	39.67	10.17	15.17	33.29
Mdn	44.5	34.0	7.0	13.1	32.6
Mo	n/a	16	6	n/a	n/a
SD	19.95	23.05	5.84	3.59	7.17
σ^2	398.06	531.52	34.15	12.91	51.37
α^3	0.17	0.66	1.06	1.06	0.52

The strength of the absolute value of *r* can be classified into five levels for analysis: very weak (*r* = 0–0.19), weak (*r* = 0.2–0.39), moderate (*r* = 0.40–0.59), strong (*r* = 0.6–0.79), and very strong (*r* = 0.8–1). However, the absolute values of *r* are arbitrary limits. The context of *r* and the results should also be considered. D1 had a very weak *r* value (*r* = 0.020883962) with D2 (Figure 4.8) and three weak *r* values with D3 (Figure 4.9), D4 (Figure 4.10) and D5 (Figure 4.11) (*r* = 0.200902789, –0.327025048 and –0.324103906, respectively). D1 might not correlate with the other four datasets. Because the sample size was relatively small, further studies could be conducted with at least sixty subjects for statistical significance to achieve different research outcomes.

D2 showed three different values of *r* from weak (Figure 4.12) to moderate (Figure 4.13). D2 and D4 presented moderate negative connections (*r* = –0.520202212) between the participants' time spent on the initial survey and number of times they touched the LPBs. This might indicate that the participants' interactions with the LPBs during the experiment were associated with the content of the initial survey.

Discussion

Based on this research, we propose a model of an empathic longevity service that emphasizes empathy-building through the utilization of tangible artefacts (LPBs) and service touchpoints (Figure 4.14). This model was modified from Suri's diagram, which illustrates that 'To understand what people think and feel (i.e. empathy), we must observe what they say and do' (2003, p. 55). It incorporates Passera et al.'s (2012) service prototyping practical framework (SPPF), which builds upon Blomkvist's and Holmlid's (2011) prototyping frameworks. This conceptual framework considers individuals' unique experiences, taking their cultural, social, mental and physical needs into account.

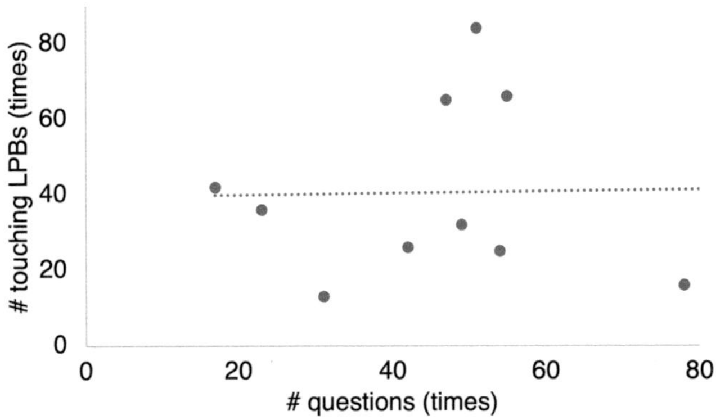

FIGURE 4.8 *The very weak correlation (r = 0.020883962) between the number of questions participants asked per session (x-axis) and the number of times participants touched the LPBs per session (y-axis).*

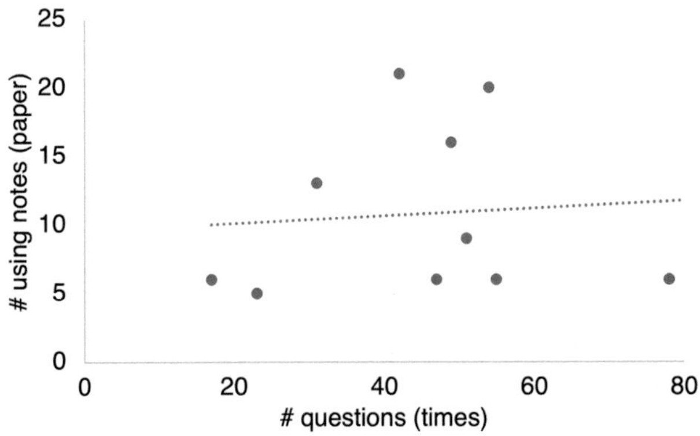

FIGURE 4.9 *The weak correlation (r = 0.200902789) between the number of questions participants asked per session (x-axis) and the number of Post-it® notes participants used per session (y-axis).*

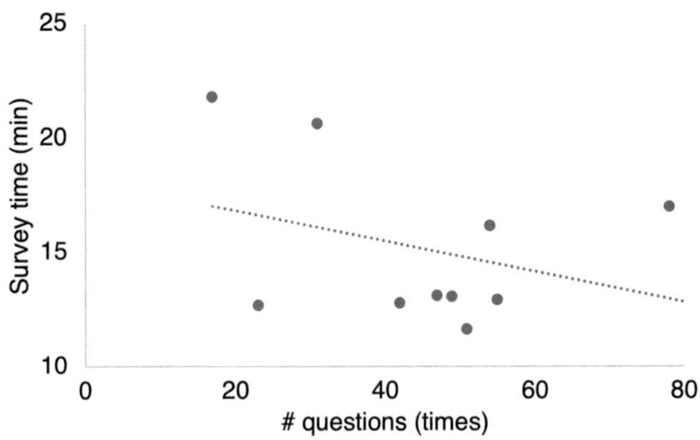

FIGURE 4.10 *The weak correlation (r = −0.327025048) between the number of questions participants asked per session (x-axis) and participants' time spent filling out the initial survey in minutes per session (y-axis).*

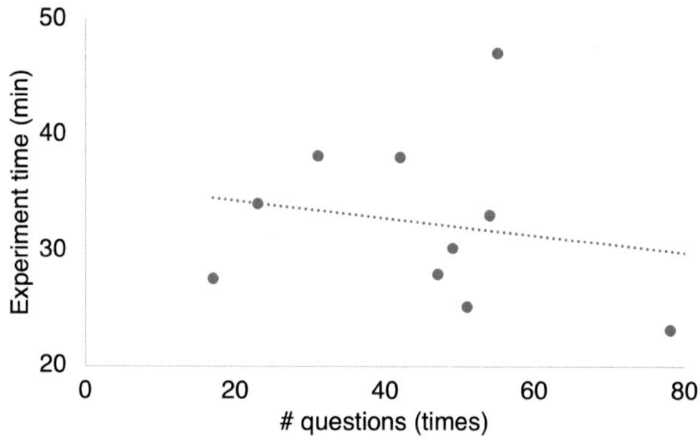

FIGURE 4.11 *The weak correlation (r = −0.324103906) between the number of questions participants asked per session (x-axis) and participants' time spent finishing the longevity coaching service experiment in minutes per session (y-axis).*

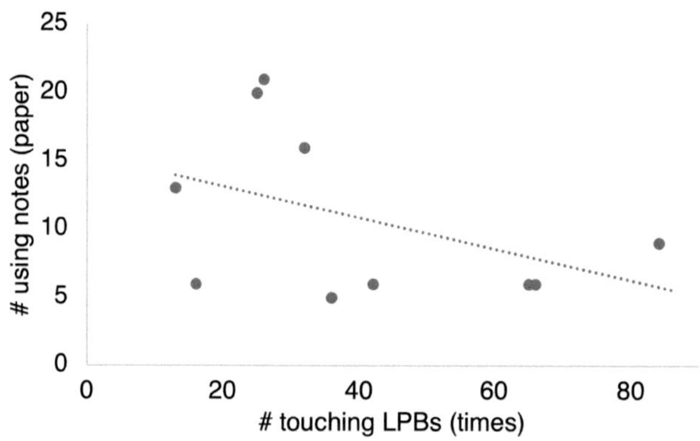

FIGURE 4.12 *The weak correlation (r = −0.341649682) between the number of times participants touched the LPBs per session (x-axis) and the number of Post-it® notes participants used per session (y-axis).*

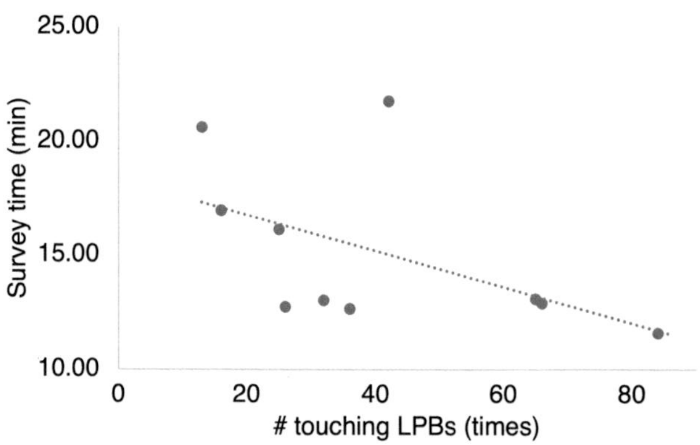

FIGURE 4.13 *The moderate correlation (r = −0.520202212) between the number of times participants touched the LPBs (x-axis) and participants' time spent filling out the initial survey in minutes per session (y-axis).*

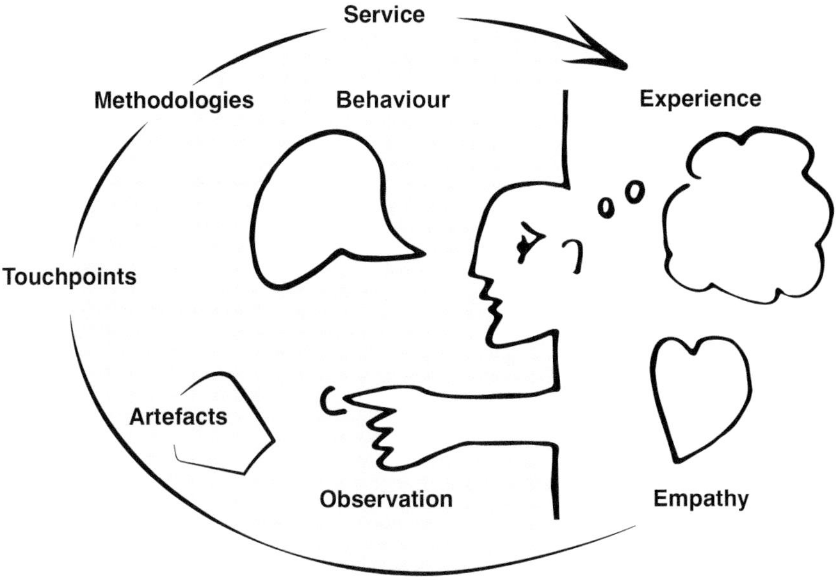

FIGURE 4.14 *An empathic service design framework. Adapted from Suri's (2001) illustration depicting the connection between behaviour, experience, observation and empathy.*

Because of the small sample size and the focus on empathy, demonstrating statistical significance when developing an argument in the early stages of the design and development process is challenging. In general, at least sixty subjects are needed to achieve statistical significance. However, in this preliminary experiment, we applied statistical analysis methods to explore how we could quantify and measure the outcomes by using the six design attributes and contribute to learning and possible opportunities for findings in the field of service design.

From design to development, the longevity service experiment focused on empathy, that is, understanding the participants' experience. Using tangible artefacts leading to empathy-driven conversations allowed participants to comprehensively understand what to consider in a longevity service. Additionally, this allows service providers to create broader, more inclusive discussions about longevity planning, considering participants' cultural and social backgrounds. Because these life-centred design touchpoints have the capacity to make participants feel understood, we believe that designing and using them can create sophisticated, empathic services that empower participants to discuss delicate topics. The questions on these tangible artefacts allowed the participants to open up and explore their options. Embracing this shift extensively broadens the scope of possibilities and encourages a deeper understanding of the human experience.

As already defined, the strength of the absolute value of *r* is categorized into five levels for analysis: very weak, weak, moderate, strong and very strong. According to the statistical analysis results (Figures 4.8 to 4.13), we note significant differences between samples associated with different 2x2 connections across various pairs from D1 to D5. Although the absolute values of *r* are set at arbitrary thresholds, the implications and context of *r* should also be considered in the analysis. Further research could investigate whether these variations are because of factors such as cultural differences, personal financial status, education level, income, family issues, decision-making processes or individual behaviours. Addressing these questions is crucial, and a more personalized and qualitative research approach (e.g. in-depth interviews and fieldwork studies) may be required to fully understand and respond to them.

Conclusion

The study's goal was to explore the following research question: 'How can we develop an empathic longevity service from prototype to eventual experience?' Drawing from preliminary experimental findings, empathy plays a central role in longevity services. First, we should harness the value of tangible toolkits, such as LPBs, to address sensitive and complex longevity challenges. By strategically leveraging the tangibility and tactility of these artefacts, service recipients (e.g. users or customers) are empowered to engage in more private and abstract conversations with longevity coaches. Second, we can improve individuals' service experiences by establishing empathic design touchpoints and processes, such as interactive and transparent tech- and touch-based service encounters that address people's pain points and satisfy their expectations. Finally, we integrate user feedback into the holistic design process to enhance user engagement. These transformative elements can impact the entire participatory design approach, from inspiration and ideation to implementation and integration.

Takeaway

In the age of longevity economics, people are actively seeking personalized services to plan for extended lifetimes. The importance of comprehending individual needs and exhibiting empathy has become more crucial than ever. The objective of this research was to investigate how to develop an empathic longevity service by incorporating empathy into a longevity service prototype. To pursue this inquiry, a controlled experiment was established encompassing tech- and touch-based service encounters. The study proposes a conceptual model for longevity services that place emphasis on empathy-building techniques through using tangible artefacts called Longevity Planning Blocks, human-centred service touchpoints, and design methods.

Acknowledgements

This book chapter was inspired by 'The Future of Service Prototyping', a 9 May 2023 online webinar hosted by Dean Satu Miettinen from the Service Design Department at the University of Lapland. The author extends heartfelt appreciation to Dean Miettinen for the invitation. The author also expresses gratitude for the constructive feedback from Professor Melanie Sarantou from the Strategic Design Department at Kyushu University and the author's PhD Committee members at the Massachusetts Institute of Technology (MIT): Deputy Dean of Engineering Maria C. Yang, Dr Joseph F. Coughlin, Professor Olivier de Weck, Professor Eric Klopfer and Professor John Ochsendorf. The author is also thankful for the valuable feedback and encouragement from his industry mentors: Gianfranco Zaccai, Co-Founder and Chief Designer of Continuum Innovation, and Michael Peng, Chief Creative Officer of Moon Creative Lab and Board Director of IDEO.org. The author also acknowledges the thoughtful contributions of his esteemed thought partners, Dr Lisa D'Ambrosio and Dr Chaiwoo Lee, Research Scientists at the MIT AgeLab, and Professor Sofie Hodara from the College of Arts, Media and Design at Northeastern University. Finally, the author thanks Marsha M. Rubin, Kjersten C. Lazar and Stephanie Emory for their invaluable contributions and professional financial planning expertise, which greatly contributed to the cocreation of this empathic longevity coaching service.

References

Batson, C. D. (2009). These things called empathy: Eight related but distinct phenomena. In J. Decety & W. Ickes (Eds.), *The social neuroscience of empathy* (pp. 3–16). The MIT Press. https://doi.org/10.7551/mitpress/9780262012973.003.0002

Blomkvist, J., & Holmlid, S. (2011, 29 May). *Existing prototyping perspectives: Considerations for service design*. Nordes 2011: Making Design Matter, 29–31 May, School of Art & Design, Aalto University, Helsinki, Finland. https://doi.org/10.21606/nordes.2011.012

Brown, T. (2019). *Change by design: How design thinking transforms organizations and inspires innovation* (Revised and updated). Harper Business.

Collins, W. (2019). *The empathy gap: Male disadvantages and the mechanisms of their neglect*. lps Publishing.
Coughlin, J. F. (2019a). Planning for 8,000 days of retirement: Advisor value in today's longevity economy. *Investments & Wealth Institute* (March/April 2019).
Coughlin, J. F. (2019b, 13 April). *Why 8,000 is the most important number for your retirement plan*. Forbes. https://www.forbes.com/sites/josephcoughlin/2019/04/13/why-8000-is-the-most-important-number-for-your-retirement-plan
Davis, M. H. (1996). *Empathy: A social psychological approach* (M. H. Davis, Ed.; 1st ed.). Routledge. https://doi.org/10.4324/9780429493898
Dolata, M., Agotai, D., Schubiger, S., & Schwabe, G. (2019). Pen-and-paper rituals in service interaction: Combining high-touch and high-tech in financial advisory encounters. *Proceedings of the ACM on Human-Computer Interaction*, 3(CSCW), 1–24. https://doi.org/10.1145/3359326
Friese, S. (2014). *Qualitative data analysis with ATLAS.ti* (2nd ed.). SAGE.
Giebelhausen, M., Robinson, S. G., Sirianni, N. J., & Brady, M. K. (2014). Touch versus tech: When technology functions as a barrier or a benefit to service encounters. *Journal of Marketing*, 78, 113–124.
Hess, J. L., & Fila, N. D. (2016). The manifestation of empathy within design: Findings from a service-learning course. *CoDesign*, 12(1–2), 93–111. https://doi.org/10.1080/15710882.2015.1135243
Houston, D. (2014). Financial security: Longevity changes everything. In P. H. Irving & R. Beamish (Eds.), *The upside of aging* (1st ed., pp. 209–224). Wiley. https://doi.org/10.1002/9781118691823.ch14
Kouprie, M., & Visser, F. S. (2009). A framework for empathy in design: Stepping into and out of the user's life. *Journal of Engineering Design*, 20(5), 437–448. https://doi.org/10.1080/09544820902875033
Kuure, E., Miettinen, S., & Alhonsuo, M. (2014). *Change through service design: Service prototyping as a tool for learning and transformation*. https://api.semanticscholar.org/CorpusID:34563450
Lee, S.-H. (2023a, 12 April). *Design for longevity: Service, systems and sustainability*. DesignWanted. https://designwanted.com/design-longevity-service-systemes-sustainability/
Lee, S.-H. (2023b, 31 May). *Design for simulation: How to project your future self?* DesignWanted. https://designwanted.com/design-simulation-how-project-your-future-self/
Lee, S.-H., Coughlin, J. F., Balmuth, A., Lee, C., Cerino, L., Yang, M., Klopfer, E., de Weck, O. L., & Ochsendorf, J. (2023, 9 October). *Designing longevity planning blocks through experimental participatory observation and interviews. Life-changing design*. The International Association of Societies of Design Research. https://www.iasdr2023.polimi.it/
Lee, S.-H., Coughlin, J. F., Yang, M. C., Balmuth, A., de Weck, O., & Ochsendorf, J. (2023, 22 August). *Design artifacts and gaming transform financial planning services: Multifaceted experimental design research*. International Design Conference (IDC) and IDSA Education Symposium. https://www.internationaldesignconference.com/
Lee, S.-H., Coughlin, J. F., Yang, M., de Weck, O. L., Lee, C., Klopfer, E., & Ochsendorf, J. (2023). Co-create financial planning services for an aging

population: Designers' perspectives. *Proceedings of the Design Society*, 3, 947–956. https://doi.org/10.1017/pds.2023.95

Lee, S.-H., Patskanick, T., Balmuth, A., & Coughlin, J. F. (2023). Applying human-centered system design to the development of a tool for service innovation. In K. Vaes & J. Verlinden (Eds.), *Connectivity and creativity in times of conflict*. Academia Press. https://doi.org/10.26530/9789401496476-006

Lee, S.-H., Yang, M., De Weck, O. L., Lee, C., Coughlin, J. F., Klopfer, E., & Ochsendorf, J. (2023). Service design in action: Transformation, consideration, and system thinking. *Proceedings of the Design Society*, 3, 3145–3154. https://doi.org/10.1017/pds.2023.315

Leonard, D., & Rayport, J. F. (1997). Spark innovation through empathic design. *Harvard Business Review*. https://hbr.org/1997/11/spark-innovation-through-empathic-design

Mager, B. (2009). Service design as an emerging field. In *Designing services with innovative methods* (p. 17). University of Art and Design Helsinki.

Mauterstock Jr., R. B. (2020). Making the transition from financial planning to longevity planning. *Journal of Financial Planning*, July 2022, ISSN 1040-3981, https://www.financialplanningassociation.org/article/journal/JUL20-making-transition-financial-planning-longevity-planning

Nielsen, J. (2010). *Usability engineering* (3rd ed.). Kaufmann.

Oxford Reference. (2024). *Empathy*. https://doi.org/10.1093/oi/authority.20110803095750102

Passera, S., Maila, R., & Kärkkäinen, H. (2012). *When, how, why prototyping? A practical framework for service development*. https://api.semanticscholar.org/CorpusID:2960014

Porcini, M. (2023). *The human side of innovation: The power of people in love with people* (1st ed.). Berrett-Koehler Publishers, Inc.

Soto, M., Miettinen, S., & Alhonsuo, M. (2023). The prototyping service experience as a tool of empathy and research: From Finland to Chile. *In Service Encounters and Meanings Proceedings* (pp. 440–457). https://doi.org/10.3384/ecp203023

Suri, J. F. (2001). The next 50 years: Future challenges and opportunities for empathy in our science. *Ergonomics*, 44(14), 1278–1289. https://doi.org/10.1080/00140130110105850

Suri, J. F. (2003). Empathic design: Informed and inspired by other people's experience. *Empathic Design, (35)*, 51–57.

Suri, J. F. (2005). *Thoughtless acts? Observations on intuitive design*. Chronicle Books.

Sustar, H., & Mattelmäki, T. (2017, June 15). *Whole in one: Designing for empathy in complex systems*. Nordes 2017: Design and power. https://doi.org/10.21606/nordes.2017.031

Tan, A. H. T., Muskat, B., & Johns, R. (2019). The role of empathy in the service experience. *Journal of Service Theory and Practice*, 29(2), 142–164. https://doi.org/10.1108/JSTP-10-2018-0221

Turkle, S. (2007). *Evocative objects: Things we think with*. MIT Press.

Vink, J., & Oertzen, A.-S. (2018). *Integrating empathy and lived experience through co-creation in service design*. https://api.semanticscholar.org/CorpusID:200073344

WHO. (2022, October 1). *Ageing and health*. https://www.who.int/news-room/fact-sheets/detail/ageing-and-health

Wu, L., Fan, A., Yang, Y., & He, Z. (2022). Tech-touch balance in the service encounter: The impact of supplementary human service on consumer responses. *International Journal of Hospitality Management*, 101, 103122. https://doi.org/10.1016/j.ijhm.2021.103122a

5

Research on Empathy-Driven Design Workshops and Social Implementation Methods

Yanfang Zhang and Masanori Fujita

Abstract

This chapter explores empathy's role in driving competence, attitude and behavioural change within design workshops and subsequent social implementation efforts by mapping empathy from workshop ideation to the social action of one project, *One Person One Flower*, which was located in Fukuoka City, Japan, during 2023. Based on longitudinal research of design workshops from 2012 until 2023 in Fukuoka, the design workshops gradually led to the social implementation of the design ideas. Through observations of stakeholders during the design workshop in 2023, empathy processes were mapped during the project's design and social implementation phases to understand the impact of the elements of empathy – attitude, competence and behaviour – on these processes. In response, this chapter will present a proposal for the practical enhancement of empathetic skills, communication proficiency and building trusting relationships. The transformative potential of empathetic design methodologies in addressing societal challenges is advocated for.

Introduction

The world faces numerous challenges, including social disparities and large-scale disasters. Simultaneously, social issues have evolved, leading to irreversible changes in consumer behaviours, habits and values. Diverse approaches are needed from individuals with varied backgrounds and values to address these continually evolving global challenges by innovating new social systems to ensure inclusivity through design. In Japan, government policies on science, technology and innovation, such as the

Science, Technology and Innovation White Paper (MEXT, 2013), which details the nation's strategies, achievements, and future goals in scientific research, technological advancements and innovation policies, the Japanese government clearly outline design as a means to create new value.

This chapter presents longitudinal research on a series of design workshops that were implemented between 2012 and 2023, here using citizen-participatory workshops (hereafter referred to as 'workshops') as design methods to address local issues creatively. Furthermore, successful initiatives have been undertaken to implement the excellent ideas generated from these workshops into society. The workshops are divided into two periods: from 2012 to 2015 and from 2016 onwards. The workshops provided platforms where individuals from diverse backgrounds could convene to address societal issues with inclusivity and diversity. From 2012 to 2015, 'Universal Design Workshops' in Fukuoka were conducted under the principle of 'friendly to everyone, everyone is friendly', serving as a cornerstone of municipal governance engaging individuals from varied backgrounds in two-day society issue-solving activities. From 2016 onwards, these workshops evolved into global events in collaboration with Global Goals Jam (n.d.), focusing on designing solutions aligned with the Sustainable Development Goals (SDGs) advocated by the United Nations. This shift aimed to address local issues on a global scale.

The incorporation of empathy into the design workshop for the social implementation process of 2023 served as the catalyst for this research. According to Manzini (2014) and Krznaric (2019), empathy is about understanding the world as others feel and sense it, hence using it to guide one's actions. Empathy holds the power to change our own lives and bring about fundamental changes in society. Existing research emphasizes the importance of design methodologies that focus on empathy (Coleman et al., 2003). For instance, it is argued that the source of design prowess lies in deep insights into what individuals think, value and perceive as problems, ranging from explicit to inarticulate. In other words, empathy towards users constitutes the essence of design (Mattelmäki & Battarbee, 2002). Tamura et al. (2014) assert that design empathy is necessary when transitioning from designing for practical functionality to designing for personal experiences and individual contexts.

Alongside these changes, which shifted from hosting design workshops to social implementation, empathy also evolved. Within the ongoing relationship between the stakeholders, individuals learn about each other's needs during the design workshops through empathetic responses, sometimes successful or not. Therefore, the objective of this chapter is to elucidate shifts in stakeholder roles and empathy and suggest novel practical methods and useful tools for design methodologies based on the findings. The research question addressed in this chapter is 'How can empathy be actively embedded in social innovation contexts to stimulate behavioural and attitudinal change?'

Thematic Literature Review

This thematic literature review will discuss design thinking for solving social issues within the impact framework of the SDGs. In addition, empathy is explored in a design context, shifting from design workshops and ideation to social implementation. Goleman (1995) defines empathy as the ability to recognize and understand the emotions of others, enabling individuals to respond compassionately and appropriately. He emphasizes empathy as a cornerstone of emotional intelligence that is essential for nurturing relationships, resolving conflicts and fostering social harmony and cooperation. Similarly, in another book, he states that empathy is crucial for effective communication and relationships (Goleman, 1998).

Brown and Wyatt (2010) discuss how design thinking can be applied to solve social issues, emphasizing empathy, creativity and collaboration in problem-solving processes. Manzini (2015) explores design's role in fostering social innovation to address global sustainability challenges, advocating for collaborative and participatory approaches. In addition, social innovation contexts are intricately connected to service design. Service design and participatory design methods are widely used in social design and social innovation contexts to address societal challenges, create social impact and generate transformational change (Alkire et al., 2020; Sangiorgi, 2011; Vink & Koskela-Huotari, 2021; Yang & Sung, 2016).

Moreover, since the United Nations proposed the seventeen Sustainable Development Goals (SDGs) in 2016 as a framework for generating societal impact, research has been conducted on the need for design methods related to the SDGs. For example, Halpern and Walther (2022) examine how the design thinking approach can potentially empower young entrepreneurs whose focus is social innovation and may help them move towards efficient and effective action in their environment. Maher et al. (2018) examine the process of designing MetaMAP, an interactive graphic tool for understanding social–ecological systems and designing well-integrated solutions. Zhang et al. (2023) explore conditions that can bridge the gap between formulating design ideas and implementing actions focused on the SDGs by utilizing design methods and models.

Empathy can be categorized into three main subdivisions: attitude, competency and behaviour (Derksen et al., 2013). In empathy, competency involves empathic skills, communication proficiency and the ability to build trustful relationships. Derksen et al. (2013) suggest that enhancing these skills can enhance overall competence. Empathy is essentially the human ability to identify with the feelings and thoughts of others. Scholars such as Wright and McCarthy (2008) and Mattelmäki and Battarbee (2002) argue that design empathy fosters a personal connection between the designer and the user, viewing the user from their own perspective rather than as an object. They emphasize the importance of maintaining one's own viewpoint while engaging with others, allowing for creative responses from that standpoint.

Alhonsuo et al. (2020) highlight the key insights from design workshops, which include learning and understanding through design methods, design thinking and dialogue and how this affected organizational culture and change. Kuure (2020) mention that service workshops are an opportunity to see design and service beyond the end result. Through them, it is possible to focus on design activities as a broader phenomenon and integrate design practices into local and relevant contexts. This kind of participatory service design workshop can be seen as a key way to designing and research with communities. In Japan, workshops often use the design thinking process (Brown, 2008), which is how the workshop used as a case study in this chapter was developed.

Longitudinal Research: From Design Workshops to Social Implementation

The present research focuses on illustrating how design workshop research has evolved over many years of practice between 2012 and 2023 (Figure 5.1). Initially, the annual design workshops generated ideas, but they gradually became initiators for social implementation. The workshops represented collaborative endeavours involving individuals with diverse expertise, aiming to devise solutions to social issues through design methodologies. They fostered the social initiatives of participants striving to nurture empathetic dialogue that generated new knowledge and artefacts. Although the specific workshop processes varied annually, they were fundamentally rooted in design thinking and structured into four main phases:

- a exploring (this phase involves theme exploration, research and problem identification);
- b responding (participants engage in ideation to propose solutions);
- c making (prototyping and visualization of proposed solutions take place); and
- d sharing (participants present their solution plans to the group).

The significance of the workshops was that participants could experience the design process in a short amount of time and put into practice the application of design methods.

Social implementation originated from design workshop ideation. Unlike workshops, social implementation represents ongoing processes characterized by changing participants, roles, locations and timelines. Throughout the workshops conducted thus far, numerous excellent ideas have emerged to address social issues, with some teams continuing to implement these proposals in society, even after the workshops. Figure 5.2 summarizes the design proposals forthcoming from the design workshops for each year. The design proposals that were also socially implemented

FIGURE 5.1 *Overview of the 2012–23 workshops. Photographs by Zhang (2019) and Nagano (2017).*

in further collaboration with the first author are indicated in dark grey. Figure 5.2 illustrates how social implementation increased to no or only one design proposal between 2012 and 2019. Thereafter, at least two design proposals were socially implemented between 2020 and 2023.

A common process was outlined by comparing the four main phases of the design workshops and data from the various workshops conducted by the researchers. From the workshop's design proposal to social implementation, a series of activities are outlined in seven steps: (a) exploring (theme exploration, research, problem identification); (b) responding (ideation of solutions); (c) making (prototyping and visualization of solutions); (d) sharing (presentation of solution ideas); (e) documenting (sharing the idea with a broader audience); (f) reconstructing (refining the proposed solution); and (g) starting (engaging in the social implementation of the solution). Social implementation involves putting the experience of design methods in the design workshop into practice. As shown in Figure 5.2, twelve meaningful projects have been socially implemented that originated from design ideas through cocreation by teams of stakeholders. The present study illustrates how the workshop progressed to social implementation.

Research Methods

Frayling's (1993) research through design (RtD) methodology was implemented as an approach to examine empathy in the design workshop

Time	Team proposals							
2012	Hakata kiten Project	The Ring	Info-rest	Hakata Shield				
2013	Double Maru Project	Plasico	Umakamon	Allergy Free Cake "Nico"				
2014	kami(i) musium for all	Amenity Lunch Box	Kizuna					
2015	Fukuoka B-ITE	Eet Japan Project	Yatai KIT	Yatai Buddy				
2016	Unbrella Yatai	Yururi Yatai	Trip Yaitai Fukoka					
2017	Toito	Wheelog!	Cazera	Baby BnB				
2018	Adventure of Underground Kingdom	City Oasis	Kawasemi Project	Know More Plastics				
2019	Wheel Chair Fashion	Project IMA	Uki Uki Park	Sustainable Our Hometown				
2020	Shikanoshima Footprint Project	Let us make a Maching bus	Design lesson of new age lifestyle	Urushi forest school	E mobycycle	Oiwayama Bamboo Base"Kizuna"		
2021	DIY Education	Project Co-en	Yaba! Mirai Farm	"sweat" turns into "dinner"	Rhizome Futures Gallery	Sendai Garappan	Garbage Racer	Waste will not be Wasted
2022	Mobile playground equipment	Rainy day in park	Login	FfF-Fab for Forest	WSH	Mitsuke you	Watering Dancers	Hatake Lab
2023	One Person, One Flower	Nakagawa Future Project	Yabakei River Step	Realization	Oh! Bento OBENTO as Media	Neighborhood Workshop URUSU	Trees for Circulation of Trees	Wild Life Supermarket
2023	New Keihoku Old Tales	Craftsmanship Nurtured by Water and Land	Contact Between Welfare & Manufacturing	Art Fusion Creation	Snap Art Project			

FIGURE 5.2 *Overview of team ideas and confirmation of social implementation among design workshops from 2012 to 2023.*

and social implementation process of 2023. Empathy was explored based on Frans Derksen et al.'s (2013) elements of empathy: attitude, competency and behaviour. The relationships among these three elements underpinning empathy were identified in both the workshop and social implementation process of 2023. Empathy maps were generated. They are typically generated for the end users of products by learning about their physical environment, demographic background and behaviour, which enables businesses to create the most suitable products for customers' needs (Ferreira et al., 2015). Empathy maps are adopted to understand people's attitudes and behaviours in an easy-to-visualize manner.

Using data from the design workshop and its subsequent social implementation project, *One Person, One Flower*, which was conducted in 2023, the first author subjectively mapped the changes in empathy in the two processes (Figures 5.2 and 5.3). Visual data from videos and photos, in addition to participant surveys, observations and notes taken during the workshops and social implementations, were collected and documented, enabling subjective mapping – here from the perspective of the workshop facilitator – of empathy across the workshop and social implementations.

The participants in both workshops and social implementation projects were included in the mapping as the key stakeholders: facilitator, student, university, citizen, designer, government and local community. Stakeholders were categorized into three levels: core, direct and indirect. Stakeholders and their specific processes vary from project to project in social implementation. Empathy maps for these stakeholders were created based on these three aspects (Figures 5.2 and 5.3). From the mapping, areas of improvement were clarified, and suggestions for better social implementation were proposed.

Design Workshops for the Social Implementation of *One Person, One Flower*

One Person, One Flower is an initiative in which each citizen, business and government member fosters flowers and greenery and fills parks, pavements, offices, homes and all other places in Fukuoka City with flora. Collaboration with Fukuoka City as a key stakeholder was essential to achieving these goals. Figure 5.3 shows the empathy levels of all stakeholders; the three vertical bars for each stakeholder indicate the attitude, competency and behaviour levels. Facilitators and social practitioners served as the primary stakeholders in this workshop. Because facilitators are responsible for planning and executing workshops, their empathy across attitudes, competencies and behaviours is consistently high. Social practitioners assume leadership roles within teams, addressing the social issues they bring to the forefront. Although their competence in empathy, such as creativity and communication skills, may vary, their attitude and behaviour towards empathy remain high. Designers also play a significant role in the workshop,

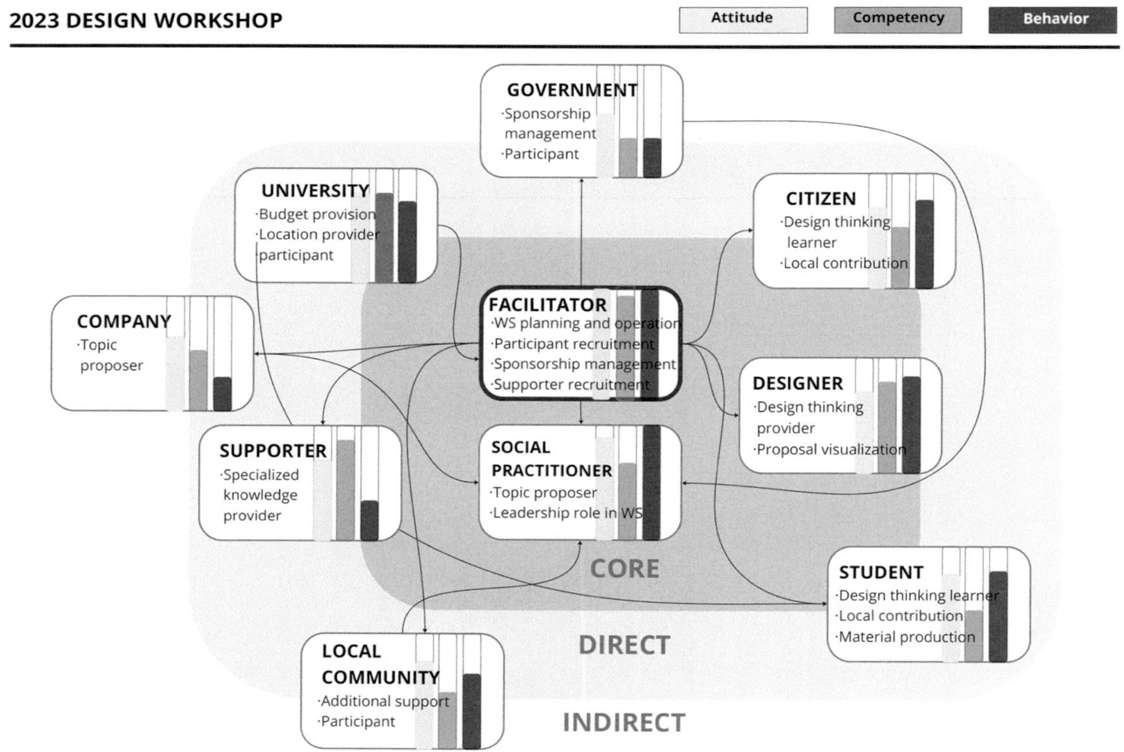

FIGURE 5.3 *Empathy map constructed by the first author for the 2023 design workshop indicating the level of involvement (core, direct and indirect) of the stakeholders and the elements that constituted empathy between them, indicated according to attitude, competence and behaviour.*

supporting design thinking processes and visualizing team proposals. Their empathy across attitude, competency and behaviour surpasses that of the other participants. Various stakeholders, including citizens, students, local communities, companies and universities, exhibit higher levels of empathy across attitudes, competencies and behaviours. Although the government may not be a core participant, it has provided sponsorship annually and has participated consistently in workshops over the years.

Figure 5.4 shows that stakeholders are changing their role from the design workshop, and the attitude, competency and behaviour levels (three vertical bars) are also changing. Core members include social practitioners, facilitators and the government. Social practitioners lead implementation activities, forming new teams and negotiating with organizations. Although empathy is generally high across attitudes, competencies and behaviours, competency may lag slightly. Facilitators provide support by utilizing design methodologies with high competency. Directly involved entities, such as designers, students, supporters and companies, exhibit empathy across all dimensions. The government has become a core member, engaging closely with other stakeholders to explore support possibilities.

Upon comparing the stakeholder empathy diagrams for both the workshop and social implementation, several key findings emerged. First, it was noted that the core members involved in social implementation differ from those involved in the workshops, with shifting positions and roles among stakeholders. Overall, compared with noncore members, core members exhibited greater levels of empathy in terms of attitude, competence and behaviour. The role of a facilitator with a background in design proves pivotal in workshops employing design methodologies and subsequent social implementation endeavours. Furthermore, effective collaboration between facilitators and social practitioners is crucial throughout the design workshops and social implementation activities.

As depicted in Figure 5.5, training can enhance competence. The significance of competencies such as empathic skills, communication skills and the ability to build trustful relationships is underscored by the design workshop. Acquiring and practising these skills during a two-day workshop is crucial for continuous improvement in social implementation efforts.

In the three key subdivisions, attitude and behaviour can be improved through long-term learning and practice. Competency, on the other hand, can be improved in a short period using effective tools. This chapter proposes methods focused on improving competence. Enhancing competence is anticipated to positively impact attitude and behaviour, thereby aiming for an overall improvement in empathy.

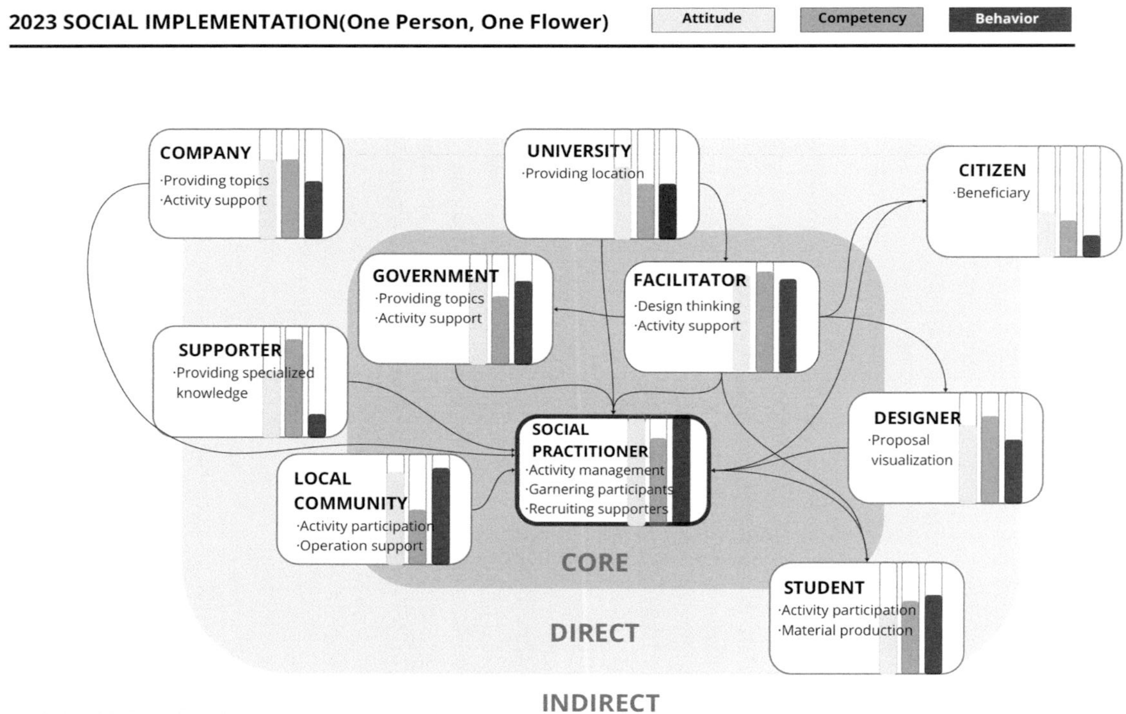

FIGURE 5.4 *Empathy map for 2023 social implementation illustrating stakeholders' empathy according to the elements of attitude, competency and behaviour.*

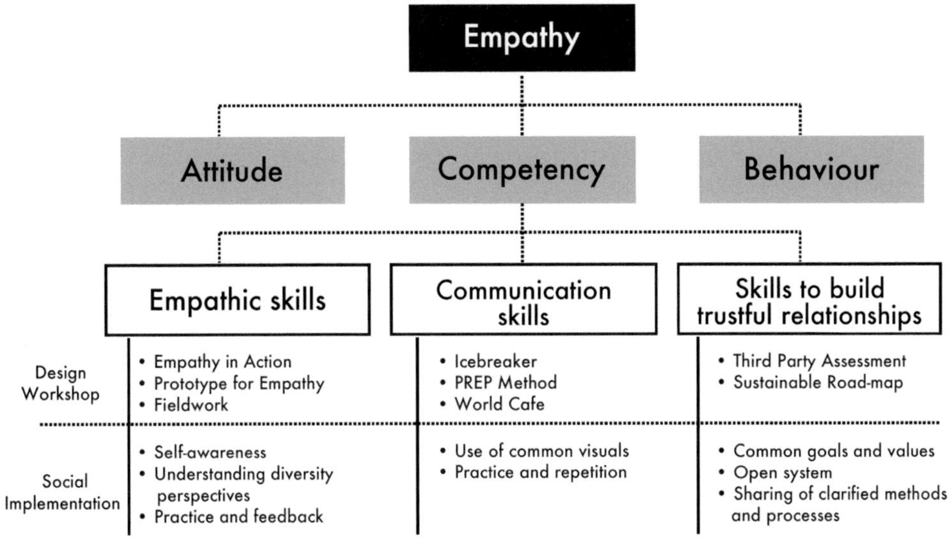

FIGURE 5.5 *Elements of empathy and competency improvement methods. Based on Derksen et al., 2013.*

Recommendations for Design Workshops

Workshops are collaborative activities conducted within limited time frames. It is recommended that framework tools be utilized to bolster competencies.

Enhancing Empathic Skills: Empathy in action involves helping individuals empathize with and comprehend a specific situation foreign to them (GGJ, n.d.). The first step in facilitating empathy for action in a design workshop would be to analyse gathered data and decide on an issue to address. Once this is complete, workshop participants and organizers should brainstorm how laypeople can easily understand and empathize with the chosen issue. A prototype solution can be submitted to trial-and-error tests, where continuous feedback is collected. Once the prototype can properly solve the issues discussed, it can be presented to the public. Throughout these steps, it is important to make a record using video, notes or photos to gather insights about the issue that can aid future workshops. Prototyping for empathy aids in understanding users' needs and problems when using a concept. The steps for this prototype begin with identifying specific aspects of a user experience that lack clarity. From there, strategies can be developed to understand potential problems a user might face. A quick prototype should be developed and tested among the design team. The results from the tests should then be used to refine the design idea. Tools such as the persona spectrum (GGJ, n.d.)

can be employed or utilized as models to aid in building empathy while designing. Fieldwork entails conducting research in the actual situations of the target individuals rather than in design studios or laboratories. This part of the design workshop involves setting objectives and planning to meet them. Necessary tools such as questionnaires should be created to help address any issues in the target area. Once these steps are complete, observations and interviews can be conducted, and the data from this should be analysed.

Enhancing Communication Skills: Icebreakers are activities conducted at the beginning of the workshop to foster rapport among participants, ease tension and initiate communication. For example, in the Silent Machine Game, participants silently move or combine their bodies to form a single machine. The audience must guess the function of their machine; this icebreaker is excellent for facilitating team bonding and communication. Another tool that aids communication skills is the PREP method (Ooshima, 2006). This framework enhances persuasiveness and ease of communication in a short amount of time by stating the conclusion first. Those involved in the project or discussion must provide reasons for citing that conclusion. Providing specific examples can help enhance persuasiveness among team members. Once the responses are complete, the final step is to summarize the conclusion. The World Cafe is an excellent tool that can facilitate discussions among a large group about multiple topics (Jung, 2019). The organizers need to brainstorm and select topics first, which can then be shared while assigning participants to groups. A topic can be given to each group, and discussions can occur within the group during the allotted time. Afterwards, a group can share and exchange ideas with other groups and incorporate what they have gathered into their own findings. The final step is for each group to share their conclusions about their respective topics with the entire audience.

Building Trustful Relationships: Third-party assessment involves inviting an external group to review the situation and providing feedback that can be beneficial during the workshop (Bamber et al., 2004). A group can first clarify their issues and then invite a group selected from a demographic of end users of their project or product. A presentation that communicates the favoured concept can be made, followed by design reviews from external groups. Another tool for building trusting relationships is the Sustainable Roadmap (GGJ, n.d.). This tool maps challenges and plans for the future and can provide a clear direction for social implementation efforts. The first step is to identify local challenges. A matrix where challenges are mapped from the present to 2030 can be created. Using a matrix can help designers identify any 'gaps', which can then be filled to create a roadmap. Then, a challenge can be selected on the roadmap, and its 'path' can be followed to generate ideas.

Recommendations for Social Implementation Activities

Social implementation activities following the workshop are iterative cocreation processes. Alongside the aforementioned frameworks, it is crucial to establish a sustainable mechanism with a comprehensive, long-term perspective to enhance competency capabilities.

Enhancing Empathic Skills: Empathic skills can be developed by enhancing self-awareness, and self-understanding is vital for improving empathic skills (Krznaric, 2019). Exploring one's emotions and values enhances one's self-awareness. In addition, understanding diverse perspectives – recognizing the viewpoints and positions of others – is essential for enhancing empathic skills. Embracing the emotions and experiences of diverse participants beyond one's own perspective can foster deeper empathy (Derksen et al., 2013). Finally, continuous practice and feedback are crucial for building empathy. These skills develop through daily interactions and regular conversations, improving as we receive feedback from others.

Enhancing Communication Skills: Two important points to enhance communication skills to achieve social implementation through the utilization of common visual tools are the incorporation of common visual aids (such as whiteboards, poster boards and markers) in group work and presentations. This facilitates the visualization of ideas and concepts and enhances communication among participants. Additionally, regular practice is essential for improving communication skills. Practising in various contexts, including role-playing, presentations and everyday conversations, is crucial.

Building Trustful Relationships: Building solid relationships based on strengthening trust involves establishing shared goals and values. Collaborating towards common objectives and respecting each other's values and beliefs strengthens the bond. Open communication can help cultivate trust by openly expressing thoughts and emotions and avoiding hidden agendas or deceit. It is important to define specifics, such as when and with whom to undertake tasks towards a goal and ensure that these are communicated to all members. In the present study, design workshops conducted to address social issues were examined from the perspective of workshop practitioners. The outstanding ideas generated from these workshops were used in social implementation. Changes in stakeholder relationships throughout the process, from design workshops to social implementation, can be clarified from an empathetic standpoint. Additionally, practical tools and methods have been proposed. This approach maximizes the power of design, facilitates solutions through diverse collaboration and adapts to evolving social conditions. Several considerations arise from the present study. First, stakeholder empathy maps for social implementation vary across projects. Although efforts have been made to extract generalizable content using the case study of '*One Person, One Flower*', analysing more social implementation projects in the future is necessary to extract common methods. Additionally, technology plays a crucial role in empathy, and changes in methods because of the Covid-19 pandemic may lead to variations in empathy during workshops. Although not addressed in the current study, analysing these changes in the future is imperative.

Conclusion and Discussion

The analysis of the empathy maps revealed that core members consistently demonstrated higher levels, with the role of facilitator being pivotal in fostering empathetic engagement and facilitating collaboration. Compared with attitudes and behaviours, competence was found to be more likely to improve in a short period. According to previous research, design workshops are an effective method for improving understanding through dialogue among participants and for participatory design practice projects (Alhonsuo et al., 2020; Kuure, 2020). In this chapter, we find that empathy can be actively embedded in social design and innovation contexts to stimulate behavioural and attitudinal change. Specifically, the analysis of design workshops illustrates how to support different stakeholders in feeling and expressing more empathy towards other stakeholders and viewpoints. This can pave the way for participants to become more inclined to act because their needs are acknowledged. As shown in the case study, core stakeholders for social implementation differ from those for workshops, along with their roles. Thus, the role of facilitators with design backgrounds was found to be important because design thinking can enhance empathy among stakeholders and encourage collaboration between stakeholders. The present study has also noted that, in empathy, competency improves faster than attitude and behaviour. As a result, tools have been proposed to aid future workshops or social implementation projects in enhancing overall empathy. This chapter advocates for the transformative potential of empathetic design methodologies in tackling societal and global challenges. To achieve such objectives, empathy must be central to stakeholder interactions.

Takeaway

Based on longitudinal research since 2012, this chapter proposes that design workshops leading to social implementation can be structured in the following four phases:

1. Exploring: This phase involves theme exploration, research and problem identification.
2. Responding: Participants engage in ideation to propose solutions.
3. Making: Prototyping and visualization of proposed solutions take place.
4. Sharing: The participants present their solution plans to the group.

The chapter further proposes that social implementation can be implemented using seven steps:

1. Exploring: Theme exploration, research, problem identification.
2. Responding: Ideation of solutions.
3. Making: Prototyping and visualization of solutions.
4. Sharing: Presentation of solution ideas.
5. Documenting: Sharing the idea with a broader audience.
6. Reconstructing: Refining the proposed solution.
7. Starting: Engaging in the social implementation of the solution.

Acknowledgements

In promoting this research, we received the cooperation of Global Goals Jam in Japan's facilitator Shinichiro Ito, Tokushu Inamura, and Koichiro Fumoto; the cooperation of Director Saneyuki Uehara, Section Chief, Daisuke Ushifusa and Nanako Yamashita, from the One Flower Promotion Department, Housing and Urban Affairs Bureau, Fukuoka City; and Mieko Kimura, Managing Director of the Enjoy Well-Being Society. We would like to express our gratitude for their contributions.

References

Alhonsuo, M., Hookway, S., Sarantou, M., Miettinen, S., & Motus, M. (2020). Healthcare design sprints: What can be changed and achieved in five days? In S. Boess, M. Cheung, & R. Cain (Eds.), *Synergy – DRS International Conference 2020, 11–14 August, Held online by Design Research Society*. 74–991. https://doi.org/10.21606/drs.2020.231

Alkire, L., Mooney, C., Gur, F. A., Kabadayi, S., Renko, M., & Vink, J. (2020). Transformative service research, service design, and social entrepreneurship: An

interdisciplinary framework advancing wellbeing and social impact. *Journal of Service Management*, 31(1), 24–50.
Bamber, C. J., Sharp, J. M., & Castka, P. (2004). Third party assessment: The role of the maintenance function in an integrated management system. *Journal of Quality in Maintenance Engineering*, 10(1), 26–36.
Brown, T. (2008). Design thinking. *Harvard Business Review*, 86(6), 84.
Brown, T., & Wyatt, J. (2010). Design thinking for social innovation. *Development Outreach*, 12(1), 29–43.
Coleman, R., Lebbon, C., & Myerson, J. (2003). Design and empathy. In J. Clarkson, S. Keates, R. Coleman, & C. Lebbon (Eds.), *Inclusive design* (pp. 478–479). Springer. https://doi.org/10.1007/978-1-4471-0001-0_29
Derksen, F., Bensing, J., & Lagro-Janssen, A. (2013). Effectiveness of empathy in general practice: A systematic review. *British Journal of General Practice*, 63(606), e76–e84.
Ferreira, B., Silva, W., Oliveira, E., & Conte, T. (2015). Designing personas with empathy map. In *SEKE* (Vol. 152).
Frayling, C. (1993). Research in art and design. *Royal College of Art Research Papers Series*, 1(1), 1–5.
Global Goals Jam (GGJ). (n.d.). *Jammer resources*. Global Goals Jam. https://globalgoalsjam.org/jammer-resources
Goleman, D. (1995). *Emotional intelligence*. Bantam Books.
Goleman, D. (1998). *Working with emotional intelligence*. Bantam Books.
Halpern, J., & Walther, C. (2022). Design thinking and the UN sustainable development goals: Design thinking and youth empowerment case study ForUsGirls (US) and start-up Africa (Kenya). In C. Meinel & T. Krohn (Eds.), *Design thinking in education* (pp. 233–253). Springer. https://doi.org/10.1007/978-3-030-89113-8_16
Jung, H. (2019). Changes in the students' consciousness through co-learning in multicultural classrooms: An analysis based on world cafe session results. In *CAJLE Annual Conference Proceedings* (pp. 128–136). https://www.cajle.ca/wp-content/uploads/2022/10/16_CAJLE2019Proceedings_JungHyeseon_Final.pdf
Krznaric, R. (2019). 共感する人, ホモ・エンパシクスへ、あなたを変える六つのステップ [The empathetic person: Six steps to change yourself into a homo empathicus] (K. Tanaka & T. Ogino, Trans.). Puneumasha.
Kuure, E. (2020). *Service design workshops in design practice*. https://urn.fi/URN:ISBN:978-952-337-235-1
Maher, R., Maher, M., Mann, S., & McAlpine, C. A. (2018). Integrating design thinking with sustainability science: A research through design approach. *Sustainability Science*, 13, 1565–1587.
Manzini, E. (2014). Making things happen: Social innovation and design. *Design Issues*, 30(1), 57–66. https://doi.org/10.1162/DESI_a_00248
Manzini, E. (2015). *Design, when everybody designs: An introduction to design for social innovation*. MIT Press.
Mattelmäki, T., & Battarbee, K. (2002). Empathy probes. In *Proceedings of the Participatory Design Conference PDC 2002, Malmö, Sweden, June 2002* (pp. 266–271). CPSR.
Ministry of Education, Culture, Sports, Science and Technology (MEXT). (2013). 科学技術・イノベーション白書 (Science, technology and innovation white paper). https://www.mext.go.jp/b_menu/hakusho/html/kagaku.htm

Ooshima, T. (2006). PREP法で簡単に身につく 論理的に「話す」技術 (Logical 'speaking' techniques easily acquired with the PREP method). Nihon Jitsugyo Publishing.
Sangiorgi, D. (2011). Transformative services and transformation design. *International Journal of Design*, 5(2), 29–40.
Tamura, D., Sakai, Y., Sawamura, M., & Shigemura, M. (2014). メンタルモデル: ユーザーへの共感から生まれる UX デザイン戦略 (Mental models: UX design strategies born from empathy for users). Maruzen Publishing.
Vink, J. E., & Koskela-Huotari, K. (2021). Social structures as service design materials. *International Journal of Design*, 15(3), 29–43.
Wright, P., & McCarthy, J. (2008). Empathy and experience in HCI. In *Proceedings of the SIGCHI Conference on Human Factors in Computing Systems, Florence, Italy, April 2008* (pp. 637–646). Association for Computing Machinery.
Yang, C. F., & Sung, T. J. (2016). Service design for social innovation through participatory action research. *International Journal of Design*, 10(1). https://www.ijdesign.org/index.php/IJDesign/article/view/2456
Zhang, Y., Sarantou, M., & Cruz, C. (2023). An education model for creating engagement in SDG action. In D. Jones, N. Borekci, V. Clemente, J. Corazzo, N. Lotz, L. M. Nielsen, & L.-A. Noel (Eds.), *The 7th International Conference for Design Education Researchers*, 29 November – 1 December 2023, London, United Kingdom (pp. 1–15). Design Research Society.

6

Using Service Design to Create Smart Jewellery Low-Fi Prototypes for Pandemic Challenges

Virve Inget, Mari Suoheimo, Manuela Quaresma and Jonna Häkkilä

Abstract

This chapter investigates the conceptual design and potential use of jewellery-like wearable devices during pandemics, such as the Covid-19 pandemic, via service design facilitation. We report three workshops involving service prototyping held in Norway, Finland and Brazil, where the participants took part in three different activities. First, they participated in an empathy-building activity that was part of the service design practice. The participants received three images depicting the Covid-19 time challenges and created notes about them. Second, they received a low-fi prototyping kit to create wearable jewellery that could address the challenges discovered. Third, a subset of the concepts was chosen for a more in-depth ideation concerning the stakeholders and services required for the functionality of the wearable device. These wearables aim to assist individuals in navigating the challenges of pandemic times and provide practical applications for everyday use. This research advances the understanding of how to concept, prototype and design human-centric aesthetically pleasing wearables.

Introduction

In recent years, several global crises have had drastic impacts on our lives. One of these was the Covid-19 pandemic, which succumbed to rapid mobility restrictions and social isolation. This challenged the means of being in touch with loved ones on an everyday basis and highlighted the important role of digital communication (Karhu et al., 2021). Currently, forecasts predict the increased likelihood of future pandemics because of global changes (Baker et al., 2022), so it is essential that we continue to explore how to better prepare for such pandemics or other transmissional diseases. The rapid identification of infected individuals is crucial for detecting pandemics such as Covid-19. Research indicates that wearable devices show promise in early infection detection by measuring physiological parameters, such as heart rate, respiratory rate and skin temperature (Mitratza et al., 2022). In the case of the transmission of seasonal flu, wearables and services can play a role in preventing it (Radin et al., 2020). In this chapter, we explore how new communication device concepts, in the form of wearable jewellery, can be utilized in a pandemic context, particularly to enhance communication between people.

Wearable devices are electronic devices designed to be worn on the body, often as accessories or embedded in clothing. These devices are equipped with sensors, processors and communication technology, allowing them to collect data, perform computations and interact with other devices or users. Jewellery wearables or smart jewellery is a type of jewellery that has been enhanced with such technology. These devices can perform a variety of functions, such as tracking fitness or health data, receiving notifications, controlling other devices or even making payments; here, the functionality is embedded in bracelets or rings (Silina & Haddadi, 2015). Wearables often connect to a smartphone or other device through Bluetooth or other wireless technology, allowing the wearer to access and control the device's features through a companion application. Different products or artefacts, such as smart jewellery, can be touchpoints (Pääkkönen et al., 2021) in a service and play an essential role in ensuring that the service works smoothly.

As wearable devices evolve into various forms, such as rings, bracelets and clothing accessories, it becomes important to address users' expectations for aesthetics, appeal and form factors. These requirements ensure that wearables blend seamlessly with users' personal styles and do not disrupt their look (El-Gayar et al., 2019). The design of smart jewellery should be based on an understanding of the artistic and functional characteristics of traditional jewellery and their societal and cultural significance as adornments for their wearers (Ahde-Deal et al., 2017; Silina & Hsu, 2018). Smart jewellery is often designed by industrial designers and engineers, with little involvement from jewellery designers or goldsmiths, not to mention regular jewellery users. To design better wearables, it is important to understand what the local people and their needs are and what service design can provide via empathy creation (Miettinen et al., 2016).

In this chapter, we investigate the possibilities of smart jewellery in the context of pandemics. We take a participatory and service design approach and focus on the following research questions:

1 How might smart jewellery, as part of a service, help address challenges during the pandemic?

2 How can service design and participatory design be used to lower the threshold for creating smart jewellery concepts?

Our research contributes to the broader discussion on wearable health and wellness technologies, particularly in their conceptualization and prototyping. It should be noted, however, that technical sensing solutions and medical aspects are outside the scope of our chapter because our focus is on participatory design and service design.

Prototyping of Empathic Service Design as an Approach

User-centredness, cocreation and empathy creation are at the core of service design (Miettinen et al., 2016). In service design, sequencing, evidence and holism are valued (Stickdorn et al., 2018). The aim of using different tools is often to increase participation, which is important, as participation can build empathy (Rytilahti & Miettinen, 2016), for example when designing experiences (Soto et al., 2023). Empathy involves placing oneself in others' shoes or exposing and understanding the needs or feelings of the user. The aim of participatory and service design is to provide voices for the user and, at the same time, to distribute power (Kindon et al. 2007; Souleles, 2017; Stickdorn et al., 2018). Participatory design has long roots in Scandinavia, dating back to the 1970s with democratic planning (Bødker & Pekkola, 2010), and some of the reasons for this success could be because power relations are generally quite low in Nordic countries. Service design has also been proven to be a useful approach when designing new products for emerging technologies (Hamid & Suoheimo, 2023). Using different service design perspectives can aid emerging technologies' product design to maintain user-centred focus, thus supporting that the users' needs are met (Lee et al., 2022; Suoheimo et al., 2023).

When developing a design artefact or service design touchpoint, one of the key stages in the process is prototyping. Prototyping allows us to analyse, test and validate solutions for the problem we are trying to address. Prototypes are understood as representations, embodiments or manifestations of concepts and ideals for possible design solutions (Blomkvist & Holmlid, 2011). Prototypes enable us to test the potential and validity of an idea from the user's perspective and in the context where the artefact will be used right at the beginning of a project. Service design prototyping aims to assess the customer experience through the composition, continuity and consistency of interactions with the service (Blomkvist, 2016). Prototyping and evaluating this whole chain of parts, elements and touchpoints are not always easy tasks for understanding the experiences of those who will interact with the service.

Service design has also been applied to develop service concepts that include interactive artefacts as a key part of the service. For instance, service concepts with interactive wallpapers have been investigated with projected visualizations simulating the artefact (Häkkilä et al., 2023), and wearable displays for connecting with one's partner have been explored with nonfunctional prototypes (Jarusriboonchai et al., 2020). A systematic literature review reveals that the human–computer interaction (HCI) field has used service design, especially for gaining a journey perspective on users' actions, and in addition to general design research methods such as interviews, customer journey maps are the most frequently used method (Lee et al., 2022).

Research Method

The methodological approach used in our research on service design and participatory design methods was applied to a case study. Low-fidelity (low-fi) prototyping refers to the use of simple, rough or unfinished prototypes to quickly and inexpensively test ideas and concepts and gather feedback during the design process (Walker et al., 2002). Low-fi prototyping can be a useful tool for designing smart jewellery because it allows designers to quickly and inexpensively explore a range of concepts and ideas and gather feedback from users early in the design process (Fortmann et al., 2016). As a case study, we took the topic of smart jewellery design to overcome pandemic challenges and used it as an example of an emphatic service concept creation for connecting people with an unconventional computing form factor. Case studies 'can include any issue which the researcher finds worth studying' (Gummesson, 2017, p. 6), and they can result in recommendations or theory generation (Gummesson, 2017). The data were collected in three workshops in Norway, Finland and Brazil (Table 6.1), consisting of views from those countries.

The presented data consist of three workshops (2022–3) in Norway, Finland and Brazil. All twenty-four participants were design students at their respective research institutes. The workshops in Norway and Brazil were entirely in person, but the workshop in Finland was hybrid, with some attending online and some in person. In total, there were eight women and three men. The average age of the participants was 30 years, ranging from 18 to 42 years. All participants completed consent forms, and the workshops were recorded. All participants in the present study had spent time in quarantine (at home) during the Covid-19 pandemic so that the context of the research was familiar to them.

The participants completed the consent form and background questionnaire. This was followed by a short expert presentation introducing the area of smart jewellery. The participants were engaged in three different activities during the workshop. The first task was a warm-up exercise with images to understand Covid-19 time challenges. In the second stage, the participants ideated smart jewellery concepts to address the challenges encountered and used prototyping materials to articulate their concept ideas in a tangible manner. In the third task, a stakeholder map was created to understand what people or entities were needed to make such a product work as a service. The data consisted of recordings, notes, Post-it® notes, pictures, stakeholder maps and templates the participants had filled in. In all three workshops, the participants were given the same three activities with the same images and templates to use.

Prior to the workshops, a low-fi prototyping toolkit was created and tested in a pilot workshop on smart jewellery concept design (Inget et al., 2022). The kit contained easy-to-use materials such as blu-tack, polymer modelling clay, paper and different kinds of stickers (Figure 6.1). Like cardboard and paper, polymer clays can be found in a variety of colours that can be further considered to obtain new colours.

TABLE 6.1 *Participatory design workshops*

Place	**Amount of participants**	**Hours of recording**	**Month and year**
Norway	5	2 h 31 min	04/2022
Finland	11	2 h 26 min	10/2022
Brazil	8	3 h 47 min	07/2023
TOTAL	24	8 h 44 min	2022–2023

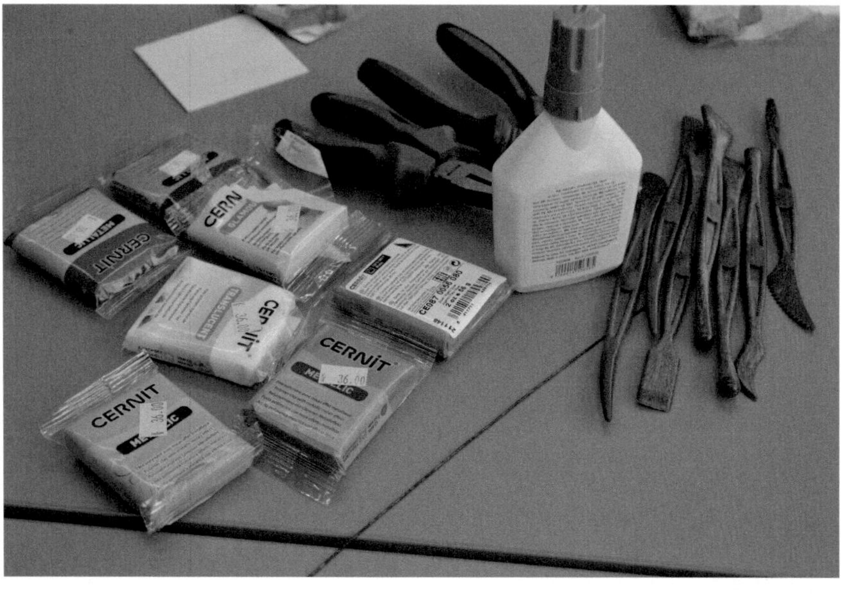

FIGURE 6.1 *Low-fi prototyping materials used during the research process.*

Results

Empathy-Building Activity via Images

In the first workshop activity, three images were placed on the wall as stimuli. One image displayed a person travelling on a bus, the second image was taken inside a grocery store and the third image presented a restaurant with a person giving an order at a desk. The participants were asked to use Post-it® notes to write down challenges that one would encounter in these specific situations during the Covid-19 pandemic and to place the Post-it® notes around the image, as illustrated in Figure 6.2. After this, the notes were discussed.

Prototyping Low-Fi Smart Jewellery

From the pandemic challenges identified with Post-it® notes, the participants could then select a specific topic to further investigate how the challenge could be met with a smart jewellery device and a service linked to it. The participants were instructed to imagine a jewellery smart device that they could use during Covid-19 isolation, for instance, to maintain contact with friends or family or to provide health-related information.

The participants then proceeded to low-fi prototyping of their concept idea. The participants explored and further developed their ideas through rough models made with paper, modelling clay and other craft supplies provided in the pre-prepared prototyping toolkit (Figure 6.3).

FIGURE 6.2 *Empathy-building activity, where workshop participants were thinking of pandemic challenges in particular situations.*

FIGURE 6.3 *A participant in Norway modelling the concept and writing it down on a template.*

Participants created concepts and models including necklaces, earrings and bracelets, which, in most cases, were targeted to aid in communication between people during a pandemic or to detect health issues. Most of the created concepts were jewellery items, but other gadgets also emerged. Figure 6.4 illustrates the four concepts. First, from the left side is a ring that could show a person's mood if a person is feeling lonely. The picture beside presents a concept from Brazil that is an earring measuring oximetry, showing if the wearer is ill. The third concept on the left side illustrates a necklace that has a screen and communicates short messages. The last concept from a student in Finland, on the right side, is a necklace that shows if the person has had recent contact with another person and if there has been a longer time since the last contact.

Figure 6.5 showcases a conceptual design from a workshop in Norway. This jewelry concept combines an earring with an in-ear headphone, drawing inspiration from students' experiences of wearing face masks while dealing with multiple items around or in their ears. The same image shows how the participants in the workshops were given stickers to vote on the concepts they liked best.

The participants shared their design ideas by explaining their concepts to each other. The participants were able to explore different design directions and consider various approaches to the challenge by experimenting with their ideas with available materials. Ideas were sketched, discussed with other

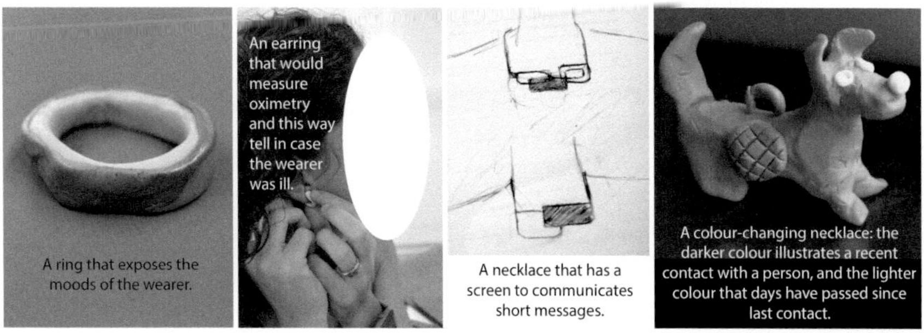

FIGURE 6.4 *Examples of different smart jewellery concepts created in the workshops.*

participants and finally shared within the group. In Finland, in the hybrid workshop, the ideas and concepts were shared, presented and commented upon online using a web-based Miro collaboration tool. In Norway and Brazil, ideas and concepts were presented in person.

Stakeholder Mapping Services

The third activity began with a voting exercise, a common activity in service design projects to identify what areas, topics or concepts are seen as the most important for future development. Each participant gave three stickers (votes), each one to a separate concept that would be elected for future development (Figure 6.6). The concepts with the most votes were elected for a stakeholder mapping activity. The purpose of the exercise was to help the participants think about how the smart jewellery product was part of a larger service. This involved considering what stakeholders needed to be involved in making and using the product. Figure 6.6 shows how workshop participants in Brazil created a stakeholder map on a table with Post-it® notes.

Figure 6.7 (right side) presents a stakeholder map on a provided template, divided into three parts: (A) the core team that develops the product and service; (B) the stakeholders that should be involved, meaning the ones involved in the development phase; and (C) the informed stakeholders, that is, the ones that are not actively taking part but need to be informed during the development process. The concept ideas being quite different resulted in the stakeholder maps being quite different from each other. The proposed service concepts often included a connected digital service that would require a connection with a smartphone service platform (e.g. Apple Music). Additionally, the development of specific apps was proposed. Interestingly, some participants brought up social media influencers as needed stakeholders to promote the product to potential users. Additionally, the participants considered the production of the product and what entities

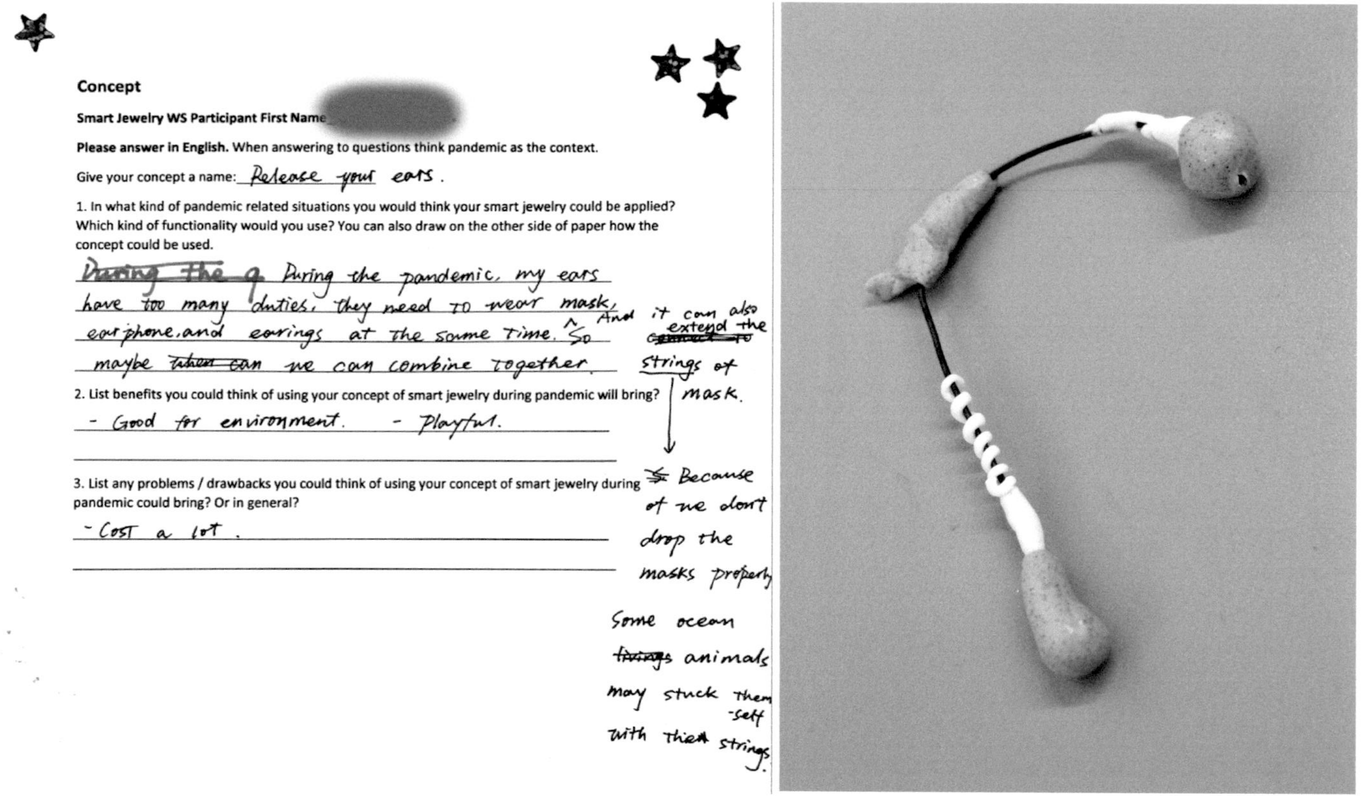
Concept

Smart Jewelry WS Participant First Name [redacted]

Please answer in English. When answering to questions think pandemic as the context.

Give your concept a name: Release your ears.

1. In what kind of pandemic related situations you would think your smart jewelry could be applied? Which kind of functionality would you use? You can also draw on the other side of paper how the concept could be used.

~~During the~~ During the pandemic, my ears have too many duties, they need to wear mask, earphone, and earings at the same time. So maybe ~~then can~~ we can combine together. And it can also extend the ~~connect to~~ strings of mask.

2. List benefits you could think of using your concept of smart jewelry during pandemic will bring?

- Good for environment. - Playful.

3. List any problems / drawbacks you could think of using your concept of smart jewelry during pandemic could bring? Or in general?

- Cost a lot.

Because of we don't drop the masks properly. Some ocean ~~things~~ animals may stuck themself with their strings.

FIGURE 6.5 *Earring concept from a workshop in Norway.*

FIGURE 6.6 *Participants in Brazil making a stakeholder map.*

were needed to make the actual product, including financing, raw materials, manufacturing and logistics.

Figure 6.7 (left side) illustrates the concept of a wearable earring with a pearl that lights up when one receives a message from their loved one. By pressing it, one can listen to it as many times as the person wishes. The stakeholder map on the right shows what kinds of partners or entities are needed to create such a product. In the core team, there are stakeholders such as jewellery designers, UX/UIs, programmers, product testing personnel, sound technicians and project managers, and the list goes on.

Discussion

Empathy and Insight through a Participatory Approach

The participatory workshops functioned well in terms of generating smart jewellery concept ideas, and the participants could place themselves in the role of designing a service for pandemic challenges. Wearable low-fidelity prototypes, crafted from easily accessible materials, such as paper and art supplies, provide a straightforward and intuitive method for generating preliminary depictions of design concepts. The participants could articulate

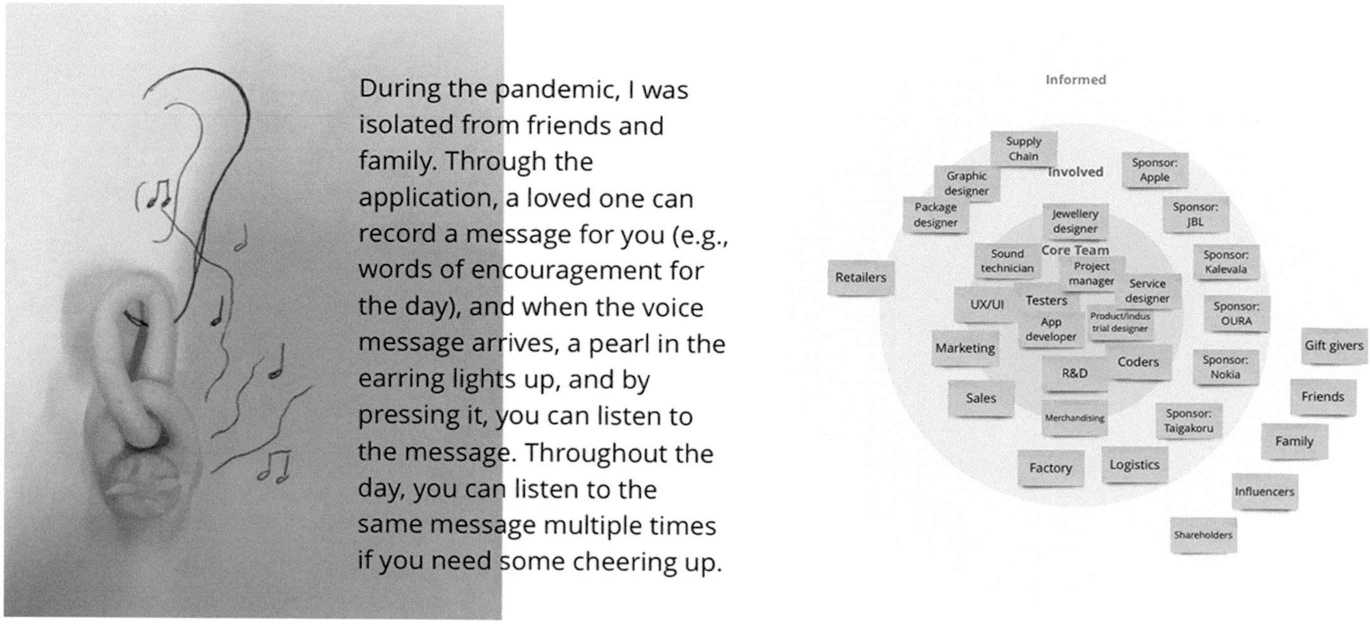

FIGURE 6.7 *Concept of pandemic time earring that sends you messages from your beloved ones (left) and the stakeholder map for creating the product and service (right).*

their thinking, share insights and discuss challenges encountered in the prototyping phase by showcasing various facets of their prototypes. They could also express their preferences for both form factors and functionality, even though they believed these might be unattainable because of technical constraints. As workshop facilitators, we observed that the participants started working quite easily on the materials provided to them. The participants asked for clarifying instructions, but in general, turning an idea into a concrete prototype with modelling clay was perceived as a simple and low-threshold activity.

When presenting their design ideas to other participants, storytelling served as an effective tool (Kim et al., 2019). In service design, storytelling promotes an inclusive and user-centric approach that facilitates communication and collaboration. The participants were able to ground their design ideas in real-world contexts and explain their solutions with real-life examples. Incorporating storytelling into the prototyping phase helps create concepts that resonate emotionally with stakeholders; it allows for simulating various user interactions and identifying potential problems in the early design phases. By the end of the workshop, the participants had a clearer understanding of the strengths and weaknesses of their design ideas.

In summary, the results of the workshop encompass idea generation, concept exploration, prototype creation, user feedback and iterative design. The workshop served as a dynamic and collaborative environment for rapidly evolving design concepts with low-fi tools and refining them based on participants' insights and feedback. We recognize that our research is limited by the small sample size of mainly female participants and methods, fitting the early concept ideation phase rather than the product development phase. Additionally, despite our research interest in smart jewellery, we recognize that this approach is not a suitable solution for many use cases, but it must be carefully selected because of its usability and practicality. However, we believe that our research provides useful insights into how to integrate hands-on participatory design methods in the area of empathy in design as well as smart jewellery. Although smart jewellery is not the solution for all use cases and is unlikely to address the challenges of pandemic times, we see that the potential of smart jewellery is still underexplored, and new concepts integrating smart jewellery as part of a service may emerge.

Empathy Viewpoints in the Cross-Cultural Approach

Although we had three participatory workshops run in three culturally different countries, we found many similarities between concepts of smart jewellery and identified challenges. This approach improved the validity of the study. There were still some issues that were different because of cultural backgrounds and country-specific demands and that are interesting in terms of empathy. We used identical stimuli to maintain a consistent workshop. However, the participants in Brazil questioned the picture of a bus that was taken in Finland because the buses were different from their city. Additionally, they reported that the government prohibited the use of air conditioning during the Covid-19 outbreak, making it difficult for local people to travel at high temperatures and humidities. The fear was that air conditioning could spread the virus inside the bus. As facilitators, this taught us that we should bring context-specific images to the workshops, which will demand knowledge of the local culture beforehand. Similarly, although one of the images was from an international restaurant chain present in all three countries, there can still be slight differences. Better images of the local context could help the workshop participants think about what their local challenges are. This was a good reminder that cultural and local sensitivity is important when planning workshops and diminishes possible challenges.

Notes on Service Design

With stakeholder mapping, a common tool used by service designers, the participants were able to identify key stakeholders and actionable steps for further development and refinement. Although the real stakeholders were not involved in the workshop, relevant stakeholders were identified. Stakeholder feedback to prototypes can later provide valuable insights into user preferences and needs as well as comments on technical development and business aspects. The activity also successfully showed how the physical low-fi prototypes were seen as part of a larger, often digital, service concept. The participants would see a need to connect wearables with apps for other digital existing or non-existent services. Additionally, thinking about stakeholders and their needs in a design process is one form of empathy-building.

According to the workshop concepts, smart jewellery has the potential for facilitating communication and being in touch during pandemics, alerting to dangerous situations, and communicating with users about important issues such as information about possible infections. The stakeholder maps from the workshops illustrated the role of physical smart jewellery in a service. With this exercise, we could see how physical products, such as wearable jewellery, are able to play a vital touchpoint role in a larger service. Touchpoints are often physical objects that a user of a service can touch on, such as taking a ticket to pay for parking or reserving a table via an app.

Bridging the Gaps in Smart Jewellery Design

Industrial designers and engineers are skilled in concepting, creating and implementing technical devices. However, the design challenge is to involve a wider audience in cocreating wearables, especially smart jewellery, who have varying levels of knowledge and skills. For instance, traditionally trained jewellery craftsmen might not have technical skills but may be experts in jewellery production techniques and understanding the historical and cultural significance of different types of jewellery. Regular jewellery users have multifaceted relationships with their jewellery (Ahde-Deal et al., 2017; Silina & Hsu, 2018). Jewellery may carry various meanings for wearers and is often associated with culture, memories and emotions. Collaborating with jewellery industry professionals and regular jewellery users can provide new perspectives and understanding for the design of smart jewellery.

To enable cocreation prototyping with craftsmen and regular jewellery users, easy-to-use, low-fidelity toolkits are essential. These toolkits should be designed to lower the participation barrier, allowing everyone to use basic skills to create concepts. The currently available toolkits often include technical components, such as ready-made sensors and circuits (Harjuniemi et al., 2023), which limit their easy modification to meet different design needs. Participants are typically required to build upon predefined measurements and forms. In contrast, using mouldable, colourful materials alongside basic moulding tools can provide greater opportunities for experimenting with different form factors and product sizes. This approach encourages creativity and flexibility in the prototyping process, making it more accessible and engaging for all participants.

One aspect of the future vision for smart jewellery low-fi prototyping is the introduction of new types of smart materials. New materials, such as flexible electronics, conductive textiles and shape-memory alloys, can revolutionize the prototyping process of smart jewellery. Researchers specializing in smart materials should collaborate closely with designers to fully leverage these advancements, ensuring that emerging materials are easy to use and can be utilized from the early phases of product conceptualization.

Conclusion

By conducting three workshops in three different countries, we were able to gather a good number of ideas on how smart jewellery could address challenges during pandemics via interaction and service design. The solutions were diverse, including concepts that were informative, alerting or for staying connected with loved ones. Low-fi with modelling clay was a successful method for lowering the threshold to create and share ideas among workshop participants. The workshops included several empathy-building activities, such as re-remembering the Covid-19 period and addressing challenges via images or making stakeholder maps and connecting wearables in a larger service. Based on our experience, we recommend using low-fi prototyping materials to construct service concepts when tangible touchpoints or physical products are considered part of the service ecosystem. This also blends the boundary between the physical product and the service connected with it because both play an important role in making a service experience. In the future, we aim to continue developing smart jewellery concepts and explore them as design probes. Moreover, we recommend that future studies investigate how different prototyping techniques can support physical–digital service creation.

Takeaway

Hands-on prototyping is a practical and essential tool for empathy-building when creating new service concepts and touchpoints, such as those utilizing smart jewellery. Having participants from different cultures and versatile backgrounds can be a better way to create empathy because the perspectives and needs can be different, even though the challenges may be the same. Low-fidelity prototyping methods that are flexible and allow creativity to flourish should be further developed to integrate emerging technologies and empathy.

References

Ahde-Deal, P., Paavilainen, H., & Koskinen, I. (2017). 'It's from my grandma.' How jewellery becomes singular. *The Design Journal*, 20(1), 29–43.

Bødker, S., & Pekkola, S. (2010). Introduction to the debate section: A short review of the past and present of participatory design. *Scandinavian Journal of Information Systems*, 22(1), 4.

Baker, R. E., Mahmud, A. S., Miller, I.F., Rajeev, M., Rasambainarivo, F., Rice, B. L., Takahashi, S., Tatem, A. J., Wagner, C.E., Wang, L. F. and Wesolowski, A. (2022). Infectious disease in an era of global change. *Nature Reviews Microbiology*, 20(4), 193–205.

Blomkvist, J. (2016). Benefits of service level prototyping. *The Design Journal*, 19(4), 545–564. https://doi.org/10.1080/14606925.2016.1177292

Blomkvist, J., & Holmlid, S. (2011). Existing prototyping perspectives: Considerations for service design. *Proceedings of the Nordic Design Research Conference 2011*. Design Research Society, 1–10. https://doi.org/10.21606/nordes.2011.012

El-Gayar, O., Nasralah, T., & Noshokaty, A. (2019). Wearable devices for health and wellbeing: Design insights from Twitter. In *Proceedings of the 52nd Hawaii International Conference on System Sciences*. ScholarSpace / AIS Electronic Library (AISeL) (pp. 3858–3867).

Fortmann, J., Root, E., Boll, S., & Heuten, W. (2016). Tangible apps bracelet: Designing modular wrist-worn digital jewellery for multiple purposes. In *Proceedings of the 2016 ACM Conference on Designing Interactive Systems* (pp. 841–852). ACM.

Gummesson, E. (2017). *Case theory in business and management: Reinventing case study research*. SAGE Publications.

Hamid, U. Z. A., & Suoheimo, M. (Eds.). (2023). *Service design for emerging technologies product development: Bridging the interdisciplinary knowledge gap* (Springer Series in Design and Innovation, Vol. 29). Springer Nature.

Häkkilä, J., Löchtefeld, M., Brun, D., & Colley, A. (2023). Service design methods for the design of smart surfaces. In U. Z. A. Hamid & M. Suoheimo (Eds.), *Service design for emerging technologies product development: Bridging the interdisciplinary knowledge gap* (pp. 235–247). Springer International Publishing.

Harjuniemi, E., Jarusriboonchai, P., & Häkkilä, J. (2023). Light it up: Designing electronic textiles with light as a design material. *Connectivity*, 304. https://library.oapen.org/handle/20.500.12657/85854

Inget, V., Suoheimo, M., & Häkkilä, J. (2022). Smart jewellery prototyping with children as part of a service design process. In *Toolkits and Wearables 2022 Workshop at CHI 2022* (pp. 1–4). University of Lapland.

Jarusriboonchai, P., Li, H., Harjuniemi, E., Müller, H., & Häkkilä, J. (2020). Always with me: Exploring wearable displays as a lightweight intimate communication channel. In *Proceedings of the Fourteenth International Conference on Tangible, Embedded, and Embodied Interaction* (pp. 771–783). ACM.

Karhu, M., Suoheimo, M., & Häkkilä, J. (2021). People's perspectives on social media use during the Covid-19 pandemic. In *Proceedings of the 20th International Conference on Mobile and Ubiquitous Multimedia* (pp. 123–130). ACM.

Kim, M., Srinivasan, D., & Zhou, X. (2019). The morphology of dignity: Service storytelling and prototypes for a service design tool. *The Design Journal*, 22(6), 793–812.

Kindon, S., Pain, R., & Kesby, M. (2007). *Participatory action research approaches and methods: Connecting people, participation and place* (Routledge Studies in Human Geography, Vol. 22). Routledge.

Lee, J. J., Yap, C. E. L., & Roto, V. (2022). How HCI adopts service design: Unpacking current perceptions and scopes of service design in HCI and identifying future opportunities. In *Proceedings of the 2022 CHI Conference on Human Factors in Computing Systems* (pp. 1–14). ACM.

Miettinen, S., Jylkäs, T., Tikkanen, H., & Jeminen, J. (2016). Service design for business: Value creation opportunities through service design research. In *Proceedings of the 20th DMI Academic Design Management Conference* (pp. 455–470). Design Management Institute.

Mitratza, M., Goodale, B. M., Shagadatova, A., Kovacevic, V., van de Wijgert, J., Brakenhoff, T. B., Dobson, R., Franks, B., Veen, D., Folarin, A. A., Stolk, P., Grobbee, D. E., Cronin, M., & Downward, G. S. (2022). The performance of wearable sensors in the detection of SARS-CoV-2 infection: A systematic review. *The Lancet Digital Health*, 4(5), e370–e383.

Pääkkönen, T., Sarantou, M., & Miettinen, S. (2021). Meaning innovations with design support: Towards transparency and sustainability in the IT field. *Design Management Journal*, 16(1), 79–92.

Radin, J. M., Wineinger, N. E., Topol, E. J., & Steinhubl, S. R. (2020). Harnessing wearable device data to improve state-level real-time surveillance of influenza-like illness in the USA: A population-based study. *The Lancet Digital Health*, 2(2), e85–e93.

Rytilahti, P., & Miettinen, S. (2016). For profit, for good: Developing organizations through service design. In Piia Rytilahti and Satu Miettinen (Eds.), *For Profit, for Good* (pp. 3–10). University of Lapland.

Silina, Y., & Haddadi, H. (2015). New directions in jewelry: A close look at emerging trends & developments in jewelry-like wearable devices. In *Proceedings of the 2015 ACM International Symposium on Wearable Computers* (pp. 49–56). ACM.

Silina, Y., & Hsu, A. (2018). Infusing meaning into social wearables: Lessons from sentimental jewelry. In *Proceedings of the 32nd International BCS Human–Computer Interaction Conference* (pp. 1–5). BCS Learning & Development.

Soto, M., Miettinen, S., & Alhonsuo, M. (2023, November). The prototyping service experience as a tool of empathy and research: From Finland to Chile. In *Entanglements and Flows: Service Encounters and Meanings* (pp. 440–457). Linköping University Electronic Press.

Souleles, N. (2017). Design for social change and design education: Social challenges versus teacher-centred pedagogies. *The Design Journal*, 20(1), 927–936.

Stickdorn, M., Hormess, M., Lawrence, A., & Schneider, J. (2018). *This is service design doing: Applying service design thinking in the real world. A practitioner's handbook*. O'Reilly Media.

Suoheimo, M., Trapani, P., & Miettinen, S. (2023). Historical perspectives on service design and technology. In U. Z. A. Hamid & M. Suoheimo (Eds), *service design for emerging technologies product development: Bridging the interdisciplinary knowledge gap* (pp. 25–44). Springer International Publishing.

Walker, M., Takayama, L., & Landay, J. A. (2002). High-fidelity or low-fidelity, paper or computer? Choosing attributes when testing web prototypes. In *Proceedings of the Human Factors and Ergonomics Society Annual Meeting* (Vol. 46, No. 5, pp. 661–665). SAGE Publications.

PART TWO

Implications for Creating Empathic Services within Sensitive Contexts

7

The Role of Cultural Sensitivity in Fostering Empathy in Service Design

Amélia Steunou and Luca Simeone

Abstract

In today's globalized world, increased cultural diversity presents both an opportunity and a challenge for service designers. Recognizing, embracing and valorizing cultural diversity enables service designers to expand their perspectives and adopt more empathetic approaches (Hao, 2019; Sarantou & Miettinen, 2022). However, designers might not always feel equipped to appreciate this cultural diversity and achieve a fine-grained understanding of how (the creation and use of) services can be affected by cultural specificities (Dennington, 2017; Walsh, 2016).

This chapter will focus on the role that cultural sensitivity – that is, the ability to effectively engage in cross-cultural interactions and communications, reflecting a development process through which individuals cultivate the aptitude to identify and value cultural differences (Chen & Starosta, 1997) – can play in facilitating the development of services that resonate deeply with specific cultural groups. This chapter describes a service design process carried out across Denmark and China and shows how four processes connected to cultural sensitivity (*recognition*, *learning*, *respect* and *adaptation*) are instrumental in fostering trust and empathy with and among actors. The case study will chronicle how cultural considerations played a pivotal role in creating user-friendly experiences, facilitating user appropriation, enhancing service attractiveness and, overall, fostering empathy.

This chapter builds on studies and conceptualizations of cultural sensitivity developed in cultural anthropology and social research (Appadurai, 1996; Escobar, 2012; Marcus, 1995; Rosaldo, 1989; Tsing, 2011) to provide analytical tools and actionable methods for incorporating cultural sensitivity within design processes, thereby enriching the empathetic fabric of services in an increasingly globalized world.

Introduction

Broadly characterized as the ability to step into another person's shoes (Keen, 2006), empathy equips designers with the capacity to appreciate the subjective experiences of users and other stakeholders affected by the design process and establish a sense of trust, fostering the creation of services that not only meet practical needs but also resonate with users' needs, desires and perspectives. As users increasingly seek personalized experiences (McKinsey, 2021), the ability of designers to empathize with diverse users becomes a critical factor in achieving service relevance.

In addition, with underlying processes such as increased cross-national mobility, migration flows and globally operating media platforms, our societies become the theatre where 'different cultural systems, representations, imaginaries converge and give rise to new transcultural practices' (Vanni, 2016, p. 7). The significance of empathy becomes more pronounced, particularly when designing for users whose cultural contexts diverge from those of the designers themselves. For service designers, this implies the need to conceive of their design process to welcome this diversity, valorize it and engage with it respectfully; it also presents designers with the delicate task of avoiding missteps in communication and interpretation.

This delicate balance underscores the importance of cultural sensitivity – that is, knowledge and acceptance of cultural diversity – in approaching and fostering empathy across cultural landscapes. Although some studies have traced the connection between empathy and cultural sensitivity (Goncu Berk, 2013), the topic remains underexplored. This chapter aims to bridge this gap by analysing a cross-cultural case study across Denmark and China.

Literature Review

Empathy appears central in service design, enabling identification and connection with others' thoughts and feelings (Fulton Suri, 2003). Two primary forms are affective, characterized by instinctively feeling others' experiences, and cognitive, involving an analytical understanding of others' situations (Callahan, 2018). However, culture profoundly shapes how these empathetic states are expressed and understood (Masuda et al., 2008; Mesquita & Ellsworth, 2001; Parson, 1993). Indeed, empathy, being a historically and culturally situated notion (Eichbaum et al., 2023), is not universally standardized but varies across cultures. Forging empathic connections across cultural boundaries becomes a challenge because such connections are generally easier to form with individuals who share a common cultural background (Surma-aho, 2022). Considering that empathy builds a third culture between two beings who, of different origins, begin to communicate (Broome, 1991), cultural sensitivity emerges as crucial in design processes.

Cultural sensitivity is a multifaceted concept defined in various ways. Some describe it as a concept and some as a mindset (Hart et al., 1980), while others see it as an ability to effectively communicate cross-culturally (Chen & Starosta, 1997). Appadurai (1996) emphasizes the importance of understanding the ways in which cultural identities are constructed and negotiated in our globalized world. To him, cultural sensitivity requires navigating the power dynamics and hierarchies that shape cultural exchange and recognizing the importance of cultural translation across different cultural contexts. According to Escobar (2012), cultural sensitivity entails recognizing, respecting and valuing the diversity of cultural practices and knowledge systems. It involves comprehending how culture shapes our interactions with the world and is crucial for challenging dominant discourses and structures that perpetuate inequality (Escobar, 2012). Similarly, Harkin and Rosaldo (1990) believe that cultural sensitivity involves understanding and respecting diverse human experiences and perspectives. Cultural sensitivity should be grounded in understanding the historical, social and political contexts that shape cultural identities and practices. According to Marcus (1995), cultural sensitivity is the recognition of the diversity and complexity of cultures and the power dynamics that can affect cross-cultural communication and interaction. Finally, Tsing (2011) highlights the need to respect and accommodate cultural differences and to recognize and engage with the complex interconnections between cultures; her perspective emphasizes the importance of examining power dynamics and inequalities in cultural interactions.

Scholars' discussions on cultural sensitivity highlight the necessity of questioning our positionality when navigating cultural differences. Positionality entails critical reflection on our position as designers and its implications for research (Knight & Deng, 2016). This introspection of our biases allows us to transcend our perspectives, immerse ourselves in others' experiences and navigate cultural complexities. Recognizing our own assumptions and preconceptions influences every aspect of research, from formulating hypotheses to selecting methods and tools (Rodriguez Schon & Celi, 2023). Table 7.1 summarizes key aspects of cultural sensitivity emerging from the literature.

Service design – eminently sociable and entwined with the real world – is imbued with cultural aspects (Akama & Prendiville, 2013). Operating within an ever-increasing globalized market and transcultural societies, service designers feel the urge to give greater relevance to cultural research. This awareness stems from the observed reality that failure rates on international assignments double in the absence of cross-cultural training (Pachter et al., 1995). Indeed, culture significantly impacts the processes of appropriation and rejection (Cohen, 2021), making users feel alienated in cases of cultural mismatches between their culture and the culture established in a service system (Morgan & Morgan, 2005). Therefore, designers require a deeper understanding of how value is created across complex transcultural

TABLE 7.1 *The four aspects of cultural sensitivity stemming from the literature*

The four aspects of cultural sensitivity	
Recognise	Be aware of cultural differences and their influence on human beings. This includes recognizing one's own cultural biases and assumptions.
Learn	Study the cultural norms, values, and beliefs specific to a given culture, and understand how cultural differences affect communication and interaction with others.
Respect	Approach every culture equally and without judgement, to avoid stereotypical representations.
Adapt	Use acquired knowledge and soft skills to work effectively in cross-cultural settings. This is done through adapting one's own behaviour and actions, and developing empathy and flexibility.

negotiations within the entire service process (Heinonen & Strandvik, 2015). The ability to run service design processes by paying attention to elements such as cultural resonance and adaptation would not only help people overcome cultural barriers but also ensure that products and services, here when embedded in the context in which they operate, are perceived as genuinely relevant and attractive (Clatworthy, 2011).

However, a notable absence of discourse exists within service design surrounding the execution of culturally sensitive studies (Dennington, 2017), and limited research links cultural sensitivity with empathy. This chapter addresses this gap by exploring the following research question: 'How can cultural sensitivity foster empathy in service design?'

Methodology

The present paper conducted an in-depth analysis of a single case study (Eisenhardt & Graebner, 2007). Case studies allow for the identification of critical insights through a fine-grained examination of one or more examples (Yin, 2018). The case is a cross-national service design project: a feasibility study for a bubble tea shop service operating in Denmark and China. The shops will employ Danish and Chinese workers and work with local products. The bubble tea shop wishes to foster a pleasing localized and culturally sensitive experience, hoping to build and strengthen the brand, boost sales and promote customer engagement. As we embarked on a service design journey to carry out user research and to model the service for these two initial bubble tea shops (in Copenhagen and Beijing), we tried to infuse key aspects of cultural sensitivity into the process. This project became a space to (1) experiment with the deliberate integration of cultural sensitivity within key service design methods and (2) reflect upon our research question.

We acknowledge that more case studies in different parts of the world are needed to validate our findings and that what we present in the next sections should be considered preliminary and provisional.

Cultural Sensitivity at Play in the Service Design Process

In our case, we crafted a tweaked version of the double diamond (Drew, 2019), which integrates culturally sensitive considerations (Figure 7.1).

Pre-project: We researched cultural sensitivity and identified the four key aspects (recognize, learn, respect and adapt). This characterization guided the subsequent phases; as a lens, it was constantly used in conjunction with key design methods to ascertain the cultural aspects of the project.

This stage marked the start of our empathetic design approach; through reflections on our backgrounds, cultural norms and beliefs, we recognized the influence of our positionality as designers. This process helped address the power dynamics within our methodologies, striving to empower the participants in the research.

Discover: In this initial exploratory phase, insights and inspirations were gathered using qualitative and quantitative research methods (desk research, benchmark analysis, a survey and photo diary interviews). The desk research focused on building a multilayered understanding of the two cultural contexts and touched upon historical notes and specific economic and social dimensions. A subsequent benchmark analysis of similar services already operating in Beijing and Copenhagen provided further insights into the different cultural aspects at play in the two contexts.

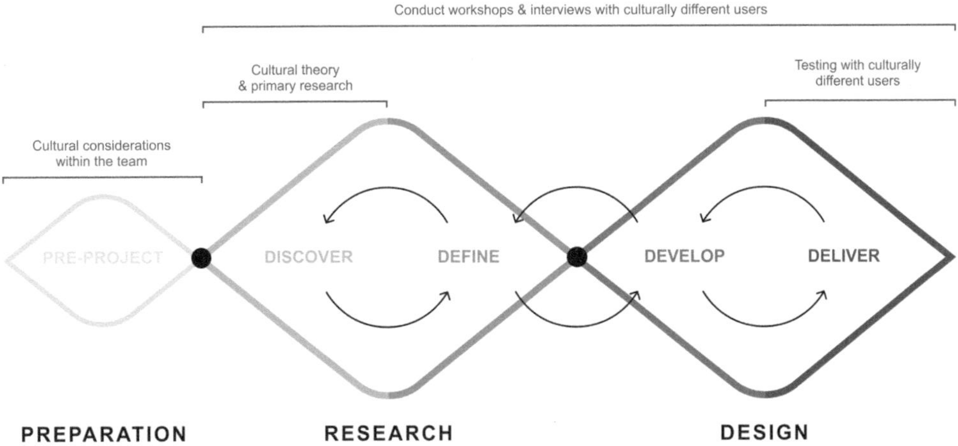

FIGURE 7.1 *Culturally sensitive double diamond (to have access to Figures at a high resolution, refer to the Miro board): https://rb.gy/k3omuu.*

Empathy with users is often hindered, particularly without face-to-face interactions. To bridge this gap, tailored approaches have been employed. One such approach involves using photo diary interviews, which allow users to document their daily experiences, thoughts and feelings in their own words and contexts. This method offers deeper insights than traditional interviews by depicting users' lives over time through their own lens. Here, the participants were asked to capture their 'bubble tea journey' through photographs by visiting existing tea shops in Copenhagen and Beijing. The participants were given 10 days to do so, followed by one-on-one interviews.

In Copenhagen (Figure 7.2), the participants' photos and subsequent interviews emphasized interest in specific ingredients, sustainable practices and individual experience. For Danish consumers, the bubble tea experience seemed to lean more towards individual enjoyment, prioritizing product quality. In contrast, the Chinese participants shared numerous pictures of group experiences (Figure 7.3), featuring friends enjoying bubble teas, the shop's interior and overall space, surrounded by a plethora of branded merchandise and promotional products. Overall, the Chinese participants adopted a more collective approach, emphasizing the importance of bubble tea in their social lives.

FIGURE 7.2 *Pictures extracted from the photo diaries of participants from Copenhagen.*

FIGURE 7.3 *Pictures extracted from the photo diaries of participants from Beijing.*

Despite potential challenges in empathizing with users – especially in the absence of face-to-face interactions – the portraits emerging from the diaries, together with deeper cultural insights from other research methods, served as a bridge to empathic understanding.

Define: This phase channelled the findings of the discover phase towards actionable tasks, where key insights were interpreted and clustered to narrow the project's scope. This phase unfolded through the following methods and processes: an interview with a bubble tea shop owner, the synthesis of findings and the creation of an ecosystem map and user group personas. These last two methods are worth highlighting in relation to their potential to foster empathy. The ecosystem map serves as a tool for visually representing the various relationships among the different actors within a service system, shedding light on their interactions and value exchanges.

The concentric rings, from user touchpoints to service providers and access providers, delineated the network of stakeholders. This mapping exercise, approached with cultural sensitivity, aimed to ensure that users' cultural contexts were considered and portrayed in the service system. The created ecosystem map (Figure 7.4) revealed the dynamics of the bubble tea service systems, highlighting diverse actors and their functional relationships. The obvious primary role of the bubble tea shop in the service proposition directed attention to the factors influencing service quality and cultural resonance.

However, as cultural considerations continued to emerge throughout the mapping process, the limitations of this map became evident. Based on desk research and prior findings, it can be hypothesized that the significance of interpersonal and hierarchical relationships in a Chinese cultural context may greatly impact interactions among actors within the system. However, capturing these dynamics within the ecosystem map proved challenging, given its nature and scale. Additionally, in Beijing, the shop is seen as an important social hub that influences the overall service experience. In the Danish context, the emerging nature of the bubble tea service poses challenges for accurate representation within this mapping framework because of the limited information available about this relatively new service. Different interpretations of value exchanges and still undefined cultural factors, including customs and regulations, added complexity to the representation. These complexities emphasize the need for a more nuanced representation of value exchanges within the ecosystem map. Hence, while providing some clarity, the map demonstrated clear limitations, and its fragility became apparent when assessing its alignment with the cultural realities of Danish and Chinese contexts.

Another key method of fostering empathy is personas. Although personas have become widely used tools in service design, they face criticism for potentially oversimplifying and stereotyping user groups, leading to irrelevant solutions that do not account for the complexity and diversity of experiences (Cabrero et al., 2016). Despite this critique, the benefits of using personas in this case study were significant; three personas each represented user groups from Copenhagen and Beijing (see Figures 7.5 and 7.6). Inclusivity was shown by defining user groups instead of specific individuals with names and ages (those personas embody user stances, e.g. 'the mindful buyer'). Abstract, low-detail drawings were chosen over photographs to avoid stereotypical representations. Moving away from quantitative gauging, a more empathetic understanding was encouraged; instead of using conventional sliders often seen in traditional personas, emphasis was placed on discussing emotions and character traits.

The personas acted as cultural narratives, presenting personal stories that were aiming for authenticity. Iteration was a fundamental aspect of persona creation, ensuring that the profiles evolved along with the cultural understanding developed throughout the project.

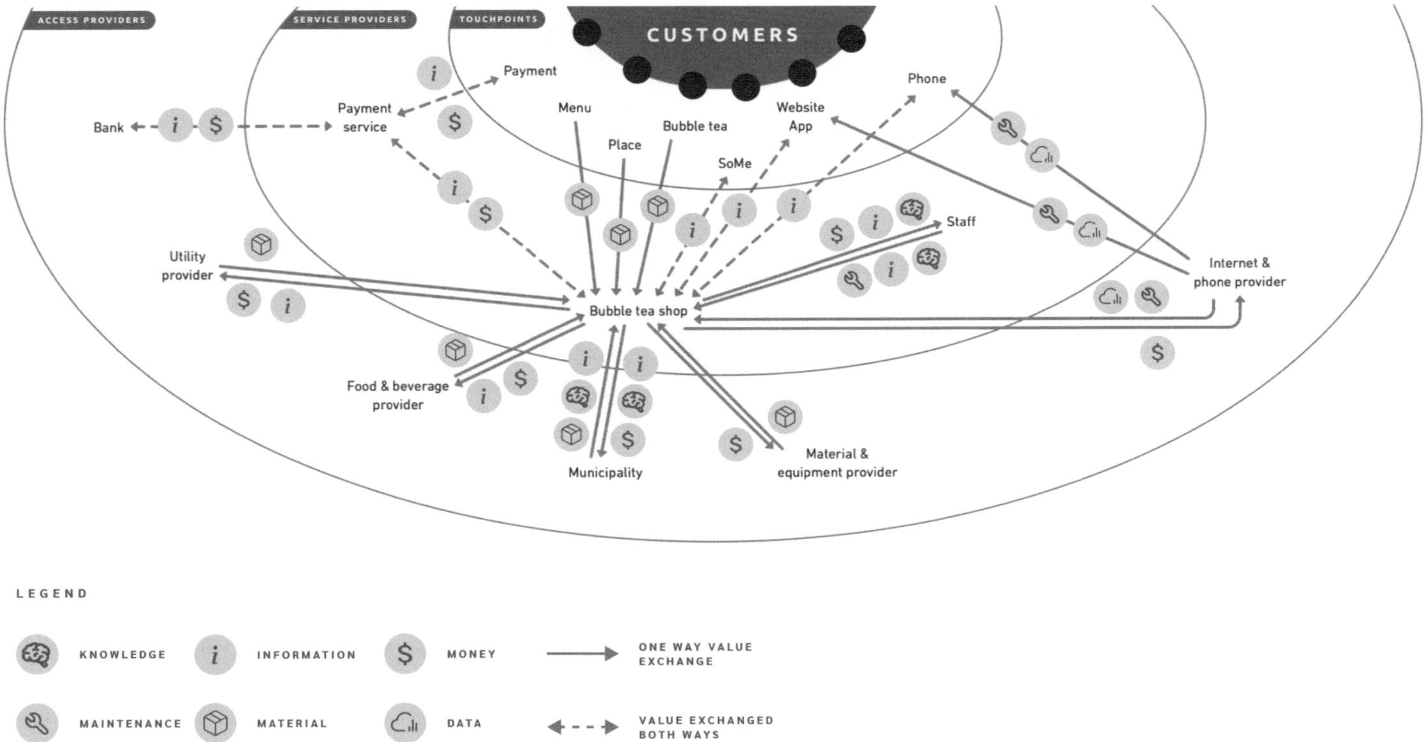

FIGURE 7.4 *Ecosystem map of the bubble tea shop.*

The Sceptic Sipper

Personality Traditional & Simple

Consumption Rarely

Positives

- Appreciate cozy and intimate atmospheres
- Enjoy authentic food and drinks
- Receive an efficient service

Frustrations

- Feeling out of place in trendy environments
- Artificial bubble tea (Sceptical of its health benefits and ingredients)
- Dealing with complex and unfamiliar menus

Story The Sceptic Sipper enjoys going to his local café where he can relax in a cosy and intimate setting. He has already tried bubble tea but is skeptical of its health benefits and ingredients. He's not one to jump at new experiences, but he's willing to give something a try if it aligns with his traditional values and beliefs. He appreciates straightforward and efficient service with great quality and authenticity, and is not a big fan of trendy or popular spots, enjoying a more low-key experience.

FIGURE 7.5 *Group persona representing the Sceptic Sipper (Copenhagen users).*

The Social Drinkers

Personality Outgoing & Caring

Consumption At least once a month

Positives

- Spend quality time together and make fun memories
- Enjoy spacious environments to sit and chat
- Get varied menu options to accommodate everyone's preferences

Frustrations

- Difficulty finding spacious environments to chat comfortably
- Limited customisation options to please everyone in the group
- Hate boring and dull atmospheres

Story The Social Drinkers are a very close group of friends who like to spend time together whenever possible. To them, the bubble tea experience is not only about the drink itself but also about the atmosphere and quality time spent with each other. They consider it a fun moment to be shared together, where they always share and care for each other. Therefore, customising their orders is important to ensure everyone is happy. They are always looking for new flavours and toppings to add to their drinks. When it comes to choosing a place, they seek a lively and fun atmosphere. They want to feel like they are part of a bustling, energetic environment where they can socialise and have a good time.

FIGURE 7.6 *Group persona representing social drinkers (Beijing users).*

Develop: In the development phase, the service concept was refined through multiple steps to ideate, develop and frame a final concept of the different service components. This phase consisted of the 'How might we?' exercise, the creation of user journeys, service blueprints, use cases, concept framing presentation and a testing workshop.

The testing workshop sought feedback from potential service users from Copenhagen and Beijing. Participants from both contexts were selected to capture an inclusive range of perspectives and ways of interacting with the prototypes of the concept presentation. The emphasis was on validating that the service concepts were commonly accepted, culturally appropriate and met users' expectations while promoting discussion and cocreation. This workshop shed light on the clarity and usefulness of the service concept and necessary improvements.

The workshop was conducted online and followed a one-to-one approach (participant and facilitator). It consisted of several activities interacting with prototypes (Figures 7.7 and 7.8) to gather user feedback regarding the brand name, logo, menu design and shop 3D visuals. For example, for prototypes geared towards Chinese users, vibrant colours and dynamic illustrations were chosen to match cultural styles. The preferences previously voiced by Danish users were incorporated when creating the different prototypes: using natural colours, cultivating a cosy atmosphere, fewer menu choices but emphasizing product quality and clear steps guiding the customers through the ordering process.

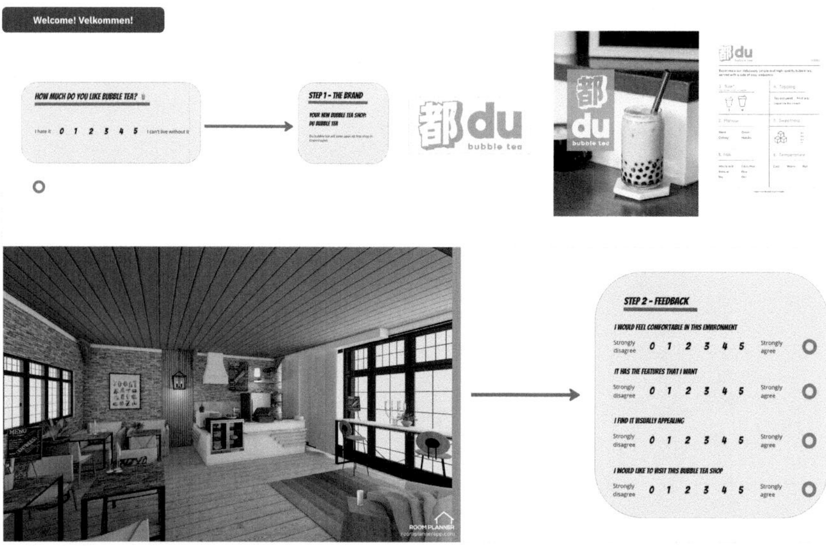

FIGURE 7.7 *Prototypes presented on a Miro board (Copenhagen version).*

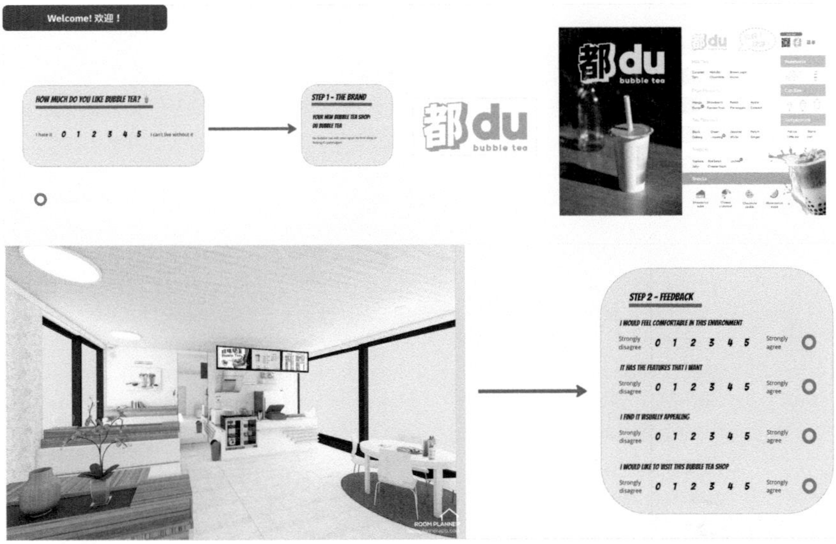

FIGURE 7.8 *Prototypes presented on a Miro board (Beijing version).*

***Deliver*:** This final stage focused on the delivery phase of the concept, including the final concept presentation, suggestions for implementation and guidelines for service designers. Distinct reports were created to provide a dedicated context for each cultural area, allowing the findings and recommendations to resonate with the respective clients. An interview with a professional service designer was conducted to gather an external viewpoint on the design process; it enriched the final deliverables with practical wisdom and industry expertise.

Discussion

Table 7.2 presents some highlights from our analysis while reconnecting to our research question: 'How can cultural sensitivity foster empathy in service design?'

The case study highlighted the importance of the process of *recognize* in laying a strong empathic foundation: Designers should cultivate an acute awareness of cultural differences and their profound influence on the entire service ecosystem (interactions with users, stakeholder engagement, deliverable development and overall service implementation). In our study, this process unfolded through multifaceted engagements with users. The immersive diary exercise provided a firsthand glimpse into the realities of users' lives and contexts of product use, compensating for the absence of in situ observation. By creating empathic visual aids, the moodboards themselves

TABLE 7.2 *How cultural sensitivity promotes empathy within our service design process*

Key aspects of cultural sensitivity as deployed in our project		**Impact on empathy within service design**
Recognize	It had to do with acknowledging cultural nuances and potential cultural differences and their implications through-out the entire project and service and was mirrored by a self-reflection process, where the designer reflected on their own positionality.	It established the groundwork for an empathetic approach to the project; recognizing cultural elements ensured that the service not only met practical needs but also contributed to a sense of cultural continuity and familiarity. Empathy became a driving force in the design process, steering decisions towards solutions that are functional and culturally resonant.
Learn	It had to do with studying cultural norms, values and beliefs, acquiring an understanding of different cultures and of how cultural values can shape users' perceptions and behaviours toward the service.	It helped us acquire a more accurate empathetic approach while synthesizing findings. It enabled the implementation of tailored strategies and decisions and more culturally resonant designs (as able to construct meanings).
Respect	It had to do with approaching cultures, being aware of own biases and preconceived notions and with a commitment to integrate cultural perspectives into the design process.	It helped identify potential cultural barriers to the adoption and usage of the service and to create inclusive and empathetic representations.
Adapt	It had to do with adapting one's behaviour and actions, as well as adapting methods and tools. Iteratively testing concepts with users as a key aspect.	It helped ensure the comprehensibility and acceptability of early ideas, representations and the overall service by tailoring the design to meet specific cultural needs and preferences.

became cultural artefacts. In alignment with the insights offered by Callahan (2018), our study posits the greater prominence of affective empathy in the development of artefacts as opposed to cognitive empathy. The photo diary exercise and subsequent interviews facilitated immersion in the emotional landscapes of the participants, which captured their emotions and aspirations. However, during the analysis of the moodboards and transcripts, our engagement gradually shifted towards cognitive empathy, deciphering the meanings behind users' narratives, thoughts and metaphors. Conclusively, our exploration of recognition highlighted the dynamic relationship between affective and cognitive empathy. Combining personal narratives and cultural insight, the photo diary interviews emerged as a bridge to empathy; it is through the clear divergent perspectives between the Copenhagen and Beijing participants that we recognized the need for nuanced approaches tailored to local expectations. However, the challenge to *recognize* lies in acknowledging potential disparities that shape user behaviours, preferences and expectations and in admitting and confronting personal cultural biases and assumptions. This exercise served as a means of self-confrontation, limiting potential unconscious biases. By assessing our positionality, we acknowledged our nonneutral role as designers, heeding Bredella's claim that 'being intercultural means acknowledging that we belong to a culture and exploring how we are shaped by our culture just as others are shaped by theirs' (2003, p. 594).

Learn represents the study of cultural norms, values and beliefs specific to the cultures in focus. Learn highlighted the importance of cultural immersion, where designers should engage in in-depth research, participate in cultural activities and foster direct interactions with users. In our project, the role of cultural knowledge became particularly evident in the preparation for the interviews and workshops. The first few minutes of an interview set the tone of the entire interaction and are crucial for establishing an effective rapport with participants. Knowledge is key to creating environments conducive to adequate communication and effective data collection. Therefore, it was possible to design and conduct a workshop embracing a relational approach that was intentionally crafted to foster a space where participants could engage in meaningful dialogue and shape the service concept. By treating the participants as experts in their own experiences and as essential collaborators, the workshops served as platforms for concept validation, ensuring resonance and shared understanding. In this collaborative approach, knowledge was created by both the designer and the user, aligning with the assertion that codesigning is 'a transformative process in revealing and unlocking tacit knowledge, moving people along on a journey to "make real" what proposed services might be like in the future' (Akama & Prendiville, 2013, p. 30). The workshops led to concrete results, notably the improvement of service prototypes, which served as forms of practical empathic modelling. Through visual prototypes, we simulated customers' specific circumstances, offering insights into their physical experiences. Our research emphasized the varying significance of personal space norms across cultures. Because of such experience prototyping, the workshops stimulated ideation, creative solutions and a dynamic exploration of viewpoints between participants and designers. *Learn* enhanced our project's empathetic reach. Our findings resonate with those of Morgan and Morgan (2005), who note that even a little involvement from representatives of the target culture during the initial stages significantly enhances the acceptance and effectiveness of digital systems within that culture. By deepening our cultural understanding, we recognize that empathy is not merely a unilateral approach. Instead, it emerges as a co-constructed and participatory process, necessitating active participation and mutual engagement, echoing the perspectives of scholars such as Broome (1991).

Respect emerged as a guiding principle that ensures an equal, unbiased and nonjudgemental approach to culture. It embodies a commitment to integrating diverse cultural perspectives and considerations throughout the design process, avoiding stereotypical representations and preconceived notions to ensure that the service resonates with the target audience. In our study, *respect* was demonstrated and developed through considerations of all tools, models and representations, particularly in the construction of personas. Often perceived as a more passive and narrative method, personas have yet to prove their effectiveness in humanizing users. Aligned with Appadurai's insights into the dynamic nature of cultural identities, our iterative persona creation approach transformed them into empathic reflection prompts. Therefore, we believe that crafting personas can be, in itself, a respectful act because carefully 'depicting the other' is the first step towards empathy. Furthermore, *respect* echoes the previous *learn* phase in its consideration of interpersonal dynamics, including the subtleties of communication and relational intricacies, such as tone of voice, spatial orientation and body language. Our approach involved reviewing the core concept definitions to ensure a clear understanding of terminology; key concepts such as 'simple', 'healthy' and 'value' were analysed, revealing the disparities between Danish and Chinese interpretations. Later interview questions facilitated the clarification of these terms in their respective cultural contexts. Moreover, it was possible to tailor approaches by respecting nuances in dialogue and cultural preferences, such as directness in Danish culture and rapport-building in Chinese culture. Recognizing the significance of cultural hierarchies was key in anticipating the participants' preferences and expectations in interviews. Cultural research informed our understanding of users' preferences and expectations, allowing us to formulate the following hypotheses: Within the context of Danish culture, characterized by a lack of emphasis on social status and hierarchy, a designer seen as an outsider would be generally well accepted. In contrast, in Chinese culture, the introduction is crucial for establishing rapport and showing respect to participants, here considering the nuances of power dynamics. Therefore, positioning ourselves as an insider – someone sharing communication patterns and fostering a sense of familiarity – was perceived as beneficial. Hence, when incorporated and shown with respect to the service concept, its designs and communication materials, cultural nuances contributed to the execution of improved empathetic processes. This approach fostered a sense of empowerment among the individuals involved, enhancing empathy within the design framework.

Adaptivity embodies a responsive approach to cultural feedback gained from target groups, and it is characterized by continual testing, tailoring and adjustments of content, design choices and personal behaviours based on ongoing cultural insights. In our case study, *adaptation* was instrumental in addressing the challenges arising from cultural nuances. Informed by ongoing dialogue with users, adjustments were made to prototypes, service features and communication strategies. These adjustments were essential for accommodating diverse communication styles, hierarchical expectations and preferences and were particularly evident during workshop sessions and the creation of the ecosystem map. This mapping exercise prompted reflection on the empathic accuracy of technical diagrams; here, empathy was compromised but was still demonstrated through recognizing our own limitations in representation. We questioned replicable models and underscored the importance of a culturally reflexive approach and the acknowledgement of the non-neutral role of designers in shaping cultural narratives. As a result, in-depth observation and direct engagement with users proved more effective in understanding user needs than simulations of customer experiences alone. Our findings underscore the significance of authentic engagement in uncovering the multifaceted nature of user experiences. This finding supports Harkin and Rosaldo's perspective (1990) on the importance of acknowledging the diversity of truths shaped by cultural contexts. By facilitating the creation of culturally specific service variations and tailored deliverables, *adapt* enhanced the comprehensibility, acceptability and overall impact of the service. It is through *adaptation* that empathic commitment can be reinforced, ensuring that the service stays attuned to the ever-changing cultural contexts within which it operates.

In summary, our exploration of the intersection of empathy and cultural sensitivity reveals profound insight: context is the foundation upon which accurate empathic understanding is built. Cultural sensitivity emerges as essential for facilitating and sustaining empathy, revealing how its absence can hamper such processes. Our design journey illustrates the transformative power of empathy. Although the path to empathic understanding may initially cause uncertainty and unsettlement, it ultimately leads to clarifying the scope and requirements of the design process, thereby fostering purposeful action. In essence, empathy thrives in contexts in which cultural sensitivity prospers. Empathy is a dynamic journey that begins with understanding, flourishes through emotional connection and culminates in compassionate action.

Takeaway

Empathy flourishes in design when grounded in active engagement and understanding. It evolves beyond a one-sided understanding into a co-constructed journey requiring designers' active involvement in cultural contexts. Effective empathy begins by accurately depicting cultural narratives, where designers acknowledge their influential role in shaping these narratives and the influence of their own cultural context. True empathic understanding embodies a continual journey of shedding biases, embracing different perspectives and adapting to cultural insights. As such, embracing cultural sensitivity is imperative for driving informed, empathetic design practices, enriching design practices and fostering services with heightened relevance and resonance.

References

Akama, Y., & Prendiville, A. (2013). Embodying, enacting and entangling design: A phenomenological view to co-designing services. *Swedish Design Research Journal*, 9, 29–40. https://doi.org/10.3384/svid.2000-964X.13129

Appadurai, A. (1996). *Modernity at large: Cultural dimensions of globalization*. University of Minnesota Press.

Bredella, L. (2003). What does it mean to be intercultural? In G. Alred, M. Byram, & M. Fleming (Eds.), *Intercultural experience and education* (pp. 225–240). Multilingual Matters.

Broome, B. J. (1991). Building shared meaning: Implications of a relational approach to empathy for teaching intercultural communication. *Communication Education*, 40(3), 235–249. https://doi.org/10.1080/03634529109378847

Cabrero, D. G., Winschiers-Theophilus, H., & Abdelnour-Nocera, J. (2016). A critique of personas as representations of 'the other' in cross-cultural technology design. In *Proceedings of the First African Conference on Human Computer Interaction* (pp. 149–154). https://doi.org/10.1145/2998581.2998595

Callahan, E. (2018). Cross-cultural empathy: Learning about diverse users in design thinking process. In *Proceedings of the International Conference on Human–Computer Interaction* (pp. 385–397). Springer. https://doi.org/10.1007/978-3-319-92270-6_32

Chen, G.-M., & Starosta, W. J. (1997). A review of the concept of intercultural sensitivity. *Human Communication*, 1, 1–16. https://files.eric.ed.gov/fulltext/ED408634.pdf

Clatworthy, S. (2011). Service innovation through touch-points: Development of an innovation toolkit for the first stages of new service development. *International Journal of Design*, 5(2), 15–28.

Cohen, J. (2021). *La culture c'est .... Organisation visuelle de l'information sur les interfaces digitales* [PowerPoint slides]. ISIT Paris, Arcueil.

Dennington, C. (2017). Service design as a cultural intermediary: Translating cultural phenomena into services. *The Design Journal*, 20(sup.1), S600–S613. https://doi.org/10.1080/14606925.2017.1353008

Drew, C. (2019, September 3). *The double diamond: 15 years on*. Design Council. https://www.designcouncil.org.uk/news-opinion/double-diamond-15-years

Eichbaum, Q., Barbeau-Meunier, C.A., White, M. et al. (2023). Empathy across cultures – one size does not fit all: From the ego-logical to the eco-logical

of relational empathy. *Health Sciences Education*, 28, 643–657. https://doi.org/10.1007/s10459-022-10158-y
Eisenhardt, K. M., & Graebner, M. E. (2007). Theory building from cases: Opportunities and challenges. *Academy of Management Journal*, 50(1), 25–32.
Escobar, A. (2012). *Encountering development: The making and unmaking of the Third World*. Princeton University Press. https://doi.org/10.1515/9781400839926
Fulton Suri, J. (2003). Empathic design: Informed and inspired by other people's experience. In I. Koskinen, K. Battarbee, & T. Mattelmäki (Eds.), *Empathic design: User experience in product design* (pp. 51–57). IT Press.
Goncu Berk, G. (2013). *A framework for designing in cross-cultural contexts: Culture-centered design process* [Doctoral dissertation, University of Minnesota]. University of Minnesota Digital Conservancy. https://hdl.handle.net/11299/155725
Hao, C. (2019). *Cultura: Achieving intercultural empathy through contextual user research in design* [Doctoral dissertation, Delft University of Technology]. https://doi.org/10.4233/UUID:88322666-9CF1-4120-BBD6-4A93438BCA74
Harkin, M., & Rosaldo, R. (1990). Culture and truth: The remaking of social analysis. *Ethnohistory*, 37(4), 445. https://doi.org/10.2307/482864
Hart, R. P., Carlson, R. E., & Eadie, W. F. (1980). Attitudes toward communication and the assessment of rhetorical sensitivity. *Communication Monographs*, 47(1), 1–22. https://doi.org/10.1080/03637758009376016
Heinonen, K., & Strandvik, T. (2015). Customer-dominant logic: Foundations and implications. *Journal of Services Marketing*, 29(6/7), 472–484. https://doi.org/10.1108/JSM-02-2015-0096
Keen, S. (2006). A theory of narrative empathy. *Narrative*, 14(3), 207–236. https://doi.org/10.1353/nar.2006.0015
Knight, W. B., & Deng, Y. (2016). N/either here n/or there: Culture, location, positionality, and art education. *Visual Arts Research*, 42(2), 105–111. https://doi.org/10.5406/visuartsrese.42.2.0105
Marcus, G. E. (1995). Ethnography in/of the world system: The emergence of multi-sited ethnography. *Annual Review of Anthropology*, 24, 95–117. https://www.jstor.org/stable/2155931
Masuda, T., Ellsworth, P. C., Mesquita, B., Leu, J., Tanida, S., & Veerdonk, E. V. (2008). Placing the face in context: Cultural differences in the perception of facial emotion. *Journal of Personality and Social Psychology*, 94, 365–381.
McKinsey & Company. (2021, November 12). *The value of getting personalization right—or wrong—is multiplying*. McKinsey & Company. https://www.mckinsey.com/capabilities/growth-marketing-and-sales/our-insights/the-value-of-getting-personalization-right-or-wrong-is-multiplying
Mesquita, B., & Ellsworth, P. C. (2001). *The role of culture in appraisal*. In K. R. Scherer, A. Schorr, & T. Johnstone (Eds.), *Appraisal processes in emotion: Theory, methods, research* (pp. 233–248). Oxford University Press.
Morgan, K., & Morgan, M. (2005). *Cultural considerations in interactive system design*. University of Bergen, Norway. http://www.metrics.it/Cultural_Considerations.pdf
Pachter, B., Brody, M., & Anderson, B. (1995). *Prentice-Hall complete business etiquette handbook*. Prentice Hall Publishers.

Parson, E. R. (1993). Ethnotherapeutic empathy – Part II: Techniques in interpersonal cognition and vicarious experiencing across cultures. *Journal of Contemporary Psychotherapy*, 23, 171–182.

Rodriguez Schon, V., & Celi, M. (2023). Pursuing positionality in design. In D. De Sainz Molestina, L. Galluzzo, F. Rizzo, & D. Spallazzo (Eds.), *IASDR 2023: Life-changing design, 9–13 October, Milan, Italy*. https://doi.org/10.21606/iasdr.2023.371

Rosaldo, R. (1989). *Culture and truth: The remaking of social analysis*. Beacon Press.

Sarantou, M., & Miettinen, S. (Eds.). (2022). *Empathy and business transformation* (1st ed.). Routledge. https://doi.org/10.4324/9781003227557

Surma-aho, A. (2022). The role of empathy in design practice – Theoretical development and empirical exploration of an integrated model. [Doctoral Thesis, Aalto University]. Aalto University. http://urn.fi/URN:ISBN:978-952-64-0953-5

Tsing, A. L. (2005). *Friction: An ethnography of global connection*. Princeton University Press. https://doi.org/10.2307/j.ctt7s1xk

Vanni, I. (2016). The transcultural edge. *PORTAL Journal of Multidisciplinary International Studies*, 13(1). https://doi.org/10.5130/portal.v13i1.4829

Walsh, T. (2016). *Online surveys in collecting cross-cultural qualitative user experience feedback* [Doctoral dissertation, Tampere University of Technology] (Publication No. 1376). http://urn.fi/URN:ISBN:978-952-15-3733-2

Yin, R. K. (2018). *Case study research and applications: Design and methods* (6th ed.). Sage.

8

The Role of Empathy in Reliving Indigenous Cultural Heritage:

Insights from Arts Ceduna, South Australia

Amna Qureshi and Sherrie Jones

Abstract

Globally, Indigenous cultures face existential threats from historical colonization, contemporary globalization and ongoing cultural assimilation. This study examines the role of empathy in reinterpretation and reliving Indigenous cultural heritage (ICH), focusing on a case study from South Australia involving a conversation with an Indigenous artist facilitated by Arts Ceduna. Empathy, encompassing cognitive, affective and behavioural components, is vital for fostering mutual respect, understanding and collaboration between Indigenous and non-Indigenous communities. This chapter explores how empathy aids in blending traditional elements with contemporary influences and addresses power dynamics, historical trauma and systemic inequalities. This study engages with the Indigenous artist community in Ceduna by utilizing qualitative interviews and participatory observation to understand how empathy influences ICH reinterpretation. The findings indicate that empathy enhances cultural preservation, facilitates the integration of traditional and contemporary elements, empowers Indigenous communities and promotes social justice. This study discusses the challenges and limitations of empathy-based approaches, such as superficial engagement and power imbalances. It proposes strategies for enhancing empathy in cultural preservation initiatives, including cultivating cultural understanding, centering Indigenous voices and agency, promoting cross-cultural benefits and facilitating dialogue and reconciliation. This research contributes to a deeper understanding of empathy dynamics in ICH and suggests more inclusive and effective strategies for upholding Indigenous cultural and intellectual property (ICIP) rights.

Introduction and Background

Indigenous cultural heritage (ICH) goes beyond just creating knowledge from people's stories: 'It includes ideas, experiences, worldviews, objects, forms of expression, practices, knowledge, spirituality, kinship ties, and places valued by Indigenous Peoples' (Indigenous Cultural Heritage: Heritage BC Online Resources, 2023, sec. Learning Centre/Reconciliation). Through active participation, storytelling offers a way for Indigenous people to recognize and value each other's stories. It involves not only retelling these stories but also reliving them over time, all while preserving ancestral knowledge. Frank (2002) affirms that Indigenous communities strive for improvement and are built on shared agreements about what is better to keep their heritage alive; in this context, values indicate not only individual preferences but also the community's collective narratives (p. 115).

Although the urgency to preserve Indigenous cultures and uphold Indigenous rights has gained increasing recognition (Wiessner, 2011), this chapter specifically emphasizes the significant role of empathy in reliving and reinterpreting ICH. This highlights the importance of empathy in interactions with Indigenous communities, particularly through artistic activities and collaborative initiatives. These engagements involve witnessing the communities as they revisit their ICH, blending traditional elements with contemporary influences as time progresses.

This chapter explores a discussion with an Indigenous artist from Ceduna, reflecting on how empathy benefits from the reinterpretation of ICH. This examination is grounded in our engagement with the Aboriginal Artist Community in Ceduna, as facilitated by Arts Ceduna – an Aboriginal Arts Centre. The chapter not only explores the importance of empathy, but it also addresses challenges and proposes strategies to enhance empathy in cultural initiatives. The role of empathy in cultural preservation requires a thorough investigation, particularly to address power dynamics, historical trauma and systemic inequalities, as explained by Wain et al. (2012); these authors emphasize that this investigation is crucial for devising effective strategies that foster respect, understanding and collaboration between Indigenous and non-Indigenous communities.

The main research question driving this inquiry is as follows: 'How can empathy effectively support ICH reinterpretation, particularly within artistic and collaborative engagements for and with Indigenous communities?' Addressing this question is crucial for understanding empathy's role in cultural preservation and its significance within Indigenous contexts. The chapter aims to deepen the understanding of empathy dynamics in ICH, here contributing to the development of more inclusive and effective strategies for supporting Indigenous communities in preserving their heritage while upholding Indigenous cultural and intellectual property (ICIP) rights. Janke and Iacovino (2012) note that the ICIP encompasses various cultural elements passed through generations, requiring communal stewardship to prevent exploitation by external entities. This communal stewardship not only safeguards cultural heritage but also strengthens the community's sovereignty and identity. By exploring the interplay between empathy and ICH, the chapter underscores the importance of culturally sensitive approaches in policymaking and heritage management.

Understanding Empathy

Understanding empathy goes beyond mere application; it entails grasping its essence and impact deeply (Hart & Ingle, 2024). Davis (2016) explains empathy as a key component of human connection, consisting of cognitive, affective and behavioral components. Cognitive empathy is the ability to understand and recognize another person's point of view, allowing people to comprehend their feelings, thoughts and experiences (Decety & Jackson, 2004). Affective empathy, on the other hand, is emotional resonance with others' sentiments, in which people share and experience the emotions of others around them (Eisenberg & Eggum, 2009). However, behavioral empathy entails taking action to help others based on empathic knowledge and transforming emotional understanding into supportive behaviors and actions (Batson, 2011).

Empathy is not merely a tool but also a vital aspect of nurturing interpersonal relationships and fostering prosocial behavior. Empathic responses play a crucial role in creating a sense of connection and mutual understanding between individuals, contributing to the development of trust and cooperation (Preston & de Waal, 2002). Thus, empathy promotes altruistic acts and compassionate behavior because individuals are more inclined to help those they empathize with more (Dovidio et al., 2006). However, it is essential to examine the distinctions of empathy critically in the context of power dynamics and historical trauma, particularly concerning Indigenous and non-Indigenous relationships. Power imbalances and historical injustices can influence the dynamics of empathy, shaping how individuals from dominant and marginalized groups perceive and respond to each other's experiences (Cunneen, 2005). Historical trauma stemming from colonialism, dispossession and systemic oppression can impact the capacity for empathy and trust within Indigenous communities (Krieg, 2009).

Preservation of Indigenous Cultural Heritage

In the context of Indigenous cultural preservation, understanding empathy requires sensitivity to historical and cultural factors that shape interactions.

Recognizing and acknowledging the historical trauma experienced by Indigenous people is essential for building empathetic connections and fostering reconciliation (Atkinson, 2002; Menzies, 2019). Wain et al. (2012) propose that 'cultural empathy can bridge cultural differences by providing a means to integrate an attitude of openness to diversity with appropriate knowledge and skills to work successfully with people from other cultures' (p. 21). Similarly, in Indigenous cultural preservation, empathy can play a pivotal role by facilitating mutual respect, understanding and collaboration between Indigenous and non-Indigenous communities. When individuals and communities empathize with Indigenous peoples, they tend to understand the significance of Indigenous cultures better, offering increased support for preservation efforts while also facilitating the bridging of cultural gaps and the cultivation of authentic relationships built on trust and respect.

Engaging Indigenous communities in artistic activities can enhance their empowerment in preserving their ICH and asserting their rights, as highlighted by Miettinen et al. (2016), Qureshi et al. (2025) and Sarantou et al. (2024). Recognizing and respecting Indigenous perspectives and traditions within these activities can establish a platform where Indigenous voices are acknowledged and honored. Gair (2012) endorses the notion that facilitating opportunities for Indigenous individuals to express their cultural identity through art grounded in empathy can nurture a deep sense of pride and ownership over their ICH. These efforts, as emphasized by Kennedy (2015), are pivotal in assisting Indigenous communities in their ongoing endeavors for cultural preservation and empowerment.

Empowering Indigenous Communities

In New Zealand, government-supported programmes aimed at educating non-Māori about Māori culture have fostered greater empathy and appreciation for Māori traditions (Spolsky, 2005; Wilson, 2001). Through such initiatives, non-Māori individuals are encouraged to engage with the Māori language and cultural practices, thus empowering the Māori people to revitalize their ICH. Another example is Australia, where empathy-based approaches such as cultural awareness training have helped raise awareness about Indigenous issues and support cultural preservation efforts among the Aboriginal and Torres Strait Islander communities (Butler et al., 2019). By educating government officials and non-Indigenous Australians about Indigenous cultures, these programmes promote empathy and understanding, ultimately empowering Indigenous communities to protect their ICH. Similarly, in North America, initiatives such as Indigenous cultural competency training have promoted empathy and an understanding of Indigenous cultures among educators, policymakers and healthcare professionals. By integrating Indigenous perspectives into mainstream curricula and policies, these initiatives empower Indigenous communities to combat stereotypes and assert their rights to their ICH (Nagy et al., 2014).

According to these examples, one cannot forget to mention the Sámi people from the Nordic regions in discussions about Indigenous cultures and their resilience. The Sámi have a rich ICH spanning thousands of years, characterized by their close relationship with nature and unique traditions. Their traditional livelihoods, such as reindeer herding and fishing, are deeply intertwined with the Arctic environment they inhabit. Despite facing historical oppression, the Sámi have revived by maintaining their cultural identity and advocating for their rights (Kuokkonen, 2000). In recent years, there has been growing recognition of Sámi rights and efforts to address historical injustices. The Sámi Parliament in Norway, Sweden and Finland represents a significant step towards self-governance and cultural preservation for the Sámi people (Stepien et al., 2015). Moreover, the Sámi have played a crucial role in environmental activism, highlighting the impacts of climate change on the Arctic and advocating for sustainable practices that protect their traditional lands and way of life. Their struggle and resilience serve as powerful examples of Indigenous resistance and the importance of preserving diverse cultures in an ever-changing world. One instance of empathic collaboration for Sámi cultural preservation involves non-Sámi researchers partnering with Sámi communities to document and protect traditional ecological knowledge (Riseth et al., 2011). Through respectful dialogue and mutual understanding, researchers have integrated Sámi perspectives into scientific studies, empowering Sámi communities and aiding in the sustainable management of their ancestral lands.

Promoting Social Justice

By acknowledging current social injustices, individuals can gain insights into the repercussions of both past and present colonial policies (Horvath, 1972). According to Griffiths et al., 'social justice is central to equitable practices across every sector in society' (2016, p. 19). This entails determining the causes of social justice and injustice, assessing the impact of those causes and putting policies in place to address those causes (Cunneen, 2005). Integrating empathy into these interactions can be a powerful tool for advancing social justice and empowering Indigenous communities, because it fosters mutual respect, trust and collaboration, facilitating knowledge co-creation and the revitalization of Indigenous cultures. O'Faircheallaigh (2009) discusses the importance of social impact assessments in resource development on Australian aboriginal land, stressing the need for aboriginal control and practical measures. Ultimately, empathy plays a vital role in the ICH sector, especially when engaging with Indigenous communities, leading to transformative change rooted in equity, equality and respect for

Indigenous knowledge and perspectives and contributing to healing and reconciliation efforts.

Methodology

In the methodology section of this chapter, the comprehensive inclusion of both the foreground setting and background narrative elucidates a specific focus on empathy – highlighting empathy's significance beyond its routine application as a tool within various services and organizations.

The Foreground Setting

The methodology utilized in this study incorporated interdisciplinary cooperation among researchers from three faculties at the University of Lapland in Finland, alongside engagement with the Indigenous community at Arts Ceduna in South Australia. The visit was conducted as part of the *Promoting Sustainable Practices for Digitalising Indigenous Cultural Heritage – Global North and South Juxtaposed* (TRUST) project, with a focus on building practices for digitalizing ICH (Qureshi et al., 2023). Before the visit, extensive preparations were made, including coordination with the Ceduna Aboriginal Corporation (CAC) and discussions with Arts Ceduna management. The research team, comprising non-Indigenous researchers, acknowledged the importance of understanding local culture and community objectives (WINHEC, 2010). Practical guidelines were provided to ensure respectful engagement during the visit, which included considerations of cultural sensitivity and ethical research practices (Francett-Hermes & Pennanen, 2019).

During the two-week visit, various activities were undertaken to facilitate interaction and engagement with the Indigenous artist community (see Figure 8.1). These initiatives embraced a participatory and culturally sensitive approach, prioritizing empathy and respect for Indigenous knowledge and perspectives.

A meet-and-greet activity was organized at Alexander's Beach, providing an opportunity for bonding and informal interaction. Subsequently, visits to the Arts Centre allowed researchers to explore Indigenous arts and craft collections and engage with local artists in a collaborative workshop to understand the narratives behind their artworks. Empathy played a significant role throughout the visit, helping establish trust and understanding with the Indigenous community. Researchers demonstrated genuine interest and respect for Indigenous culture and traditions, fostering an environment that was beneficial for opening dialogue and knowledge exchange. In addition to informal interactions, a community information seminar was held towards the end of the visit, addressing topics such as understanding ICIP rights, documenting and digitalizing artistic processes and ethical considerations in

FIGURE 8.1 *Meet-and-greet activity (top left and right); workshop with the Indigenous artist community (bottom left); seminar held in cooperation with the Arts Centre (bottom right). Photograph by Amna Qureshi (2023).*

sharing Indigenous artwork online. Moreover, the importance of relational accountability and the mutual benefits of the research to the local community were emphasized (Francett-Hermes & Pennanen, 2019; WINHEC, 2010). Despite facing challenges such as technical issues and extremely hot weather conditions, the seminar provided a platform for meaningful discussion and collaboration between researchers and the Indigenous artist community.

Background Narrative

Apart from participating in the workshop and seminar, conducting further interviews and engaging in informal discussions with Indigenous artists formed the foundation for adopting the *yarning* technique, providing a meaningful means of connecting with the Indigenous artist community. Rooted in Indigenous Australian traditions, *yarning* goes beyond mere conversation (Atkinson et al., 2021; Bessarab & Ng'andu, 2010; Dean, 2010). It embodies a culturally rich form of communication that emphasizes storytelling, deep listening and mutual respect. Through *yarning* sessions, the researchers created inclusive spaces where community members felt comfortable sharing their experiences, perspectives and aspirations. These sessions became pivotal moments for building connections and understanding within the research process. This empathic engagement fostered a sense of validation and empowerment among community members who felt that their voices were being heard and valued within the research process.

Furthermore, empathy serves as a bridge for cross-cultural understanding, enabling researchers to navigate sensitivities and complexities within the Indigenous context (Kennedy, 2015; Sikes, 2006). As researchers empathized with the lived realities of the Indigenous artist community, they gained insights into the challenges, aspirations and cultural distinctions shaping community life. This understanding laid the groundwork for meaningful collaboration and co-creation of knowledge between researchers and Indigenous artists.

Empathic Perspectives: Insights from an Indigenous Artist

During an hour-long conversation, Sherrie Jones, a 48-year-old Indigenous artist at Arts Ceduna, conveyed a profound sense of empathy through her work (see Figure 8.2).

Through qualitative analysis of the transcribed interviews, Jones's Indigenous cultural background, artistic techniques and motivations were explored. Thematic coding and interpretation were employed to elucidate how Jones's artistic expressions embody cultural reinterpretation and empathy towards her ICH, her ancestors and her community. Later, this sentiment was meticulously distilled into a 10-minute video summarizing the essence of Jones's journey as an Indigenous artist, her rich ICH and her commitment to preserving and disseminating Indigenous wisdom (Arts Ceduna, 2023). Throughout the discussion, Jones also demonstrated empathy towards the struggles faced by her ancestors, acknowledging their hardships as integral to her ICH.

Born in Perth, Western Australia, and raised in Carnarvon until the age of 10, Jones's upbringing laid the foundation for her artistic expression. Inspired by her father, who was a landscape artist, she delved into painting at a young age. Her family, representing the Mirning, Wirangu, Yindjibarndi and Yamitji groups, imparted ancestral stories that she weaves intricately into her artwork. As a mother of four, Jones hopes that her artwork would serve as a guiding light for her children, carrying forward her ancestral stories and ICH. She believes that passing down these traditions is essential for preserving CH and fostering empathy and understanding among different communities.

> Most of my family has passed on now, and that is a way of staying connected with my ancestors as well. The stories that they have passed down to me, I will pass them on to the next generation, like my children and then so on and on.

Jones's role as an inspirational arts worker at Arts Ceduna showcases her multitalented and experimental nature. She not only practices various art forms but also shares her expertise through teaching. From painting

FIGURE 8.2 *Conversation with the Indigenous artist Sherrie Jones, Ceduna, SA. Photograph by Amna Qureshi (2023).*

and pottery to basket weaving and felt making, Jones mirrors her deep connection to her culture. Through her art, she depicts stories from the country, intricately mapping landscapes, healing and ceremonies. Jones emphasizes the vital role of water in her artwork, illustrating her empathy towards all life and reliving her ICH. Through her paintings, she captures water's cyclical nature, reflecting its significance in sustaining life and connecting humanity to the natural world.

> I do a lot of paintings about water and life because, without water, we have got no life, and that is for every being, the planet, and humanity. The water does not stay still. It evaporates, and then it comes back again…it is like a circle. It goes back to the same circle of life again. It just gradually turns and turns and does not stop!

This portrayal not only preserves cultural narratives but also fosters empathy towards the environment, ensuring that these values endure for future generations. Moreover, the significance of obtaining permission and guidance from elders and community members when depicting sacred stories or symbols reflects deep-seated respect for cultural protocols and demonstrates how empathy plays a crucial role in navigating sensitive cultural landscapes. The artist's commitment to honouring traditional knowledge and protocols exemplifies a form of empathetic engagement with her ICH.

> Some stories are sacred, which I am not allowed to paint but can be only done by the elders; I am not allowed to paint them. In addition, if I decide to do anything like this, then I need to ask for permission from

> my aunties and elders of the family. There are different ranks of women in the family, and they are not just making up stories. They are passed down through generations. I sometimes tell the same stories and then I also paint the stories, but interpret the way that I would tell them. However, I stay in the traditional storyline, not overlapping or leaving my boundaries and always giving respect to my ancestors. This is a part of me. I have nothing to hide. I just see myself as an artist and tell the truth about our culture and tradition. I have true respect for my elders. In addition, as an artist, I like to learn different things because one should never stop learning. As a child, I was curious and asked questions such as why the earth was created, and then I learned the process of how we interpret things as Indigenous people.

Jones's artistic storytelling serves as a vital form of cultural preservation, translating oral traditions into visual narratives and reliving them with changing times. Through media, such as basketry and felting, she expresses her deep connection to water and nature, emphasizing its importance as a source of life. In addition to preserving Indigenous culture, her work promotes empathy, a connection to the natural world and global preservation. She believes that her ICH is keeping it alive.

> The baskets tell stories from the land, sea and sky starting from the water hole. The tight weaving shows connections, and it tells a story about how water ripples, changing from light blue to dark blue, with different tones and shades.

Throughout the conversation, Jones consistently demonstrates empathy and care for her community. She emphasizes the significance of organizations such as the Arts Centre in reaching out to marginalized artists within the community and offering them opportunities for fulfilling employment. She notes that, in addition to enhancing living conditions, such initiatives could provide artists with fair job prospects and dignified incomes: 'This Arts Centre is a place where I can be myself, express myself, let go of my worries, share stories and train other artists'. Jones acknowledges the challenges artists often face and suggests that they could benefit from participating in various activities and learning opportunities held at such organizations. She also highlights the importance of transferring traditional and contemporary methods between generations to foster new connections and learning experiences within the Indigenous community:

> Learning new skills and blending them with my traditional art makes me happy. Felting is one of the oldest crafts in the world, but it is new to me. When I started using it and applied my stories to it, I felt as though I was able to connect various crafts, preserving both myself and the others. I enjoy what I do.

Reflecting on Jones's perspective, one can also understand the significance of the galleries, libraries, archives, and museums sector (GLAM, n.d.), which can provide services, particularly in handling sensitive topics, because they can support artists in preserving their culture while exploring new avenues of expression.

Strategies for Enhancing Empathy in Indigenous Cultural Heritage

To mitigate challenges and foster genuine empathy in ICH, several strategies can be implemented (Kennedy, 2015; Reitsma et al., 2013; Wiessner, 2011). These include (1) cultivating cultural understanding, (2) centring Indigenous voices and agency, (3) promoting cross-cultural benefits and (4) facilitating dialogue and reconciliation. Ongoing evaluation and adaptation are essential to ensure the effectiveness and relevance of these strategies in diverse cultural contexts.

Immersive learning experiences, engagement with Indigenous knowledge holders and honest communication with community members are crucial. Acknowledging and critically evaluating one's positionality, prejudices and privileges while undermining colonial frameworks are essential for understanding Indigenous cultures, histories and worldviews.

Cultural heritage initiatives should prioritize Indigenous views, voices and agency. Valuing Indigenous knowledge, co-designing projects with communities and aligning outcomes with community goals are vital for developing empathy and respecting Indigenous perspectives.

Efforts should provide tangible benefits to Indigenous communities, focusing on capacity building, co-producing knowledge, sharing resources and advocating for Indigenous rights and sovereignty. This approach fosters trust and mutual benefits.

Empathetic methods facilitate meaningful conversations and reconciliation between Indigenous and non-Indigenous groups. Establishing safe environments for challenging discussions, recognizing past wrongs and addressing historical injustices through social justice efforts are essential for building understanding and unity.

Challenges and Limitations

Empathy-based methodologies can risk superficial engagement, exacerbate power imbalances and reinforce biases without critical self-reflection (Wiessner, 2011; Williams, 2020). Navigating cultural differences and translating empathy into concrete outcomes, such as policy changes and community empowerment, are significant challenges. Despite efforts to be empathic, effectively translating this empathy into meaningful action remains difficult, representing only a small contribution to the broader goals of Indigenous cultural preservation and social justice.

Conclusion

The present study has shown that empathy plays a crucial role in reliving, reinterpreting and preserving ICH and fostering mutual respect, understanding and collaboration. However, to be effective, empathy-based strategies must include cultivating cultural understanding, centring Indigenous voices and agency, promoting cross-cultural benefits and facilitating dialogue and reconciliation. These approaches must be continuously evaluated and adapted to address challenges, such as superficial engagement, power imbalances and systemic inequalities. Ultimately, although empathy is essential, it must be paired with critical reflection and culturally sensitive strategies to truly empower Indigenous communities and advance social justice in cultural preservation and revitalization efforts.

Takeaway

The current study highlights the essential role of empathy in reliving and revitalizing ICH. Empathy fosters mutual respect, understanding and collaboration between Indigenous and non-Indigenous communities, facilitating the integration of traditional and contemporary cultural elements. It empowers Indigenous communities and promotes social justice. However, the present study also acknowledges challenges, such as superficial engagement and power imbalances. To overcome these challenges, strategies such as cultivating cultural understanding, centring Indigenous voices, promoting cross-cultural benefits and facilitating dialogue and reconciliation are proposed. Ultimately, the present research underscores the need for empathy-driven, inclusive and effective cultural preservation strategies that uphold ICIP rights.

Acknowledgements

The project received strategic funding from the University of Lapland. The ethics committee of the University of Lapland approved the research, following the ethical outlines of the Finnish National Board on Research Integrity (TENK). Additionally, we are grateful to SA Australia for funding the workshop at Arts Ceduna.

References

Arts Ceduna. (2023, 11 September). *Stories from the Ceduna Aboriginal Arts & Culture Centre, South Australia.* [Video]. YouTube. https://www.youtube.com/watch?v=tILyehNOgdw

Atkinson, J. (2002). *Trauma trails, recreating song lines: The transgenerational effects of trauma in Indigenous Australia.* Spinifex Press.

Atkinson, P., Baird, M., & Adams, K. (2021). Are you really using yarning research? Mapping social and family yarning to strengthen yarning research quality. *AlterNative: An International Journal of Indigenous Peoples*, 17(2), 191–201. https://doi.org/10.1177/11771801211015442

Batson, C. D. (2011). *Altruism in humans.* Oxford University Press.

Bessarab, D., & Ng'andu, B. (2010). Yarning about yarning as a legitimate method in Indigenous research. *International Journal of Critical Indigenous Studies*, 3(1), 37–50.

Butler, T. L., Anderson, K., Garvey, G., Cunningham, J., Ratcliffe, J., Tong, A., Whop, L. J., Cass, A., Dickson, M., & Howard, K. (2019). Aboriginal and Torres Strait Islander people's domains of wellbeing: A comprehensive literature review. *Social Science and Medicine*, 233, 138–157. https://doi.org/10.1016/j.socscimed.2019.06.004

Cunneen, C. (2005). Colonialism and historical injustice: Reparations for Indigenous peoples. *Social Semiotics*, 15(1), 59–80. https://doi.org/10.1080/10350330500059130

Davis, M. H. (2016). *Empathy: A social psychological approach.* Routledge.

Dean, C. (2010). A yarning place in narrative histories. *History of Education Review*, 39(2), 6–13.

Decety, J., & Jackson, P. L. (2004). The functional architecture of human empathy. *Behavioural and Cognitive Neuroscience Reviews*, 3(2), 71–100. https://doi.org/10.1177/1534582304267l

Dovidio, J. F., Piliavin, J. A., Schroeder, D. A., & Penner, L. A. (2006). *The social psychology of prosocial behaviour.* Psychology Press.

Eisenberg, N., & Eggum, N. D. (2009). Empathic responding: Sympathy and personal distress. In J. Decety & W. Ickes (Eds.), *The social neuroscience of empathy* (pp. 71–83). MIT Press.

Francett-Hermes, M., & Pennanen, H. (2019) Relational ethics in Indigenous research: A reflexive navigation of whiteness and ally positionality. *Dutkansearvvi dieđalaš áigečála*, 3(2), 125–148.

Frank, A. W. (2002). Why study people's stories? The dialogical ethics of narrative analysis. *International Journal of Qualitative Methods*, 1(1), 109–117. https://doi.org/10.1177/160940690200100102

Gair, S. (2012). Feeling their stories: Contemplating empathy, insider/outsider positionings, and enriching qualitative research. *Qualitative Health Research*, 22(1), 134–143. https://doi.org/10.1177/1049732311420580

GLAM, Cultural Heritage. (n.d.). *Wikipedia.* https://en.wikipedia.org/wiki/GLAM_(cultural_heritage)

Griffiths, K., Coleman, C., Lee, V., & Madden, R. (2016). How colonisation determines social justice and Indigenous health: A review of the literature. *Journal of Population Research*, 33, 9–30. https://doi.org/10.1007/s12546-016-9164-1

Hart, T., & Ingle, M. (2024). The (deep) end of empathy. *Journal of Humanistic Psychology*, 64(2), 211–231.
Horvath, R. J. (1972). A definition of colonisation. *Current Anthropology*, 13(1), 45–57. http://dx.doi.org/10.1086/201248
Indigenous Cultural Heritage: Heritage BC Online Resources. (2023, March 29). Heritage BC. https://heritagebc.ca/learning-centre/reconciliation/first-peoples-heritage/
Janke, T., & Iacovino, L. (2012). Keeping cultures alive: Archives and indigenous cultural and intellectual property rights. *Archival Science*, 12(2), 151–171.
Kennedy, R. (2015). *Designing with indigenous knowledge: Policy and protocols for respectful and authentic cross-cultural representation in communication design practice*. Swinburne University of Technology.
Krieg, A. (2009). The experience of collective trauma in Australian indigenous communities. *Australasian Psychiatry*, 17(sup1), S28–S32. https://doi.org/10.1080/10398560902948621
Kuokkonen, R. (2000). Towards an 'Indigenous paradigm' from a Sami perspective. *Canadian Journal of Native Studies*, 20(2), 411.
Menzies, K. (2019). Understanding the Australian Aboriginal experience of collective, historical and intergenerational trauma. *International Social Work*, 62(6), 1522–1534. https://doi.org/10.1177/0020872819870585
Miettinen, S., Sarantou, M., & Akimenko, D. (2016). Collaborative art and storytelling as an empowering tool for service design: South Australian case study. In P. Rytilahti & S. Miettinen (Eds.), *For profit, for good: Developing organizations through service design* (pp. 74–80). University of Lapland Press.
Nagy, J., Sepsi, E., & Vassányi, M. (Eds.). (2014). *Indigenous perspectives of North America: A collection of studies*. Cambridge Scholars Publishing.
O'Faircheallaigh, C. (2009). Effectiveness in social impact assessment: Aboriginal peoples and resource development in Australia. *Impact Assessment and Project Appraisal*, 27(2), 95–110. https://doi.org/10.3152/146155109X438715
Preston, S. D., & de Waal, F. B. M. (2002). Empathy: Its ultimate and proximate bases. *Behavioural and Brain Sciences*, 25(1), 1–20. https://doi.org/10.1017/S0140525X02000018
Qureshi, A., Linkola, A., & Girardi, D. (2023, 29 June). *From Global North to South and back: A research visit to the Indigenous Community in South Australia* [Blog]. Faculty of Law, University of Lapland, Finland. https://fromlaplandwithlaw.blogspot.com/2023/06/from-global-north-tosouth-and-back.html
Qureshi, A., Wilson, D., & Sarantou, M. (2025). Indigenous communities re-interpreting and preserving cultural heritage through narratives while navigating the digital age. In I.-A. Linkola-Aikio, P. Keskitalo, R. Ballardini, & M. Sarantou (Eds.), *Digital Indigenous cultural heritage* (1 ed., pp. 121–143). Palgrave Macmillan, Cham. https://doi.org/10.1007/978-3-031-76941-2_7
Reitsma, L., Wallace, J., & Rodgers, P. (2013). Exploring respectful design directions for indigenous communities. In *2013 International Conference on Culture and Computing* (pp. 131–132). IEEE. https://doi.org/10.1109/CultureComputing.2013.31
Riseth, J. Å., Tømmervik, H., Helander-Renvall, E., Labba, N., Johansson, C., Malnes, E., Bjerke, Jarle W., Jonsson, C., Pohjola, V., Sarri, L. E., Schanche, A., & Callaghan, T. V. (2011). Sámi traditional ecological knowledge as a guide

to science: Snow, ice and reindeer pasture facing climate change. *Polar Record*, 47(3), 202–217. https://doi.org/10.1017/S0032247410000434

Sarantou, M., Qureshi, A., Jones, S., & Gunter, S. (2024). 'I will now be more aware': The intersection of human-computer interaction and indigenous artmaking in living cultural heritage. In A. Bramwell-Dicks, A. Evans, M. Winckler, H. Petrie, & J. Abdelnour-Nocera (Eds.), *Design for Equality and Justice*: INTERACT 2023 IFIP TC 13 Workshops, York, UK, 28 August – September 1, 2023, Revised Selected Papers, Part II (pp. 294–298). Springer. https://doi.org/10.1007/978-3-031-61698-3_31

Sikes, P. (2006). Decolonizing research and methodologies: Indigenous peoples and cross-cultural contexts. *Pedagogy, Culture & Society*, 14(3), 349–358. https://doi.org/10.1080/14681360600892017

Spolsky, B. (2005). Māori lost and regained. In A. Bell, R. Harlow, & D. Starks (Eds.), *Languages of New Zealand* (pp. 67–85). Victoria University Press.

Stepien, A., Petrétei, A., & Koivurova, T. (2015). Sámi Parliaments in Finland, Norway, and Sweden. In T. H. Malloy, A. Osipov, & B. Vizi. (Eds.), *Managing diversity through non-territorial autonomy: Assessing advantages, deficiencies, and risks* (pp. 117–138). Oxford University Press. https://doi.org/10.1093/acprof:oso/9780198738459.003.0007

Wain, T., Sim, M., Hayward, C., Coffin, J., Mak, D., & Rudd, C. (2012). Creating cultural empathy and challenging attitudes through indigenous narrative project. *eCULTURE,* 5(1). https://ro.ecu.edu.au/eculture/vol5/iss1/3

Wiessner, S. (2011). The cultural rights of indigenous peoples: Achievements and continuing challenges. *European Journal of International Law*, 22(1), 121–140. https://doi.org/10.1093/ejil/chr007

Williams, V. R. (2020). *Indigenous peoples: An encyclopaedia of culture, history and threats to survival.* ABC-CLIO, Santa Barbara.

Wilson, C. (2001). Decolonizing methodologies: Research and indigenous peoples. *Social Policy Journal of New Zealand*, 17, 214–217. https://link.gale.com/apps/doc/A81762115/AONE?u=anon~17c8aba8&sid=googleScholar&xid=9ec35178

WINHEC. (2010). *World Indigenous Nations Higher Education Consortium. Research standards.* WINHEC Research and Journal Working Group (1st ed. adopted 26 August 2010).

9

Creating an Empathy Network: Insights into a Service Design Intervention in a High School in Japan

Moe Shimomura, Leon Loh, Yanfang Zhang and Noriko Takano

Abstract

This study explores new meanings and knowledge of empathic service design through a case study conducted in a Japanese high school. Using action research, the design researchers collaborated with teachers to codesign educational programmes and materials, incorporating feedback from questionnaires and classroom observations. The development process included three phases: a pilot project, expansion to more participants and iterative refinement based on feedback. Through long-term practical involvement, a chain of empathy networks formed, starting between design researchers and teachers and expanding to students. Design researchers initially opened their space to learn from teachers, listening to their concerns and desires to embark on a challenging educational journey. This triggered the teachers to open up their space to learn and listen to the students. Furthermore, students shared their thoughts with teachers and provided feedback to the design researchers on how to contribute to the chain.

Introduction

Currently, the main educational trend is a pedagogical shift towards more student-centred, project-based and inquiry-based forms of learning activity

(Carvalho & Goodyear, 2018) to foster twenty-first-century skills (Binkley et al., 2012). This is good for students entering the workforce, who need those skills to survive in a world characterized by volatility, uncertainty, complexity and ambiguity. However, the needs of teachers should also be explored. When education is perceived as a service, the following questions arise: Should teachers, as service providers, be sufficiently equipped with the necessary pedagogical strategies, and teaching and learning strategies, for student-centred and inquiry-based instruction?

Currently, Japanese teachers are ill-equipped to conduct student-centred and inquiry-based instruction, especially in local high schools, where traditional and conventional educational methods still strongly influence teaching practices. Teachers need to provide education that is currently appropriate when teaching students with self-directed learning and twenty-first-century competencies. However, Japanese high school teachers face challenges in creating new learning experiences. One of the reasons for this is that much of what is implemented in traditional classrooms is constrained by inflexible and slow-changing teaching practices that are affected by the national curriculum, physical school environment and grouping of students into age-based classes (Carvalho & Goodyear, 2018).

The authors of this chapter belong to the Sustainable Development Goals (SDGs) (Griggs et al., 2013) Design School at Kyushu University, Faculty of Design. In 2019, teachers from a public high school asked the authors to jointly develop an educational programme that applies the design process to meet the contemporary demands of current education. The SDGs Design School is an organization of design researchers who research and develop education programmes and strategies that incorporate design-based approaches into teaching and learning. One of the key objectives of the SDGs Design School is to propagate the value of design education towards the SDGs. To achieve this, the SDGs Design School provides educators with services that include programmes and tools necessary to implement the education.

Education is a service (Ng & Forbes, 2009). Education has been surprisingly neglected in service design research (Carvalho & Goodyear, 2018). Through the lens of service design research, the authors will provide insights into educational service design using a social implementation project at a public high school where design methodology intervenes in the field of education. The aim of this chapter is to identify the key elements for a new model of empathic service design for conventional closed communities based on reflections from the cocreation of an educational programme implemented in a Japanese public high school.

Two key research questions are answered in this chapter to identify the key elements that form the empathic service design model.

1 What kind of empathy will emerge between design researchers and high school teachers when cocreating an education programme as a form of service?

2 How can empathy be mediated between design researchers and teachers through the development of design education programmes?

The value of this research extends beyond the traditional teacher-centred educational setting; it seeks to provide insights that are applicable to cases where there are imbalanced relationships within the service, such as parenting services between parents and children or health services between doctors and patients.

In the literature review, we first delve into the cultural context of Japanese high schools. In addition, we define empathy within the context of service design and explain why the school context provides a good breeding ground for empathetic development through design. The subsequent chapter introduces the programme framework, which is followed by the findings from the inquiry-based educational programme development. Afterwards, we discuss the empathy fostered between design researchers and high school teachers and its impact; then, we present the conclusion.

Thematic Literature Review

This thematic literature review presents a discussion of the cultural context in Japan and the role of empathy in service design. The first theme elaborates on the relationship between high school teachers and students in the Japanese cultural context. In addition, the second theme discusses the potential for empathic service design to intervene in closed communities, such as schools.

Cultural Context in Japan

In Japanese dictionaries, the word '*sensei*' (commonly translated as 'teacher' in English) is defined in various ways (Iwanami Shoten Publishers, 1991):

1 A person who was born before.
2 A person with superior academic integrity. A person whom one studies under. Additionally, an honorific title for that person.
3 A school teacher.
4 An honorific title for a person in a leading position (e.g. doctor, lawyer).
5 As a term of endearment to address another person in a friendly or teasing manner.

In educational contexts, *sensei* is often used for second, third or fourth meanings. Thus, a teacher is not only someone who teaches learning and skills but also someone who is older and respected in a hierarchical relationship

(Akasaka, 2019; Hasuo, 1980; Morita & Yamada, 2013). In many Asian countries, including Japan, there is an ingrained culture of respect for elders and treating those in a position to teach with politeness and respect (Kang & Hong, 2008; Li, 1998; Miyoshi & Yoshino, 2005).

Because of this cultural background, education in Japan has long been teacher centred, with teachers imparting their knowledge to students (Matsuyama et al., 2019). In teacher-centred education, teachers make most of the decisions in the classroom, control instructional activities and emphasize the acquisition of basic skills and rote learning in their instruction (Pakarinen et al., 2024; Stipek & Byler, 2004). Teachers follow their own agenda, and there is typically no room for students' initiative or the individualization of instruction in the classroom (Pakarinen et al., 2024). The current teachers are of the generation that received this kind of education when they were students.

In this cultural context, teachers and students are unlikely to be in equal positions. It is difficult for students to gain the skills to think critically about what the teacher says and to discuss it with the teacher unless they have received special education. Additionally, the culture and education of discussion are not deeply rooted in Japan (Maruno, 2005). Japanese culture is characterized by a preference for prioritizing others' feelings rather than asserting individual opinions, avoiding conflict and aligning one's opinions with those of others (Nara Institute of Science and Technology, n.d.; Ogasawara, 2019). Students commonly feel uncomfortable and embarrassed about expressing different opinions in front of others during discussions.

In Japan, high school is generally a three-year process; most students are between 15 and 18 years old. The current Curriculum Guidelines for High Schools (Ministry of Education, Culture, Sports, Science and Technology, 2018) emphasize improving the learning process and represent the growing demand for student-centred education (Takahata et al., 2019). The SDGs Design School approach can be described as an educational method that introduces the design process into students' learning to solve complex problems, which is not yet widespread in general Japanese high schools (Loh et al., 2020). The advantage of this method is that students define the identified everyday life challenges and acquire the ability to think and express their ideas freely while seeking solutions.

Empathic Service Design and Education

Empathy is a person's ability to identify with and understand another person's feelings, ideas and circumstances (Sleeswijk Visser & Kouprie, 2008). Society can better address many harms by building networks of empathy (Kadri, 2020). One of the processes of empathy in design practice requires a structured investment of time (Kouprie & Sleeswijk Visser, 2009). Direct contact with users helps promote empathy (Kouprie & Sleeswijk Visser, 2009; McDonagh-Philp & Bruseberg, 2000). Designers' service design interventions are frequently constrained by limited time and budgetary resources (Takeyama, 2017) because of the projects' profit-driven nature. Service stakeholders, including designers, leave once the designed services are delivered. Therefore, once detached from the project, it is unlikely to enter the intervention chain again afterwards. In this respect, many services are developed and disappear quickly (Ehrenfeld, 2001). Although these services are more likely to initiate change, they are also less likely to foster empathy. Moreover, the stability of traditional and closed communities, such as schools, has the potential to foster empathy because it allows designers to iterate and immerse themselves over time. This is a key process for applying empathy in design (Kouprie & Sleeswijk Visser, 2009).

Outline of the Programme Framework

This chapter has applied a qualitative approach to create a case study that presents a new educational programme incorporating the design process into an inquiry-based cross-disciplinary study under the *SDGs Challenge Project* in High School F in Japan. Programme development began in 2019 with a pilot workshop programme of four sessions involving a class of seventeen first-year high school students.

In 2020, the workshop programme was expanded to three classes, each consisting of first-year, second-year and third-year students. The class size was maintained at approximately twenty students. Because of the Covid-19 pandemic, tablet devices were introduced, and classes were conducted in a hybrid online and offline format.

From 2021 onwards, the programme was expanded and implemented in an inquiry-based subject for all third-year students. In 2023, the programme was implemented for 314 high school seniors and 317 high school sophomores. The programme for sophomore students will continue into their third year of study in 2024.

Since 2019, design researchers have observed lessons and made notes on instruction, teacher–student interactions and student learning progress (Shimomura et al., 2023; Zhang et al., 2022). Figure 9.1 illustrates the number of teachers and design researchers involved in the research against the number and duration of the sessions while the types of data collected during the sessions are also clarified.

The project employed the action research method, which is a cycle of planning, action and fact-finding from the results to address specific problems directly and enable practical solutions to be implemented quickly (Carr, 2007; Lewin, 1946). The SDGs Design School codesigned new educational programmes and materials with high school teachers to enable them to integrate action research into their classroom practice. The main role of the SDGs Design School was to propose educational programmes that applied the design process to meet the needs of high school teachers and

	Number of students	Number of teachers	Number of design reserchers	Duration	Types of data
2019	Year 1, 17	1	3	4 sessions, 50 minutes each	-Students' responses in the questionnaire -Teachers' responses in the questionnaire -Observation notes of the design researchers
2020	Year 1, 20	1	3	5 sessions, 50 minutes each	
	Year 2, 17	1	3	5 sessions, 50 minutes each	
	Year 3, 28	1	3	3 sessions, 3 hours each	
2021	Year 3, 314	27	4	19 sessions, 100 minutes each	
2022	Year 3, 310	27	4	19 sessions, 100 minutes each	
2023	Year 2, 317	23	4	19 sessions, 100 minutes each	
	Year 3, 314	23	24	19 sessions, 100 minutes each	
2024 planned	Year 2, All	23 planned	24 planned	19 sessions, 100 minutes each	

FIGURE 9.1 *Programme achievements and progress.*

the literacy of high school students, provide design expertise, train teachers to facilitate teaching and design processes, observe classroom participation, propose suggestions for improvement and conduct educational research. The main role of high school teachers was to conduct classes, provide guidance to students and manage school operations related to class implementation. Teachers were responsible for all basic classroom interactions with students, including explaining the learning content and answering questions from the students. Additionally, they shared general pedagogical knowledge, curriculum knowledge, knowledge of learners and their characteristics and knowledge of educational contexts (Shulman, 1987). At the end of the programme, all participating teachers and students responded to a questionnaire to evaluate the programme. After the questionnaires had been compiled, the design researchers and teachers jointly reviewed the education programme for improvement.

Findings

The research design of the educational programme consisted of three phases as shown in Figure 9.2, following Lewin's action research methodology (Carr, 2007; Lewin, 1946). First, Phase 1 involved implementing a pilot project with a specific class. While creating the initial educational tools, accumulated expertise was gained through trial and error during classes. In Phase 2, the programme expanded to involve all third-year classes. The expanded scale posed challenges for

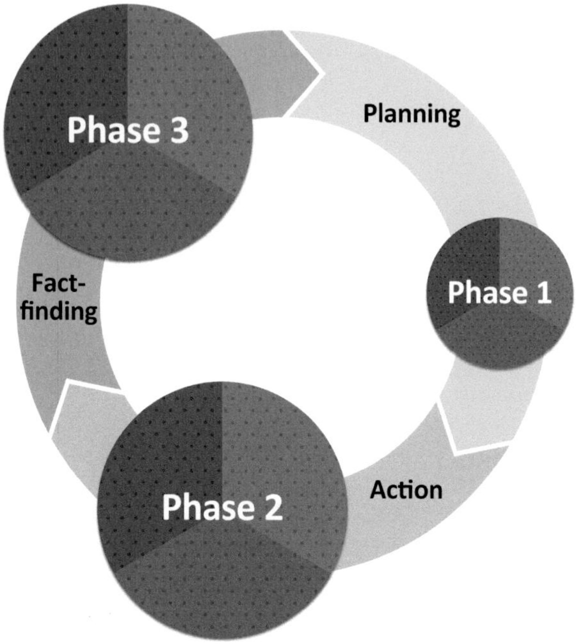

FIGURE 9.2 *Cycle of educational programme development.*

teachers because design researchers cannot fully support all classrooms. The design researchers realized that the designed materials from the pilot phase could not be applied quickly and successfully. This motivated them to invest time. In Phase 3, building on reflections from trial and error during programme expansion, the design researchers improved the programme along with teachers to provide students with deeper learning. Each phase is explained from the perspective of the intervention of the design researchers.

Phase 1: Pilot Phase

This phase involved prototyping and obtaining feedback on the educational content and materials used by teachers and students to develop a new educational programme from scratch. The first programme was conducted in 2019 using the subject 'Industrial Society and Humanity' for four 50-minute lessons. The participants were seventeen first-year high school students enrolled in the high school's own global management programme. The educational content consisted of understanding social issues, creating concepts for solutions, prototyping solutions and presenting the results using the posters created. Eight educational tools were developed collaboratively with the design researchers, high school teachers and students (Shimomura et al., 2022): a teaching plan, lesson slides, Worksheet 1: Problem identification, Worksheet 2: Survey of case studies, Worksheet 3: Idea validation, checklist for selecting ideas, comment cards and student questionnaire (Figure 9.3).

Because teaching in 2020 was conducted during the Covid-19 pandemic, strategies for in-person lessons for group discussions used an online whiteboard instead of allowing students to work closely using Post-it® notes and large papers. The teacher created a worksheet booklet based on earlier prototyping. This booklet provided students with information on the general design process and unpacked the design process for them to follow during the programme (Loh et al., 2021).

To solve complex issues related to SDGs, it is important to develop twenty-first-century skills to work towards solutions actively rather than focusing on traditional, teacher-centred education that aims to teach students to arrive at the correct answer. Observing novice students learning the design process and observing and collaborating with teachers unfamiliar with educating students to elicit discussion and lead them to optimal solutions can help design researchers understand the difficulties.

In this phase, the teachers and design researchers exchanged knowledge and developed a mutual understanding by translating each other's

FIGURE 9.3 *Worksheet of the SDGs Challenge Project. Photograph by M. Shimomura (2022).*

FIGURE 9.4 *Pilot project of 2019. Photograph by M. Shimomura (2019).*

terminology. Design researchers utilized their design expertise for students and provided guidance to teachers. In return, teachers could build a feasible programme by sharing their knowledge and curriculum structure from the field of high school education. This process was not unidirectional. The educational tools were created through dialogue and verification, in which design researchers created a draft worksheet and teachers improved the draft into a form more appropriate for communication to students. In developing the educational programme, the importance of following the format of the teaching plans generally prevalent in the educational field was confirmed so that the teachers could understand the design method. The teachers appreciated that the programme followed the familiar format of a teaching plan, making it easier to understand and implement the lessons.

In Phase 1, those teachers who had been in close communication with the design researchers directly conducted the classes. The design researchers were able to observe the lessons and communicate their findings and areas for improvement to the teachers immediately after the lesson in preparation for the next lesson. Therefore, the teachers were able to receive feedback and detailed advice as appropriate (Figure 9.4).

Phase 2: Expansion and Failure

From 2021 onwards, the programme was developed to involve more than 300 students from all classes at the same grade level using the Period for Inquiry-Based Cross-Disciplinary Study teaching framework. The classes were team-taught by two to three teachers from different specialties. All groups of students presented their work at a final presentation and were critiqued by external experts from local companies and universities (Figure 9.5). The programme and teaching materials developed on the basis of the feedback obtained in Phase 1 served as a template for what followed.

In Phase 2, supporting the teachers conducting the classes was a major challenge because the programme was expanded to include the entire school year. Teachers ranged in experience from beginning teachers to experienced teachers with decades of experience, and their understanding of the programme varied. The lack of a sample of final outcomes to explore in response to issues identified by the students, lack of knowledge and experience of the design process and lack of experience teaching a classroom format that requires mutual discussion and facilitation with students were major psychological barriers for the teachers. For these reasons, the teachers expressed anxiety about teaching the classes while not knowing the final student learning outcomes.

Several guidance sessions were held in advance to address these teachers' concerns. After the programme content, educational objectives and evaluation methods were explained, the sessions were organized to address concerns and questions. A questionnaire was administered after each class to collect teachers' questions, difficulties and requests, which the design researchers and programme's coordinating teacher answered as appropriate. Additionally, design researchers took observation notes and went around each classroom during classes to answer any questions from teachers on the spot and to provide classroom support.

FIGURE 9.5 *The presentation was critiqued by external experts. Photograph by M. Shimomura (2023).*

Another challenge was related to students' understanding of learning. Students unfamiliar with the design process struggled to generate ideas connected to the SDGs. One reason for this was that, when students heard of the SDGs, they tended to focus on major social issues in distant countries, such as hunger in Africa, drinking water in India or Black Lives Matter in the United States. As a result, the students ran into the barrier of being unable to conduct research beyond searching on the internet and in the literature and, therefore, were not able to empathize with the problem situation. The scope of experience of high school students was narrower than that of design researchers. Because they had only a shallow understanding of the problems to be addressed, they could not imagine specific scenarios and had difficulty coming up with concrete ideas about who was affected by the problems and how to solve them.

Phase 3: Deepening Learning and Programme Refinement

Fact-findings	Instead of starting the programme with an explanation of the SDGs, it would be better to start with issues thar are familiar to students in their daily lives.
	Ther explanation of the design process was written in text, but for student who have never studied design before, it would be better to explain it in easier-to-understand terms and use diagrams to make it easier to understand.
	I would like to consider the allocation of time in the programme. In particular, there was not enough time for the problem finding process, so it needea to be longer.
	I want more freedom to write on the worksheets, free space is neededd for notes and illustrations.
	There is not enough evidence for the information the students discovered thruogh Internet research. Teachers need to teach students research methods other than searching Internet-based literature.
	I would like to be able to share worksheets with others when discussing in groups. Let's encourage the use of digital tools.,
	How about showing examples of how to use the methods and ideas created by senior students?
	Most sutudents did well in converging during the process of defining the problem, but not so well in diverging.

FIGURE 9.6 *Findings of the joint programme improvement workshop.*

After the 2021 programme, the design researchers and teachers held a joint programme improvement workshop to reflect on the implemented programme the following year, as shown in Figure 9.6.

By starting the programme with an explanation of the SDGs, the students attempted to address major social issues in distant countries that they often saw and heard about in the media in the context of the SDGs. However, many students without real experience with these issues had no knowledge or experience beyond the information they could find on the internet and could only vaguely define the issues.

As such, in the following year, the programme started with the process of defining problems that would allow students to examine familiar everyday problems and empathize with them based on their own experiences. Two concrete changes were made: the steps of the process of finding problems were scaffolded in more detail, and the persona method was introduced.

The programme was developed through an annual cycle of research action. Generally, design researchers or designers make proposals or conduct research in response to client requests within limited time and space. Continuing to operate and improve the programme (service) is the responsibility of the stakeholders in the field. It is essential for empathic service design to create a framework for relationships that enables stakeholders to have continuous ownership and make ongoing improvements.

Discussion

The present study has implications for opening up the closed community shell (Figure 9.7) in Japanese education by the intervention of empathetic stakeholders. As Kouprie and Sleeswijk Visser (2009) argue, investing time in iteration and immersion is necessary to apply empathy to design. Empathetic stakeholders are distinct from others because they are involved in an iterative chain of giving and receiving empathy, not only in fulfilling service needs. Figure 9.7 shows a chain of empathy networks (Kadri, 2020) developed through mutual learning between design researchers and teachers. The design researchers initially opened their space to learn from teachers, listening to their concerns and desires to embark on a challenging educational journey. Subsequently, the design researchers equipped teachers with tools to achieve their goals, such as knowledge of the design process, educational materials and an outline of guidelines. In other words, the design researchers indirectly supported teachers and students – the touchpoints of the service – from the backstage of the service to create a richer empathy network. In turn, the teachers opened up their space to learn and listen to students through the programme. Furthermore, the students shared their thoughts with the teachers and provided feedback to the design researchers via a questionnaire.

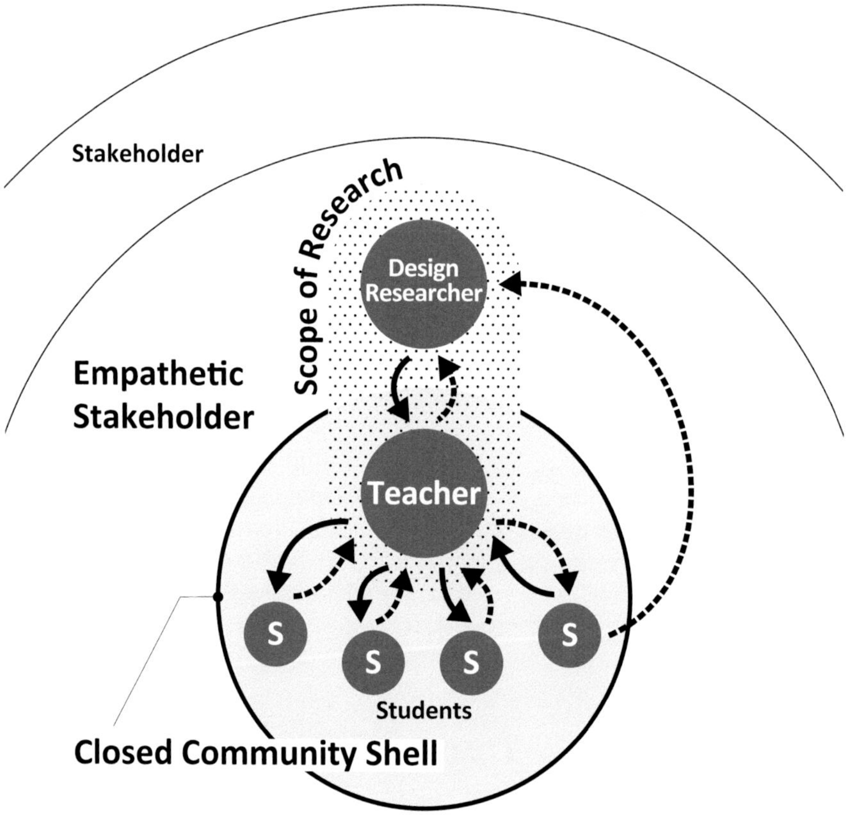

FIGURE 9.7 *Chain of the empathy network that came about through mutual learning.*

The learning arising from design researchers entering a high school setting to understand through practical experience how teachers behave in the classroom, how students react, how teacher–student dialogue proceeds and why it is difficult for teachers to undertake discussion with students is valuable. The research results indicated that real-life experiences fostered empathy between design researchers and teachers. In the same way that students' lack of empathy with social problems results from a lack of deep understanding and emotional turmoil caused by actual experience, it is important for design researchers to acquire a multifaceted understanding of the situations and emotions of people through actual experience. This section discusses how design researchers develop empathy with teachers based on their own observations during long-term, practical involvement.

First, it was important for the design researchers to develop an understanding of the students. They were able to gain knowledge of learners and their characteristics through long-term, continuous classroom observation (Figure 9.8). A longitudinal study confirmed that students' interests were strongly influenced by the news often reported in the media at the time. When considering the SDGs, the students tended to define issues as social problems in far-away countries, even though many social issues existed around them. During the learning process, the students often stopped thinking further once they believed they understood the issue after a quick internet search. Because they had never directly encountered such problems in their lives, the students were often unable to empathize with the people facing the problem or understand the problem itself in depth and could not envisage concrete means of solving the problem. The scope and experience of the students were narrower than those expected by design researchers who do not usually have daily contact with high school students. Understanding the students with whom teachers interacted helped the design researchers better understand what knowledge was important for providing teachers, what tools to make available and where to set the literacy level of their educational programmes.

Second, the results provided an understanding of student–teacher engagement. Against the cultural background of Japanese high school education, as previously stated, how can teachers teach their students new educational programmes that incorporate the design process? This was the

FIGURE 9.8 *Students in discussion. Photograph by M. Shimomura (2023).*

main concern and challenge for the teachers participating in the programme. As described earlier, in Japan, teachers are in a superior position to students and tend to be aware that their opinions have too much influence on students. Teachers in Japan do not typically have extensive experience in student-centred education. Therefore, the teachers in the current case were very cautious about how they approached students and how they guided and facilitated discussions during the programme. This was indicated by several teachers' responses in the questionnaire, in which they limited their intervention in discussions and did not feel confident in their facilitation: 'When (the teacher) gave advice, the students jumped on it or stuck to it, so it was difficult to get them to think independently'; 'Many students tend to seek answers rather than engage in the process of thinking, resulting in more situations in which discussions do not occur'; 'It was difficult to determine how deeply I should delve into discussions. The students who spoke up in the groups were always the same, and mediating that dynamic was challenging. It was a tough time for quieter students, I believe. Regarding facilitation, I did my best. It was challenging to strike a balance and not express my opinions too much'; 'I felt that intervening too much could make it seem like my thoughts were dominating the discussion, so I struggled with how much intervention was appropriate' (Figure 9.9).

In Japanese education, the teacher's opinion has been considered absolutely correct, so students feel obliged to follow the teacher's opinion.

FIGURE 9.9 *Teachers and students in the classroom. Photograph by M. Shimomura (2023).*

Therefore, the key to facilitation lies in the teacher asking questions to guide the students instead of expressing their own opinions. Student-centred education requires teachers to move from the podium to the student level and work from the same perspective and direction. By doing so, teachers can develop a good rapport with students.

Traditional teacher-centred education may have been characterized by one-way teaching from a teacher to students and a focus on right and wrong answers. However, going forward, teachers are transitioning to becoming more like mountain guides aiming for the summit together with students, hence acting as co-learners. The following signs of this shift were indicated by some teachers' comments in the post-programme questionnaire: 'I was conscious of discussing each session from the same perspective as the students' and 'I felt like I was participating with the students rather than extracting their opinions (in facilitation)'.

Finally, the authors described empathy through direct contact between design researchers and teachers, which builds on the two understandings described above. With an understanding of the characteristics of the students and the correlates of how teachers relate to them, design researchers provided adequate equipment for anxious teachers and showed them the way forward. Here, the teacher was the person to whom the design researchers delivered the service. To promote education that incorporates the design process, the SDGs Design School aimed to understand the realities of teachers who are conducting classes. The design researchers gained a multifaceted understanding of teachers from practical experience and diverse perspectives through classroom observations of student behaviour and teacher–student interactions, surveys with participating students and teachers to understand their difficulties and the implementation of practices. This process deepened empathy towards teachers. Although it may be possible to develop an education programme solely through discussions with teachers, it would not have been possible to deepen empathy towards teachers to the extent achieved without engaging in such a comprehensive education programme.

Additionally, codesigning with the teachers who managed and coordinated the programme was important. As shown in many previous cases, a key element of successful service design starts with enthusiastic key people and an initiative of opening and diversification to propagate the impact (Fassi & Manzini, 2022; Manzini, 2022; Sunaga, 2019). Supporting these key persons is important for the sustainable development of the service in the initial stage and is one of the roles of the designer. In this programme, a key teacher managed the project, explained the programme to the supervising teachers and acted as a bridge to the design researchers. Individual teachers have various characteristics, and their levels of experience vary widely. Some teachers excel in dialogue with students, while others may not. Open and equal codesign with teachers who understood the nature of education well led to the successful development of the programme.

One limitation was that the research focused on one high school. Therefore, future research should expand the number of target schools and accumulate further findings. In addition, this chapter was considered against the background of the educational context in Japan, and therefore the situation differs from other cultures.

Conclusion

Empathy between design researchers and teachers through the development of educational programmes has created a broader network of empathy. The most significant impact observed in the current case was that the empathy that started with design researchers and teachers has expanded, circulating from teachers to students and from students back to design researchers, as shown in Figure 9.7. The intervention of the design researchers was the catalyst for this change. It would be difficult to achieve this change by extending the conventions and methods of the education field and teachers' accumulated experience and mindset. However, by adopting the design process, a framework that differs from the standards that have been taken for granted in school education up to that point, teachers were able to take on the challenge with little opposition, saying, 'I do not truly understand it, but let us give it a try first'. One of the characteristics of the design process is its iterative nature. The process involves trying again and again, going back to the beginning while working towards the goal. This willingness to fail and try again in the empathy network likely encouraged educators to adopt a 'let's give it a try' mindset. Teachers are now taking the initiative to improve the programme annually. Design researchers are gradually handing over the initiative to teachers who have reached the stage where they can proceed through this cycle autonomously. This case may provide a useful roadmap for developing progressive, grounded social change.

Takeaway

The intervention of the design researchers was the catalyst for creating a chain of empathy networks. As the current study revealed, the initial step in transforming a closed community shell, such as a school, is to engage individuals who possess space for change. Empathy networks are propagated through a process of prolonged and practical observation, attentive listening and action within the community.

Acknowledgements

The authors express appreciation to the teachers and students for their contributions to this study. This work was supported by JSPS KAKENHI (grant number JP21K02527).

References

Akasaka, S. (2019). Gakkyu keiei no imi to kadai [Meaning and challenges of classroom management]. *Journal of the Japanese Association of Classroom Management*, 1, 1–4.

Binkley, M., Erstad, O., Herman, J., Raizen, S., Ripley, M., Miller-Ricci, M., & Rumble, M. (2012). Defining twenty-first century skills. In P. Griffin, B. McGaw, & E. Care (Eds.), *Assessment and teaching of 21st century skills* (pp. 17–66). Springer. https://doi.org/10.1007/978-94-007-2324-5_2

Carr, W. (2007). Philosophy, methodology and action research. In *The quality of practitioner research* (pp. 29–42). Brill. https://doi.org/10.1163/9789087903190_004

Carvalho, L., & Goodyear, P. (2018). Design, learning networks and service innovation. *Design Studies*, 55, 27–53.

Ehrenfeld, J. (2001). Designing 'sustainable' product/service systems. In *Proceedings Second International Symposium on Environmentally Conscious Design and Inverse Manufacturing* (pp. 12–23). IEEE. https://doi.org/10.1109/ECODIM.2001.992308

Fassi, D., & Manzini, E. (2022). Project-based communities: Lessons learned from collaborative city-making experiences. *CoDesign*, 18(1), 4–15.

Griggs, D., Stafford-Smith, M., Gaffney, O., Rockström, J., Öhman, M. C., Shyamsundar, P., Steffen, W., Glaser, G., Kanie, N., & Noble, I. (2013). Sustainable development goals for people and planet. *Nature*, 495(7441), 305–307.

Hasuo, N. (1980). 'Social' exchange in the classroom 'teacher–pupil relationship'. *The Journal of Educational Sociology*, 35, 146–157.

Iwanami Shoten Publishers. (1991). Sensei. In *Kojien* (4th ed., p. 1467). Iwanami Shoten.

Kadri, T. E. (2020). Networks of empathy. *Utah Law Review*, *2020*(4), 1075–1120. https://doi.org/10.26054/0D-QSHR-9WVA

Kang, N. H., & Hong, M. (2008). Achieving excellence in teacher workforce and equity in learning opportunities in South Korea. *Educational Researcher*, 37(4), 200–207.

Kouprie, M., & Sleeswijk Visser, F. (2009). A framework for empathy in design: Stepping into and out of the user's life. *Journal of Engineering Design*, 20(5), 437–448.

Lewin, K. (1946). Action research and minority problems. *Journal of Social Issues*, 2(4), 34–46.

Li, J. (1998). On Confucian educational ideology and traditional Chinese culture: In the fields of educational sociology and history. *Bulletin of the Graduate School of Education, The University of Tokyo*, 38, 125–134.

Loh, W. L., Shimomura, M., & Zhang, Y. (2020). Unlocking creative minds to engage SDGs through design education in Japanese high school. In *Proceedings of the 22nd International Conference on Engineering and Product Design Education* (E&PDE 2020, n.p.). 22nd International Conference on Engineering and Product Design Education, E&PDE 2020, Herning. https://doi.org/10.35199/EPDE.2020.16

Loh, L., Ogo, M., Zhang, Y., Takano, N., & Shimomura, M. (2021). Readjusting design education for Japanese high school students during the COVID-19 pandemic: A case study to investigate students' attitude in using online whiteboard for group discussions during design activities. In G. Bruyns & H. Wei (Eds.), *With design: Reinventing design modes* (pp. 1527–1545). Springer Nature Singapore.

Manzini, E. (2022). *Livable proximity: Ideas for the city that cares*. EGEA S.p.A.

Maruno, S. (2005, June 10). *Kyoushi no 'diskassyon kyouiku' ginou no kaihatsu to kyouiku shien shisutemu dukuri* [Developing teachers' 'discussion teaching' skills and creating an educational support system]. https://api.lib.kyushu-u.ac.jp/opac_download_md/13253/p295-p323.pdf

Matsuyama, Y., Nakaya, M., Okazaki, H., Lebowitz, A. J., Leppink, J., & Van Der Vleuten, C. (2019). Does changing from a teacher-centered to a learner-centered context promote self-regulated learning: A qualitative study in a Japanese undergraduate setting. *BMC Medical Education*, 19(1), 152. https://doi.org/10.1186/s12909-019-1550-x

McDonagh-Philp, D., & Bruseberg, A. (2000). Using focus groups to support new product development. *Institution of Engineering Designers Journal*, 26(5), 4–9.

Ministry of Education, Culture, Sports, Science and Technology. (2018, March). *High school curriculum guidelines (2018)*. https://www.mext.go.jp/content/20230120-mxt_kyoiku02-100002604_03.pdf

Miyoshi, Y., & Yoshino, R. (2005). A comparative study of work value of the East Asian people: Japan, China, Taiwan and South Korea. *Kodo Keiryogaku (The Japanese Journal of Behaviormetrics)*, 32(2), 173–189.

Morita, J., & Yamada, M. (2013). Questionnaire on the teacher–student relationship as a factor in classroom management. *Annual Report of Study of Education*, 32, 23–37.

Nara Institute of Science and Technology. (n.d.). *What is Japanese and Japanese culture*. https://www.naist.jp/ciss/coming/culture.html

Ng, I. C. L., & Forbes, J. (2009). Education as service: The understanding of university experience through the service logic. *Journal of Marketing for Higher Education*, 19(1), 38–64. https://doi.org/10.1080/08841240902904703

Ogasawara, Y. (2019). *Send your child to an 'uncomfortable place': Learning 'experiential survival strategies' from diverse careers ahead of time*. President Inc.

Pakarinen, E., Imai-Matsumura, K., Yada, A., Yada, T., Leppänen, A., & Lerkkanen, M. K. (2024). Child-centered and teacher-directed practices in two different countries: A descriptive case study in Finnish and Japanese grade 1 classrooms. *Journal of Research in Childhood Education*, 38(1), 30–49.

Shimomura, M., Loh, W. L., & Zhang, Y. (2022). Co-creation of education tools for SDGs Design School based on a pilot program in collaboration with Fukuoka Municipal Fukusho High School. *Cocreationology*, 4(1), 1–11.

Shimomura, M., Loh, L., Zhang, Y., & Takano, N. (2023). Requirements for teaching materials for high school students who are unfamiliar with the design process. In *Proceedings of the International Conference on Engineering and Product Design Education (E&PDE 2023)* (pp. 325–330). 25th International Conference on Engineering and Product Design Education, E&PDE 2023, n.p., Barcelona. https://doi.org/10.35199/EPDE.2023.55

Shulman, L. S. (1987). Knowledge and teaching: Foundations of the new reform. *Harvard Educational Review*, 57(1), 1–23.

Sleeswijk Visser, F., & Kouprie, M. (2008). Stimulating empathy in ideation workshops. In *Proceedings of the Participatory Design Conference* (pp. 174–177). Indianapolis.

Stipek, D., & Byler, P. (2004). The Early Childhood classroom observation measure. *Early Childhood Research Quarterly*, 19(3), 375–397. https://doi.org/10.1016/j.ecresq.2004.07.007

Sunaga, T. (2019). *Design no chie: jouhou design kara syakai no katachizukuru e* [Wisdom of design: From information design to shaping society] (1st ed.). Film Art, Inc.

Takahata, Y., Murata, S., Nishino, K., Yamada, T., Kyakuno, T., Kameda, K., Aoyama, H., Tanaka, M., & Tsudaka, E. (2019). A practical research on teaching methods for active learning (task-oriented research) in high school: Especially focusing on global human resource development. *Kwansei Gakuin University Researches in Higher Education*, 9, 29–44.

Takeyama, M. (2017). Development of a design method for transforming the context of service. *Journal of Serviceology*, 3(4), 4–9.

Zhang, Y., Loh, L., Takano, N., Shimomura, M., & Cruz, C. (2022). Identifying issues in teaching the problem-finding stage faced by instructors: A case study on SDGs challenge project in high school. In *Proceedings of the 24th International Conference on Engineering and Product Design Education (E&PDE 2022, n.p.)*. London South Bank University. https://doi.org/10.35199/EPDE.2022.58

10

Empathy and Distance Learning

Anu Kylmänen

Abstract

This chapter is based on a narrative produced in teaching, The Carefree Room Project (2020–1), which originated from the desire to promote conditions that increase and support students' well-being in special situations (e.g. distance learning). The goal was to create new opportunities to promote open dialogue and a culture of cooperation through the means of art and science. The double diamond (Design Council, 2019), known for service design, served as the research process. The topics were approached with visualized empathy and narrative means. The group's strengths and interests were mapped and harnessed as an asset for the project. The method helped the group's external, open discussion and thus promoted the creation of the trust necessary for the joint planning of the participants. Along the way, the Carefree Room project grew into community workshops and an art exhibition.

Introduction

The Carefree Room Project was a student-oriented design and innovation project whose purpose was to produce information on how, by increasing personal strengths and using empathy and emotions as aids, reflective and debating connections could be created in special situations. In addition, the purpose was to stimulate discussion and find opportunities to utilize methods based on empathy in changing work environments. The starting point was human well-being and mental health. The general aim of the study was to promote the appreciation, understanding and acceptance of university students themselves, each other and diversity. The goal was to find a support channel for community activities and a way to release emotions in a permissive environment through art.

The research questions of this study are:

1 How can you visualize and show your own and others' emotions when working remotely?
2 How can visual empathy be used to recognize, share and understand the emotions of others?

In the present study, the double diamond technique was applied in the service design process, and storytelling was used as the data collection method (Kosonen, 2018). The research material collected during the workshops is presented as an example case of a Carefree Room, where emotions are described with empathic visualization using different methods. In this study, learning and research were social processes (Shepard, 2000) taking place with students and the teacher through communal, empathetic and visual work methods. The learning environment was experimental and dialogic in nature, where knowledge was discussed and solved in the context of problems.

FIGURE 10.1 *Marjaana Sivonen's work 'Give your dreams wings'. Iron wire, candy wrappers, cellophane, thread. Photograph by Olivia Saari (2021).*

Theoretical Framework

During their time as students, student groups form their own social space based on meeting people and relationships. The central elements of human relationships are the small actions of everyday life. Activity, trust, tenacity, flexibility, acceptance of differences, consideration of others, encouragement and commitment, and self-respect are the cornerstones of a lasting human relationship (Määttä & Uusiautti, 2014, pp. 69–73). The elements of interpersonal success listed above serve as the cornerstones of the design of this project, which I approach through emotions, empathy and stories, which form the theoretical framework.

Storytelling

Every person has a story. Storytelling is a fundamental way to express ourselves and connect people. Stories can be an integral part of communication (Kosonen, 2018). The story we tell is part of building our own identity (Lawler, 2015). Through the story, we try to connect with other people (Liao, 2013). Storytelling unfolds as a process that includes a beginning, middle and end. Narration also affects the temporal order, the logical relationships of events, the backgrounds of events and anticipation. Each participant is responsible for how the story or narrative develops through their own interaction. The interpretation of the narrative defines the events and experiences of a person's life (Vuokila-Oikkonen, 2002). The strengths, feelings and memories of the participants in the Carefree Room group were compiled into stories. Various art-based methods were combined with the stories, the aim of which was to help participants understand and create methods based on empathy in changing study environments, thus increasing collective well-being (Leavy, 2020).

Emotions

Emotions play an important role in our lives and in achieving our goals. Emotions create behaviour in us and motivate us to be ready for action. The emotions evoked by products are motivated to exaggerate things, so emotion-based design finds these contradictions and suitable users, as well as multiple design areas (Ozkaramanli & Desmet 2012; Van Gorp & Adams 2012). Emotional states affect people multidimensionally in everyday circumstances; they indicate the direction of social interaction but also manifest as bodily sensations. Emotions also partly determine how we experience and interpret events in the environment (Nummenmaa, 2010).

The emotional reaction involve everything, including body function and behaviour. In our interpersonal interactions, the emotional expressions we observe in others also affect our own emotions. One good moment can turn a bad day into a better one. Checking emotional contagion means optimizing reactions to suitable situations and making interactions between people in an easier way (Nummenmaa, 2010). Repeated, positive emotional experiences are related to people's general well-being (Lyubomirsky et al., 2005). Emotions and taking them into account played an important role in group guidance throughout the process.

Empathy

When we can perceive our emotional states and their underlying factors, it is easier to understand our behaviour and try to influence it. This understanding of one's own emotions can develop affective empathy. It also develops compassionate empathy that comes from helping people cope with their challenging emotions and circumstances. Empathy is the only skill for understanding another person's emotions and usually everyone can naturally experience and feel empathy (Saarikivi, 2018; Sarantou & Miettinen, 2022).

In practice, empathy can stimulate participation because it requires one's detachment from one's current life world and, at the same time, allows one's life to open up (Ashworth & Lucas, 2000). Empathy is an essential part of social cognition and can be defined as the ability to understand and share the inner life of another person. It allows us to observe the actions of others in the same way as when we ourselves participate in them (Brink, 2011, pp. 1–2) or react to the inner emotional states of others (Patnaik, 2009).

The stories in this research material are structured with the help of visual imagery, which has an empathetic touch. The purpose of visualization is to make the shared story more understandable and visible so that the other person's feelings can be understood and interpreted later if necessary. In this text, I call this habit 'visualized empathy' because it best describes this activity. By working and discussing together in the Multisensory Board workshop and the Community Clay Workshop, a safe and accepting interaction space was created, optimizing the opportunity for each participant to react to a new situation from their own perspective while also providing a better understanding of the thoughts of others in that situation.

Double Diamond as a Research Process

In a Carefree Room, the target of the double diamond characteristic of service design was a university student in a challenging situation. The flow of the process was structured in four different areas of the double diamond (Design Council, 2019). In this process, the purpose of the story was to tell and visualize the student story remotely as a collaboration of the whole group using computer platforms as an aid. Each phase of the project, from design to implementation, was divided according to the double diamond into four areas: discovery, definition, development and testing. In practice, this was realized in such a way that by unravelling (finding) the story of the fictitious (definition), the group members were able to develop different ways (develop) to help a person in a challenging situation (testing).

With the help of the story, prevention, life management, peer support and information sharing emerged as the main themes of the project. These themes were approached through a design process, the core of which was the merging of research and art into concrete action. In this project, we tested art-based methods for finding and solving challenging situations. These methods include storytelling and emotion-based design (Leavy, 2020; McNiff, 2008). Combining art and research restores the value of bodily and sensory knowledge of experience. It is typical to strive for uniqueness in art and generalizability in science. Science and art commonly look at reality in a process-like manner and strive to visualize the relationships between the whole and its parts (Liljefors, 2020; Sava, 1998). The process emerges from long development work to collect observations (Bardy, 2012) and a functional community through the senses and materials (Hiltunen, 2009).

Methods Based on Visualization and Empathy Used in the Carefree Room Project

In this research, I present the methods using the example case of a Carefree Room, with a special emphasis on thinking based on empathy and the visualization of feelings arising from it (Lawler, 2015), for which I use the term *visualized empathy*. In the fall of 2020, a group of thirteen students (with an age range of approximately 20–26 years) from the University of Lapland's Faculty of Arts and Design set out to think about students' well-being during their studies. The Carefree Room Project Group was chosen as the University of Lapland's student group for 2022.

The teacher led the group's discussion to reflect on their own strengths and interests, and the discussion continued on the challenges of a university student's life. After that, he divided the topics according to the interests of the students. With the help of the stories, everyday situations requiring change were identified, and possible suggestions for improvement were sought. The group initially worked remotely through digital collaboration platforms.

We developed visual remote empathy; that is, we tried to describe our emotions or the emotions of another person by writing and drawing while connected remotely. Through these means, we tried to understand and share the inner life of another person. Here, empathy served as an enabler of connections and mutual understanding between students, which was born through the attention and presence of others and created motivation to create understanding based on the subject's and own previous experiences (Saarikivi, 2018).

Visualization supports and promotes communal interaction. Figure 10.2 is an example of the material from which answers to the research question were outlined: (1) How can you visualize and process your own and others emotions when working remotely?

Next, I will give a few examples that answer the research question: (2) How can visual empathy be used to recognize, share and understand the emotions of others? The Carefree Room Project lasted from 7 September 2020 to 31 May 2021. One of the communal models in the Carefree Room was an art-illustrated person called Sara's story, which was based on a remote conversation where students identified with the challenges of university student life. The dialogue, along with Sara's story and its analysis, revealed alternative ways to express the idea that understanding a challenging situation deepened the participants' commitment to the project. The group in the Carefree Room created an imaginary character who studied for five years at the University of Lapland.

The teacher started the discussion and encouraged the students to think about the topic through their own strengths and interests. The group

FIGURE 10.2 *Veeti Tukiainen made visual notes about the meeting on the 26th of October 2020: 'What the emotions of a challenging situation look like'. Photograph by Anu Kylmänen (2020).*

participants developed stories about this person, which were combined into a common and single story framed by a fictional character named Sara. The group created a timeline of Sara's life in which Sara's character became a challenging situation. We thought about how the situation could have been prevented and how Sara could have escaped the challenging situation. Solutions were developed for Sara's story, where the group used the story and illustrations, that is, visual empathy, to empathize with another person's (Sara) situation.

During the discussions, the group drew a picture of Sara together (Figure 10.3). The character's appearance took shape when the ideas were just being developed. The participants' joint observations, research information, ideas, imagination and real-life experiences formed one multidimensional and comprehensive story. As a result, the story served as a method for identifying even difficult topics through discussion. Sara's character provided a safe means to address the more difficult aspects of the stories that had previously been overlooked. Each member of the group had their own role in creating the story of 'Sara'; that is, everyone

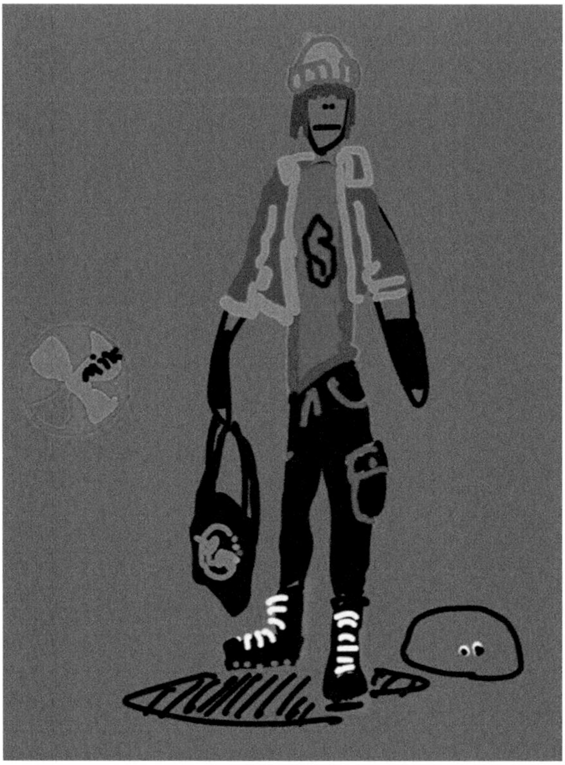

FIGURE 10.3 *A drawing representing Sara's character. Print screen: Anu Kylmänen (2020).*

contributed to this story according to their own previous experiences and strengths. At this stage, the whole group was committed to the same topic, and the motivation to continue in the next stages was good. According to the teachers' observations, the confidence of the group members in participating in the discussions was encouraging, and the sense of community was strong.

The answers were collected on a common Miro board by writing and drawing, and if desired, further discussion was carried out. The discussion involved the use of emojis, which supported the visualization of the emotional states of the participants. In the end, Sara's story crystallized into the core concepts of the Carefree Room: safety, prevention, peer support, community and release of emotions. With the help of these concepts, two workshops were built in which emotional reactions were worked on using different materials and methods. It was possible to participate in the workshops on site or remotely.

The *Community Clay Workshop* was organized on 15 February 2021. The workshop was open to all students and staff at the University of Lapland. The purpose was to offer an opportunity to try working with clay as a way to relieve stress or other challenging emotions. During the workshop, each participant received a piece of clay to shape a sculpture depicting their own emotions. The group members experienced that stress can be alleviated by practising a conscious presence and activating different senses. In particular, by activating a sense of touch during the workshop, an

FIGURE 10.4 *My emotions visualized with clay. Photograph by Annu Kuivas (2021).*

FIGURE 10.5 *Carefree room group making the sensory boards 'Sense the Wall'. Photograph by Olivia Saari (2021). Mixed technology, 120 cm × 160 cm; Fiia Sandqvist, Anu Kylmänen, Tanja Heape, Leena Nieminen, Annu Kuivas, Olivia Saari, Veeti Tukiainen, Santeri Kallankari, Pirjo Vuohtoniemi.*

open discussion was held, and the participants found the workshop to be a very relaxing experience (Figure 10.4).

The Wall Workshop consisted of the idea of making a multisensory board, where each participant made a 40 cm x 40 cm sensory board using the technique of their choice. The work consisted of twelve tiles, each of which was made to represent the emotions of its maker. The purpose of the multisensory board called Sense the Wall was to create a space to focus on the present moment, experience the different emotions of others and possibly be able to identify with them. In this case, touch played a particularly important role. The sensory wall was intended as a multisensory experience that made the viewer see, touch and, thereby, feel their own emotions through the emotions of another (Figure 10.5).

The results of the Carefree Room Project were gathered together, and thus, a community art exhibition was created, which was exhibited in the gallery of the University of Lapland in Kopio from 2 to 18 March 2021. The works in the exhibition showed how visualization can be used in different ways to process and show one's own emotions while working remotely. The works also illustrated how visual empathy can help to identify and possibly also understand the emotions of others and show one's own emotions while working remotely.

The exhibition created a space where every visitor had the opportunity to receive emotional and empathetic peer support through art based on the feelings of others. The processes of the workshops progressed according to a double diamond in both (discover) the terms that serve as the core concept (define) of safety, prevention, peer support, community and liberation, which were developed in two different workshops, The Community Clay Workshop and Sense the Wall, which were implemented in practice and tested again at an art exhibition (test).

Research Instruments and Materials

The group visits after the workshops also served as information on the discussions in which the held workshop was discussed, and the next visit was planned based on the reflection. The material was saved on the Miro collaboration platform, meeting notes and group emails and other message exchanges (i.e. mainly WhatsApp message service conversations) as well as photo recordings of the workshop task. In the task, Adobe Connect software was used as a discussion platform, and the ideas were saved on the visual collaboration platform Miro. The research materials involved the operation of the Carefree Room Project. Other data collection methods included activity observation, feedback discussion and surveys (Huovinen & Rovio, 2007).

Data Analysis in Qualitative Research

In qualitative research, the analysis of the material often proceeds analytically. In the present study, the material was examined according to the double diamond method in chronological order, and the key themes of the material were specified. In the analytical phase, the material was classified and structured according to the thematic areas to be discussed. Conceptuality and core themes emerged from the research material (Kiviniemi, 1999).

I applied this theory to the visual data, meaning that I collected all the material on the Miro collaboration platform. This material was visible to the group throughout the process. The material to be analysed progressed from reflection and reflective discussion during the activity, through the acquisition and examination of the materials, towards the comprehensiveness of understanding the process and expansion and refinement of the theory. The goal was to describe the research process to the recipient as a truthful, structured whole. In the group's joint discussions, I observed the course of the discussion about the group's thoughts. The observation took place as a physical presence with the help of remote connections in the existing environment and in concrete interaction situations with other people. After the exhibition and project, I surveyed the emotions and opinions of the participants via a feedback survey. Visitors had the opportunity to fill out a short form about the emotions evoked by the exhibition. The emotions were listed on the form, and each visitor could circle the emotions they felt were appropriate. In addition, there were two short open questions in the form, which we used to find out one of the goals of the study, i.e. whether we managed to create a safe space to release one's emotions in a permissive environment through art: (1) Did the exhibition help you survive? How? and (2) What other ways could there be to survive everyday life? We received twenty-four responses.

Findings

Based on the feedback from the Carefree Room participants, the most common feelings of the workshop were calmness, fun and relief: 'Despite the corona situation, I was pleasantly surprised by how smoothly our group was able to function, even though most of the meetings took place remotely. The tasks went smoothly despite the size of the group, the group's motivation was good, and the team spirit was visible both in the remote and on-site meetings' (Participant 3 of Carefree Room Project 15.2.2021).

The exhibition visitors felt that interest (14 times), joy (12 times), aesthetic appreciation (12 times) and calmness (12 times) were the most common emotions. An example item used to answer question 1 was 'The exhibition calmed and inspired me; I got ideas for my own coursework' (Respondent 19). This feedback suggests that the method also worked in practice. This research combines the possibilities of art and design with the perspectives of human-centred research methods. The methods helped build empathic understanding not only among the participants, but also between the participants and researcher. This laid the foundation for the success of the project, where a trusting and committed atmosphere prevailed throughout the project. The method presented here enables ways in which the goal is to involve remote workers in building mutual understanding. We managed to arouse discussion and find opportunities in the utilization of methods based on empathy in changing working environments.

'Art offered a way to express emotions: where words end, artistic possibilities only begin. Making by hand also enabled a concrete approach to dealing with emotions. Working together increased the sense of meaning and community' (Participant 2 of the Carefree Room Project 15.2.2021).

At the heart of the stories was the combination of research and art, with the help of different methods for concrete action. Methods were used, for example, for describing one's own emotions, especially for observing and understanding the emotions of others with the help of different materials. The double diamond clarity of the research process and narrative approach to the topics of the workshops aroused interest in new art-based methods but also offered concrete opportunities for handling various challenging topics. In this project, the process works well with art-based methods, even in remote work. Sara's story led to a discussion about empathy, where it was easier to process and talk about one's own emotions through the emotions and experiences of others. Because of the empathic perspective, the participants felt that the conditions were safe to reflect on their own and others' emotions, and they were motivated and open to new ways of working and communicating. A sincere and encouraging atmosphere was perceived as a key factor in finding new ways of working. This was greatly influenced by the activity of the entire group.

> A Carefree Room is a place where we sail together in the emotional waves of life, find a lighthouse and return to a safe harbour, even in the middle of storms. In that place, there is live gentleness, empathy and shared joy. The room is filled with trust and assurance that no one will be left alone. You do not always need words; sometimes, a small smile and a considerate meeting with the other are enough. In the carefree room, everyone will be accepted exactly as they are. (Participant 4 of Carefree Room Project 15.2.2021)

The sense of security created during the project was an integral part of the goal of creating the common well-being of the participants and developing art-based methods for situations, such as the cases presented here. The project's working group felt that recognizing one's own strengths bolstered the group and increased the motivation and self-esteem of the group members, which also promoted understanding of other people's actions and emotions. The visualization and stories stimulated discussions and motivated the workshop participants to develop different solutions for solving the topics. Anticipation, clarity and the instructor's genuinely empathetic approach created a sense of security in the group. Working together and understanding the well-being of the participants increased the group's commitment and deep understanding of what the joint discussion was about.

Conclusion and Discussion

The starting point was people's well-being and mental health, and the goal was to find a support channel for communal activities and a way to release emotions in a permissive environment with the help of art. An open discussion, a common topic that was handled with artistic means using different materials, and the double diamond process created a clear and safe atmosphere for the entire project. A sense of security and trust was an essential part of achieving the set goals.

In the workshops, action models related to design and art were implemented, where all participants moved forward with the collective power of the group, acting creatively and discussing. The goal was to create new opportunities to promote open dialogue, together with design and the culture of making, here through art-oriented means. An empathic approach was used to help this group, where the emotions of others and one's own emotions were considered through storytelling and visualization. The present study sought answers to two research questions using different visualization and materialization tools and methods (Koh et al., 2011; Shneiderman et al., 2013) to help visualize and understand user experiences (Prendiville et al., 2017; Segelström, 2013). In working life, designers use modelling quite commonly as a design method (Patrício et al., 2011).

Emotions have their own special meaning for a person's positive resources, but also for learning and competence results. In addition, positive emotions and experiences of joy, success, ability and appreciation contribute to many prerequisites for a happy life and create a strong foundation for coping with life's challenges (Määttä & Uusitalo, 2008). The goal was to support all participants in their strengths and inspire them to see familiar things from new perspectives. Repeated, positive emotional experiences are also considered to correspond to people's general well-being (Lyubomirsky et al., 2005). Positive emotional experiences were associated with increased creativity and more flexible behaviour and thinking patterns.

The workshops aimed at everyone's personal success and new insights. It was noteworthy that the instructor did not give ready-made development models but was involved in taking the development idea forward together with the group. In a carefree room, the freedom to express one's opinion and the courage to succeed but also to make mistakes were emphasized throughout the project (Fredrickson, 2003).

This method helped increase the confidence needed for joint planning, and the planning process progressed successfully to the end (Beaulé & De Coninck, 2023; Beaulé & Viinikainen, 2022). This method can especially enable the front-end area of the planning process (Almqvist, 2017). An example of this is Sara's story, through which the participants found the subject areas to be developed, that is, challenging situations.

My conclusion is that the research answers the two research questions. Based on the research material, emotions can be shared and understood without personal contact through different remote methods and platforms, for example, with the help of a story supported by empathetic visualization. Trust emerged as the most important core theme of the Carefree Room, with which cooperation and well-being in the group could be developed even remotely. Trust was built with the help of stories that had an empathic approach, the other person's feelings were considered, and an effort was made to truly understand them. Visualized empathy was especially helpful here.

Takeaway

The purpose of the project was to produce information on how, by increasing personal strengths and using empathy and emotions as aids, reflective and debating connections could be created in special situations. The visualized empathy of the story helped everyone participate in the process, strengthened and increased open discussion, and encouraged the social inclusion of different individuals. In addition, the purpose was to stimulate discussion and find opportunities to utilize methods based on empathy in changing work environments. Storytelling works as a method that guides thinking, where empathy works as a specialty.

References

Almqvist, F. (2017). The fuzzy front-end and the forgotten back-end: User involvement in later development phases. *The Design Journal*, 20(sup.1), S2524–S2533. https://doi.org/10.1080/14606925.2017.1352765

Ashworth, P., & Lucas, U. (2000). Achieving empathy and engagement: A practical approach to the design, conduct and reporting of phenomenographic research. *Studies in Higher Education*, 25(3), 295–308.

Bardy, M. (2012). Taide, tiede ja myötätunto. *Synteesi,* 3, 29–47.

Beaulé, C. I., & De Coninck, P. (2023). *Cooperation by design: Exploring the dynamic of trust in an Arctic collaborative project.* https://arcticyearbook.com/images/yearbook/2023/Scholarly_Papers/14_Beaule_DeConinck_AY23.pdf

Beaulé, C. I., & Viinikainen, M. (2022). Building trust with storytelling: Dialogues and encounters in the Arctic. In *Empathy and business transformation* (pp. 104–118). Routledge. https://www.taylorfrancis.com/chapters/edit/10.4324/9781003227557-10/building-trust-storytelling-caoimhe-isha-beaul%C3%A9-mari-viinikainen

Brink, C. O. (2011). *Horace on poetry: The Ars Poetica.* Cambridge University Press.

Design Council. (2019). *The double diamond.* https://www.designcouncil.org.uk/our-resources/the-double-diamond/.

Fredrickson, B. L. (2003). The value of positive emotions: The emerging science of positive psychology is coming to understand why it's good to feel good. *American Scientist*, 91(4), 330–335.

Hiltunen, M. (2009). *Yhteisöllinen taidekasvatus: Performatiivisesti pohjoisen sosiokulttuurisissa ympäristöissä* (Väitöskirja). Lapin yliopisto. http://urn.fi/URN:NBN:fi:ula-20111141039

Huovinen, T., & Rovio, E. (2007). Toimintatutkija kentällä. In H. L. T. Heikkinen, E. Rovio, & L. Syrjälä (Eds.), *Toiminnasta tietoon: Toimintatutkimuksen menetelmät ja lähestymistavat* (2nd ed., pp. 94–113). Kansanvalistusseura.

Kiviniemi, K. (1999). Toimintatutkimus yhteisöllisenä prosessina. In H. L. T. Heikkinen, R. Huttunen, & P. Moilanen (Eds.), *Siinä tutkija missä tekijä: Toimintatutkimuksen perusteita ja näköaloja* (pp. 63–83). Atena.
Koh, L. C., Slingsby, A., Dykes, J., & Kam, T. S. (2011, July). Developing and applying a user-centered model for the design and implementation of information visualization tools. In *2011 15th International Conference on Information Visualisation* (pp. 90–95). IEEE.
Kosonen, K. (2018). *Finding one's own way in design. Reflections on narrative professional identity* [Doctoral dissertation]. Aalto University.
Lawler, S. (2015). *Identity: Sociological perspectives.* John Wiley & Sons.
Leavy, P. (2020). *Method meets art: Arts-based research practice.* Guilford Publications.
Liao, T. (2013). Designing wellbeing through storytelling. In K. V. Turkka Keinonen (Ed.), *Designing for wellbeing* (pp. 72–86). Aalto ARTS Books.
Liljefors, M. (2020). Knowledge worlds apart. *Movement of Knowledge: Medical Humanities Perspectives on Medicine, Science, and Experience*, 24, 205–232.
Lyubomirsky, S., King, L., & Diener, E. (2005). The benefits of frequent positive affect: Does happiness lead to success? *Psychological Bulletin*, 131(6), 803–855.
Määttä, K., & Uusiautti, S. (2014). *Rakkaus; Tunteita, taitoja, tekoja.* Kirjapaja.
Määttä, K., & Uusitalo, T. (2008). Ihmisen vahvuudet auttavat selviytymään. In *Kasvatuspsykologian näkökulmia ihmisen voimavarojen tueksi* (pp. 7–9). Lapland University Press.
McNiff, S. (2008). Art-based research. In J. G. Knowles & A. L. Cole (Eds.), *Handbook of the arts in qualitative research: Perspectives, methodologies, examples, and issues* (pp. 29–40). SAGE Publications.
Nummenmaa, L. (2010). *Tunteiden psykologia.* Tammi.
Ozkaramanli, D., & Desmet, P. M. (2012). I knew I shouldn't, yet I did it again! Emotion-driven design as a means to motivate subjective well-being. *International Journal of Design*, 6(1), 27–39.
Patnaik, D. (2009). *Wired to care: How companies prosper when they create widespread empathy.* FT Press.
Patrício, L., Fisk, R. P., Falcão e Cunha, J., & Constantine, L. (2011). Multilevel service design: From customer value constellation to service experience blueprinting. *Journal of Service Research*, 14(2), 180–200.
Prendiville, A., Gwilt, I., & Mitchell, V. (2017). Making sense of data through service design – Opportunities and reflections. In D. Sangiorgi & A. Prendiville (Eds.), *Designing for service: Key issues and new directions* (pp. 225–236). Bloomsbury Publishing.
Saarikivi, J. (2018). Finnic and other Western Uralic borrowings in Permian. *Uralica Helsingiensia*, 11, 269–355.
Sarantou, M., & Miettinen, S. (Eds.). (2022). *Empathy and business transformation.* Taylor & Francis.
Sava, I. (1998). Taiteen ja tieteen kietoutuminen tutkimuksessa. In M. Bardy (Ed.), *Taide tiedon lähteenä* (pp. 103–121). Stakes.
Segelström, F. (2013). *Stakeholder engagement for service design: How service designers identify and communicate insights* [Doctoral dissertation]. Linköping University Electronic Press.
Shepard, L. A. (2000). *The role of classroom assessment in teaching and learning.* CSE Technical Report.

Shneiderman, B., Plaisant, C., & Hesse, B. W. (2013). Improving healthcare with interactive visualization. *Computer*, 46(5), 58–66.
Van Gorp, T., & Adams, E. (2012). *Design for emotion*. Elsevier.
Vuokila-Oikkonen, P. (2002). *Akuutin psykiatrisen osastohoidon yhteistyöneuvottelun keskustelussa rakentuvat kertomukset. Oulun yliopisto.* Universitas Ouluensis. Series D. Medica, 704.

11

Service Prototyping with Empathy and Nature-Centric Thinking

Mariluz Soto

Abstract

This chapter focuses on service prototyping as a collaborative environment that can create an optimal atmosphere to enhance creative action with others. The chapter explains people's connection with their memories in a natural environment and expresses their feelings in a collage to get the body involved. There is value in starting service prototyping workshops with a natural simulated connection, which contributes to nudging a participant's connection with themselves; this means a conscious stop, and focusing their thoughts, emotions and bodies in a simulated environment. The key research question of this article is 'How can we integrate empathy and nature into service prototyping?' The main findings are the importance of the body as a vehicle to become emotionally involved and the opportunity from the service prototyping laboratory to create an empathetic atmosphere through natural connection, here defining a collaborative starting point in the same creative mood for all participants.

Introduction

In capital cities, rush and stress are the most common expressions used to define daily feelings; sometimes, an overwhelmed mood is synonymous with productivity and is considered the consequence of being a good worker. Nature seems to be something cosmetic or functional to human beings. It is likely that these words are judgements coming from a Chilean woman who has lived in the capital city for almost her entire life. In addition, it is true that my beliefs shape the way I think, highlighting certain aspects. This may be part of my belief system, influenced by my culture.

I decided to start this chapter with the previous paragraph because it is the basis of what drives my research. A relevant area in my motivation is to identify that there are opportunities from service prototyping to influence people's experiences. However, it is essential to focus on the initial point of service prototyping because this is an instance of collective simulation of a moment of experience, and it is essential to establish a common state among participants. The premise is that defining this initial mood would allow participants to begin the service prototyping process with an optimal flow by generating a change between the state in which they arrive – considering that participants may arrive after spending a long time in overwhelming public transportation, traffic congestion in the streets, family arguments or other specific problems. These situations influence the way people are predisposed to others and, therefore, impact their interactions.

Service prototyping is usually an important part of the service design process. It has the value of visualizing the outcomes in a concrete manner, but with a service prototyping laboratory it is also an opportunity to simulate environments to transform the experience connecting with other realities (Camburn et al., 2017; Miettinen, 2011; Stickdorn et al., 2018). Currently, a way to simulate different environments considers technology to be a virtual reality or other high-tech method, but the screen-simulation environment allows participants to feel themselves in different places with just a few elements. Additionally, using the service prototyping environment could transform the perception of the process, especially those stages that need a boost in creativity.

Service prototyping is a group of people working together to visualize different interactions during a service experience, searching for improvements. The exchanges in this cocreation environment provide an opportunity to simulate experiences and simultaneously contribute to creating or strengthening the community spirit (Soto, 2021). Service prototyping could help bond work teams or groups of people with a common purpose.

This chapter explains the connections between service prototyping (and cocreation), empathy and nature as an integrative perspective to change the process of service prototyping. The main question of this chapter is as follows: 'How can we integrate empathy and nature into service prototyping?' In addition, we delve into the possibility of a positive impact on people's well-being, not only for users but also for all participants in service prototyping. To comprehend the current understanding of service prototyping and empathy, this chapter considers a set of interviews with service designers and one workshop to test the possibility of improving the construction of a place for reconnecting the living from a creative perspective. The results will be added to the methodology of service prototyping as a tool to create an atmosphere of willingness to process and creative openness.

Service Prototyping, a Collaborative Environment

Service designers have a relevant role in creating innovation, and doing and reflecting are part of the learning process (Knight, 2013). Miettinen (2011) considers service design to comprise practical and academic design areas with a focus on research and methodological and strategic emphasis to develop new knowledge. First, a service is a channel through which to achieve a goal and is the path to completing a task; the steps of a service can vary, but from the user's perspective, the service must be easy to follow (Downe, 2020). Second, service design is based on many disciplines that merge concepts and tools (Sangiorgi & Prendiville, 2017), with a human-centred approach focusing on cultural understanding (Dennington, 2017). Third, service prototyping, one of the common stages of the process that service design follows, is addressed in the last parts of the process, usually before the testing moment, emphasizing its characteristics of doing quick and unfinished mock-ups (e.g. design thinking six steps from Nielsen Norman Group (Gibbons, 2016), design thinking from d.school at Stanford or Double Diamond from Design Council with the Framework for Innovation).

Prototyping is commonly defined as 'the act of making ideas concrete in a provisional, speculative way' (Ku & Lupton, 2022, p. 38), which usually refers to making visible the idea for a product; however, for services, the concreteness of the idea is slightly different and requires people to act on the ideas. Service prototyping, according to the Service Innovation Corner (SINCO) (2013), is a collaborative, creative and innovative environment (Rontti et al., 2012). As a multidisciplinary area, it contributes to innovation by creating or making visible ideas through design thinking (Lim et al., 2019). Services create value; therefore, their impact is different in different fields (Jung Bae & Seong Leem, 2014). This value creation requires the integration of the experiences of different actors and stakeholders. Service prototyping, as a collaborative environment where people related to the service participate, captures emotions from their experience, transforming them into a valuable advantage to personalized processes and obtaining more accurate and reliable results, which benefits from the proposals that arise from this cocreation environment (Miettinen et al., 2014). The environment and predispositions of the participants are essential for visualizing their experiences and work.

Prototyping mainly involves stating ideas in a concrete way to visualize the shape of something. Service prototyping simulates the ideas, using the body and its context as a tool of the process. According to Changaris (2021), there are three core principles of embodied study and design research: (1) the embodiment is situated, which recognizes the relevance of place and memory and history; (2) bodily/systemic states such as behaviour could change the outcome; and (3) the simulation shapes the understanding of the brain environment and the person's response in line with it (p. 3). Therefore, these three principles explain what happens during service prototyping in a simulation environment, such as SINCO, the relevance of the body simulating different service moments and identifying, at the same time, accurate opportunities to improve the service.

The body, gestures and emotions are vital factors for service prototyping because they trigger many visible expressions that are relevant communication actions for participants' interactions and exchanges. Emotions have a clear impact on that interaction; therefore, they are strong and valuable drivers of the decision-making process (Lerner et al., 2015). An emotional response can be activated by a product (Desmet, 2002), such as the desire to purchase a new product from a famous brand; or, from a service perspective, it can be activated by a service experience (Miettinen et al., 2014).

Service prototyping, in addition to its relevance to the service design process, emphasizes what happens in this environment. The prototyping of services in a face-to-face environment focuses on people's experience, both in the power of tangibly visualizing an experience and in the bodily expressions of the participants and its contribution to generating connections in that exchange or fostering a community spirit.

Empathy in Service Prototyping

Emotions and empathy appear naturally in this environment. Empathy interweaves the diverse expressions, behaviours and emotions of people, creating connections that are based on sharing and exchanging those aspects that constitute us as a community and society (Riess, 2017). Empathy is an essential skill for service designers because it is a component of quality in design (Kouprie & Sleeswijk Visser, 2009), and they must manage the interaction process where emotions are involved, hence enhancing the connections and creative exchange between participants of the process to achieve innovative outcomes (Kuure et al., 2014). Considering that service prototyping has a bodily component during the simulation of service moments or touchpoints, those emotions related to service experiences emerge as much as empathy because it is a collaborative and cocreative environment for working together to identify opportunities to improve the service.

The cocreative scenario involves different challenges, roles, skills and purposes, all of which make every cocreative situation unique because it depends on each person at that specific moment (Soto, 2021). Because of the complexity of the projects and the specific requirements of communities, every creative collaboration occurs; hence, empathy is also a key concept in this case. Horvath and Carpenter (2020) define 'co-creation as a knowledge process that employs creativity through arts-based methods as an alternative way to listen to the voices of marginalized communities and involve them in generating shared understanding of their neighbourhoods and (in)justices in the city' (p. 4). According to Gupta and Vajic (2000), an experience occurs when a client – or user – has a sensation or acquires any knowledge as an outcome of an interaction with a service. The experience is also related to knowledge and learning for the symbiotic connection between how people perceive the experience and how they understand it. This understanding is the combination of previous interactions and new interactions in which two factors give people functional knowledge. The concept of interacting appears here as linking the different moments that make up an experience, hence including the various factors involved (e.g. people, concrete actions, responses, options, the environment or physical or virtual place in which everything happens, which takes into consideration the characteristics that define that place). Moments are not just something created from one person to other (unidirectional); all people that participate in those moments are shaping the moment (Heath & Heath, 2017).

Experiences that are unique, memorable and sustainable over time are considered successful because they are desirable to repeat, develop and promote with others (Pine & Gilmore, 1998). Emotions and knowledge are part of this definition of experience and, whether they are positive or not, they define the experience in the present moment; they can change the memory of previous experiences, and they are part of future experiences. However, emotions can change depending on each person, even triggering different responses for the same person at different times. At the same time, emotions in people–product interactions depend on product characteristics and the world behind each person, such as needs, goals and skills (Smith & Lazarus, 1990). Because emotions impact attitudes, opinions and affective beliefs about an object, person or phenomenon (Eagly & Chaiken, 1993), which are also applicable to interactions with a service, experience has a greater impact than just the present moment. In other words, just by living an experience, which people do many times during the day, there is a change in people's understanding of their own lives and the relationships they establish through interactions. This constant voluntary and involuntary embodied learning is invisible but leaves traces. According to Maturana and Varela (1994), 'We cannot separate our history of actions – biological and social – from how that world appears to us' (p. 10); therefore, everything we experience has a consequence regarding each person's knowledge and memory and in their encounters with others.

Living an experience is also related to time, which involves the anticipation of events and a trajectory that must ultimately reach a culminating point as well as emotional involvement throughout the process and a need for uniqueness that makes a difference from the ordinary (Dewey, 1993). The relationship between time and experience appears to be a side characteristic, but it is relevant because it defines different milestones in people's lives, creating a sense of the life journey, contributing to establishing margins of comparison between the different moments lived and allowing the creation of connections between learning and linked emotions.

Methodology

This chapter connects service prototyping with empathy and creates a simulated natural environment to be tested in a workshop in a Chilean service prototyping laboratory. After the workshop, a conversation with participants was held, which focused on their experiences, to understand their processes and meanings (Creswell & Poth, 2007), which were closely connected with a phenomenological perspective (Giorgi, 1997). The purpose of the workshop was to obtain an understanding of the experience of the people, from their own voices and considering the specific context in which each one was inserted (Muratovski, 2016), and to connect from there with the existing theory (Leavy, 2017).

The workshop was conducted in Laboratory DES, the Chilean service prototyping laboratory, which has a structure based on participatory embodiment experience through a service simulation based on role-playing. The structure of the workshop combines storytelling with images of nature and creates collages from their experiences (see Figure 11.1). The purpose of

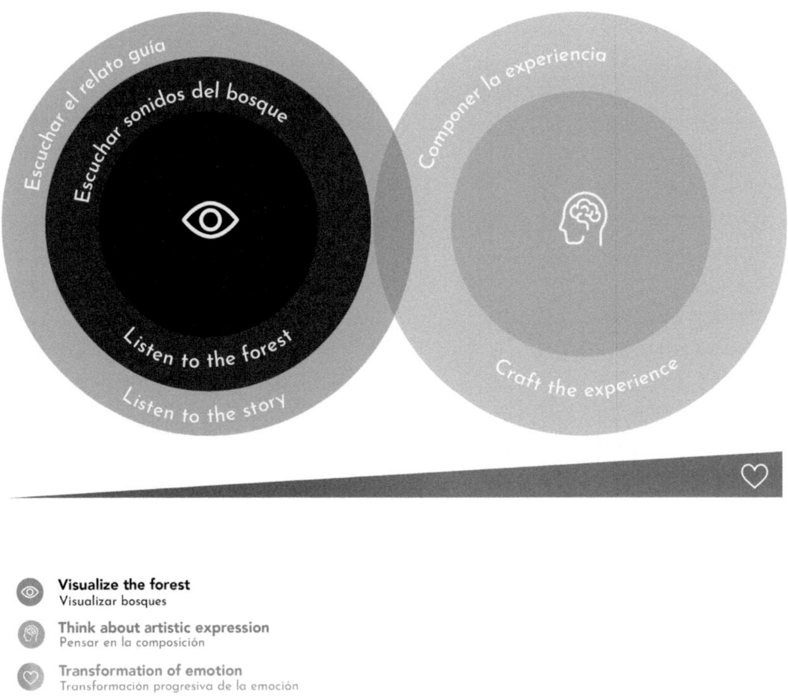

FIGURE 11.1 *Visualization of the experience. Designed by Teresita Muñoz (2024).*

the workshop was to transform the experience into a new image, focusing the participants' attention on a creative process and being conscious of the meaning of their senses (see Figure 11.2). The workshop considered a 10-minute presentation about empathy with a few questions (such as 'Do you know what empathy is?', 'Have you heard the expression "being in another's shoes"?' 'What do you think about that?') in a wider sense but integrating perspective, increasing the awareness of the disposition to connect with oneself and the environment. This is the first step towards understanding how to cocreate with plants, hence reflecting on empathy between living beings.

The experience in front of the screens watching the forest was guided by a two-minute story focused on a common memory of experience in the forest and describing bodily sensations. The story was written by the author using her observation and experience in the forest – a memory of walking in a natural environment and her bodily sensations. The purpose of the story was to give the participants certain details that would allow them to connect with the projected image. This resource was used as a conductor of the imagination so that the participants followed the awakening of sensations through the story (Lupton, 2017) and the activation of their memories of walking in a forest. Stories are part of humanity; they can preserve cultural knowledge (Beaulé & Viinikainen, 2023), make sense from different perspectives, build a common understanding and connect with other scenarios or places.

In addition, some elements, such as aromatic scents, forest sounds and temperature management, were added to the screens.

Below is the story:

> At this moment, you are in front of a forest. A vine climbs the trunks, affirmed by the roughness. It wraps itself, hugging the bark.
>
> Leaves with different greens move gently with the breeze that sneaks through the trees.
>
> Between the fog and sun rays, some drops fall after the rain. One by one, they fall on a leaf, sliding down the centre of the leaf, joining with another drop and falling together with more force onto the ground.
>
> A small river can be heard in the distance, creating a harmony of sounds similar to rain. As it passes, it moistens the roots that rise from the ground, which cross different surfaces and move, interweaving below and above the earth.
>
> You can almost feel the moist aroma of the earth. When you breathe deeply, the cold, clean air fills your lungs with rain, small drops of glistening water, resting on the many shades of green.
>
> You can almost feel the wet roughness of the tree bark on your hands. Your hands remember that coldness when walking through the forest.
>
> The raindrops generate visual layers and small sparkles that accompany our gaze.

FIGURE 11.2 *The moments of the experience. Photographs by the author (2024).*

The birds sing from the trees, announcing the eastern sun that covers all the high treetops. They hop from one branch to another, looking for food that the rain discovers and removes.

Walking in a forest is entering a different world, secret, silent and full of aromas that remain anchored in our memory.

Breathe deeply so that the forest stays in our bodies for a long time.

The purpose of the story was to accompany a connection process between listening, viewing the projected image and smelling the aromas to lead to a state of stillness. This approach allows one to create a break between the actions previously carried out and the subsequent activities to be carried out. The story is a component of the experience itself; the experience focuses on creating separation to establish an emotional state more predisposed to creativity. According to the above, it focuses on establishing moments with peaks to elevate a pleasant sensation (Heath & Heath, 2017) through a narrative that marks an arc of the experience (Lupton, 2017), emphasizing that it involves different emotions that allow the experiential flow to advance positively.

Learning from the Workshop

The focus of the workshop was on testing the process of creating a new moment as a starting point for any service prototyping activity. This starting activity aimed to transform the mood of the participants, as I said at the beginning of this chapter, especially for service prototyping carried out in large cities, which tends to be centred more on productivity, immediacy and competition.

The main findings are as follows:

- According to the size of the service prototyping laboratory, the number of participants must decrease or the projections must be wider. In Figure 11.2, one of the images shows five participants in a corner; those behind or with one of their sides close to the screen had to resort mostly to their memory to avoid losing concentration.
- It is recommended that the voice that tells the story be recorded to avoid confusion when reading.
- Because it is an activity that should last a maximum of 20 minutes, the images and collage materials should be delivered ready to select and paste to avoid the time it takes to select the magazine and define the cuts.
- People's immersion in the forest is a personal journey that should focus on a friendly scene that transports them to pleasant places. An opportunity for improvement could be the incorporation of individual VR devices.
- More elements can be added to the activity to make the experience more immersive.

Service Prototyping and the Body

Prototyping is one of the key steps in the service design process. It is a cycle to put ideas in common in a concrete manner, facilitating collaboration between participants – usually stakeholders – through the externalization of ideas (Blomkvist et al., 2011). The collaborative and creative features of service prototyping have the potential to reach more than solution development; they also modify people's perceptions through experience. This change in perception is quite abstract because it happens in a completely invisible way, impacting people's understanding of the world and personal experiences; consequently, experience becomes embodied knowledge and learning. Prototypes can trigger emotional aspects in the early stages of the process through the visualization of personalized prompts because they reveal emotional responses or reactions that provide valuable information to the process and to decision-makers (Miettinen et al., 2014).

Service prototyping involves three areas – head, heart and body – in people's interactions (Soto et al., 2023) and people who interact with themselves. According to Tolosa (2008), the head symbolizes intellectual learning, or the way in which the belief system, judgements and shared reasoning are modified. The heart contains all the feelings and emotions linked to the experience that is expected to improve in an environment of innovation and the experience of being with others while creating. Additionally, emotions provide a creative state that can vary depending on the conditions of the place and interactions with others. The body is the most visible; it is represented by all the movements made by people or by participants in a simulation of services, and it also considers proximity and expressions and even gestures that allow – or block – interactions between people. The body can also absorb learning and integrate it as knowledge.

In service prototyping in a simulation environment such as the Laboratory DES and SINCO, the trilogy is integrated into its maximum expression, and the simulation of the service experiences requires specifying each of the 'scenes' or moments through the generation of a story that allows for the uniting of everything as a choreography. This coordination is given by the shared experiences and solutions proposed in a collaborative way, and the improvements of a service represent the expectations of having results appropriate to the interaction of the users directly impacting their emotions. Choreography is a designed step and process with a creative emphasis on organizing movements from all those involved (Beiswanger, 1962), and at the same time, it is an orchestration of various elements so that the creative flow moves appropriately between the participants.

Prototyping the Atmosphere

Prototyping has been results oriented as a visualization of ideas to share quickly and easily with others and meets a specific objective within the service design process. Face-to-face meetings between people have additional value, especially after Covid-19-related quarantines and health isolation, because people experience convergent emotions and behaviours by sharing their experiences. Therefore, this prototyping environment has the possibility and opportunity to modify previous understandings or self-perceptions, and this change is new knowledge housed in the same three areas: what we recognize as part of the experience, the emotions that arise and the concrete actions associated. At the same time, prototyping acquires additional value, connecting the process in a sensitive manner to understand our surroundings and cocreating in a horizontal condition.

The purpose of the atmosphere is to provide optimal conditions for individuals to collaborate with others in a service prototyping environment. The prototyping of services and any participatory activity also requires being designed as an experience and, therefore, divided into various moments that stimulate collaborative work. The activity proposed in this chapter is an initial dynamic that would allow that atmosphere to be created by pretending to be a break between the previous actions of the participants (e.g. public transportation at peak times or traffic congestion or discussions of various kinds) to promote collaborative work. This emphasis is closer to the vision of step-out than to the actions of icebreakers or warming up. Step-out is a technique of the Alba Emoting (AE) method, in which emotions are induced through breathing. Step-out is the time to return to calm; it is to take a step back to return to a neutral point to receive – or, in the case of AE, to induce again – an emotion (Bloch, 2008). The atmosphere is also a story, travelling through the imagination and sensations to invite people to respond accordingly to themselves and others (Lupton, 2017). If the entire experience of prototyping services was thought of as a film, this first activity is the first scene – the one that determines whether we continue watching the film. Therefore, it must be designed as a set of factors that influence its realization, such as the choreography, the scenography and the emotional flow through which the participants will travel.

Conclusion

In the Western world, especially in large cities, fast-paced life and the overvaluation of productivity have caused people to move from one place or topic to another without even taking a slight pause to absorb the learning or experience they have acquired. For the same reason, there is a permanent search for 'recipes' to carry out any activity, something quick to read and learn. Fortunately, not everything is structured or measurable in the same range.

In service design, various texts and articles have been published with very good tools and methods to develop some or all the formal stages to design or improve services. I agree with Holmlid and Evenson (2007) regarding the lack of teaching on how to use all the tools that have already been developed and published. Therefore, the appropriate use of these devices depends fundamentally on the personal skills of the designers.

Generating a space of personal calm and the impulse of togetherness will enable the mental and emotional space to share experiences, opinions and comments with others. As service designers, it is essential to think of these moments as opportunities to connect with participants and give them a pleasant experience; thereby, they can contribute to the purpose that brings them together.

Takeaway

Service prototyping is a privileged environment in which to connect with people and together create proposals that can improve an experience. Therefore, it is essential to create an appropriate atmosphere for this moment of connection, creativity and togetherness. The lifestyle of large Western cities has distanced us from how vital everything living is, which, like all interactions with other people and living beings, gives us so much satisfaction. This chapter has proposed an instance to begin prototyping services focused on reconnecting with our calm, with our body, with our emotions and with others.

References

Beaulé, C. I., & Viinikainen, M. (2023). Building trust with storytelling. In M. Sarantou & S. Miettinen (Eds.), *Empathy and business transformation* (pp. 104–118). Routledge.

Beiswanger, G. (1962). Chance and design in choreography. *The Journal of Aesthetics and Art Criticism*, 21(1), 13–17.

Bloch, S. (2008). *El alba de las emociones* [The dawn of emotions]. Uqbar Publishers.

Blomkvist, J., Segelström, F., & Holmlid, S. (2011). Investigating prototyping practices of service designers from a service logic perspective. In *Proceedings of the Nordic Academy of Management Conference (NFF 2011)*, Stockholm, Sweden.

Camburn, B., Viswanathan, V., Linsey, J., Anderson, D., Jensen, D., Crawford, R., & Wood, K. (2017). Design prototyping methods: State of the art in strategies, techniques, and guidelines. *Design Science*, 3, e13. https://doi.org/10.1017/dsj.2017.10

Changaris, M. (2021). Embodied research design: A transformational paradigm in scientific discovery. In J. F. Tantia (Ed.), *The art and science of embodied research design: Concepts, methods and cases* (pp. 3–18). Routledge.

Creswell, J. W., & Poth, C. N. (2007). *Qualitative inquiry and research design: Choosing among five approaches*. SAGE.

Dennington, C. (2017). Service design as a cultural intermediary: Translating cultural phenomena into services. *The Design Journal*, 20(sup.1), S4285–S4294. https://doi.org/10.1080/14606925.2017.1352765

Desmet, P. (2002). *Designing emotions* [Doctoral dissertation, Delft University of Technology].

Dewey, J. (1993). *Demokratie und Erziehung*. BeltzReferences Verlagsgruppe.
Downe, L. (2020). *Good services*. BIS Publishers.
Eagly, A. H., & Chaiken, S. (1993). *The psychology of attitudes*. Wadsworth Group.
Gibbons, S. (2016, July 31). *Design thinking 101*. Nielsen Norman Group. https://www.nngroup.com/articles/design-thinking/
Giorgi, A. (1997). The theory, practice, and evaluation of the phenomenological method as a qualitative research procedure. *Journal of Phenomenological Psychology*, 28(2), 235–260. https://doi.org/10.1163/156916297X00103
Gupta, S., & Vajic, M. (2000). The contextual and dialectical nature of experiences. In J. A. Fitzsimmons & M. J. Fitzsimmons (Eds.), *New Service Development: Creating Memorable Experiences* (pp. 33–51). SAGE. https://doi.org/10.4135/9781452205564.N2
Heath, C., & Heath, D. (2017). *The power of moments*. Simon & Schuster.
Holmlid, S., & Evenson, S. (2007). *Prototyping and enacting services: Lessons learned from human-centered methods*. Proceedings from the 10th Quality in Services conference, QUIS.
Horvath, C., & Carpenter J. (2020). *Co-Creation in theory and practice: Exploring creativity in the global and south*. Bristol University Press.
Jung Bae, D., & Seong Leem, C. (2014). A visual interactive method for service prototyping. *Managing Service Quality: An International Journal*, 24(4), 339–362. https://doi.org/10.1108/MSQ-12-2013-0281
Knight, J. (2013). The experience design framework: Supporting design thinking in the service domain. In S. Miettinen & A. Valtonen (Eds.), *Service design with theory* (pp. 169–176). Lapland University Press.
Kouprie, M., & Sleeswijk Visser, F. (2009). A framework for empathy in design: Stepping into and out of the user's life. *Journal of Engineering Design*, 20(5), 437–448.
Ku, B., & Lupton, E. (2022). *Health design thinking*. Cooper Hewitt.
Kuure, E., Miettinen, S., & Alhonsuo, M. (2014). Change through service design: Service prototyping as a tool for transformation and learning. In Y.-K. Lim, K. Niederrer, J. Redström, E. Stolterman, & A. Valtonen (Eds.), *Proceedings of DRS 2014: Design's Big Debates* (pp. 469–482). Umeå Institute of Design, Umeå University.
Leavy, P. (2017). *Research design: Quantitative, qualitative, mixed methods, arts-based, and community-based participatory research approaches*. The Guilford Press.
Lerner, J. S., Li, Y., Valdesolo, P., & Kassam, K. S. (2015). Emotion and decision making. *Annual Review of Psychology*, 66(1), 799–823. https://doi.org/10.1146/annurev-psych-010213-115043
Lim, C., Kim, K. H., Kim, M. J., Kim, K. J. (2019). Multi-factor service design: Identification and consideration of multiple factors of the service in its design process. *Service Business*, 13, 51–74. https://doi.org/10.1007/s11628-018-0363-z
Lupton, E. (2017). *Design is storytelling*. Smithsonian Design Museum.
Maturana, H., & Varela, F. (1994). *El arbol del conocimiento*. Editorial Universitaria.
Miettinen, S. (2011). Service prototyping in action. *Touchpoint*, 3(2), 64–65.

Miettinen, S. A., Rontti, S. J., & Jeminen, J. (2014). Co-prototyping emotional value. In E. Bohemia, A. Rieple, J. Liedtka, & R. Cooper (Eds.), *Proceedings of the 19th DMI International Design Management Research Conference: Design management in an era of disruption* (pp. 1228–1246). Design Management Institute. https://doi.org/10.13140/2.1.4764.6721
Muratovski, G. (2016). *Research for designers*. SAGE.
Pine, B. J., & Gilmore, J. H. (1999). *The experience economy*. Harvard Business Review Press.
Riess, H. (2017). The science of empathy. *Journal of Patient Experience*, 4(2), 74–77.
Rontti, S., Miettinen S., Kuure E., & Lindström A. (2012). A Laboratory Concept for Service Prototyping – Service Innovation Corner (SINCO). Conference Proceedings ServDES (pp. 229–241); 8–10 February; Espoo, Finland.
Sangiorgi, D., and Prendiville, A. (Eds.). (2017). *Designing for Service: Key Issues and New Directions*. Bloomsbury.
Service Innovation Corner. (2013, March 31). *Project info*. http://sinco.fi
Smith, C. A., & Lazarus, R. S. (1990). Emotion and adaptation. In L. A. Pervin (Ed.), *Handbook of personality: Theory and research* (pp. 609–637). The Guilford Press.
Soto, M. (2021). *Emotional skills for service designers in co-creation practices* (Publication No. 300) [Doctoral dissertation, University of Lapland]. Acta Electronica Universitatis Lapponiensis. https://urn.fi/URN:ISBN:978-952-337-242-9
Soto, M., Xue, H., & Tsekleves, E. (2023). Perspectiva de cabeza, corazón y cuerpo para la salud y el bienestar desde el diseño [Head, heart and body perspective for health and well-being from design]. *Base 7 Diseño bienestar y salud, diciembre*, 26–36.
Stickdorn, M., Hornes, M., Lawrence, A., & Schneider, J. (2018). *This is service design methods*. O'Reilly Media.
Tolosa, M. (2008). *A twist in communication*. Fundacion de la Comunicología.

12

More-Than-Human Empathy in Arts-Based Material Research Processes

Jenny Bergström, Heidi Pietarinen and Satu Miettinen

Abstract

This chapter explores the role of the creative process in building a more empathic relationship between humans and the environment. Using a theoretical framework including more-than-human empathy, a material-based approach and creative processes, the authors present two case studies.

In 'The Secret Life of a Material', design students work with renewable gluten plastic, letting the material guide their creative process. 'The Biggest Bread Soup in the World' presents the artist's immersive experience with Lake Tjörnin in Reykjavik, observing and interacting with its natural environment.

These studies show how, when designers and artist–researchers engage in creative practices with nonhuman agents as coauthors, this can lead to unexpected outcomes and enhance empathic connections between humans and the natural world. This chapter highlights how an art-based research (ABR) process can deepen our understanding of and empathy towards nonhumans, encouraging collaborative practices that integrate human and natural agencies.

Introduction

Related to discussions around the human–environment relationship, this chapter focuses on and asks questions about the role of the creative process in creating a more equal relationship and empathy building between humans and nature, or natural materials, as authors in an arts-based process.

Severe global environmental changes suggest that the Earth may have entered a new human-dominated geological epoch called the Anthropocene (Lewis & Masalin, 2015). Humans may have significantly impacted the global future of the planet, possibly changing life as we know it for both humans and nonhumans. Posthumanism positions people as equal to other life forms (Nayar, 2018), suggesting that they cannot destroy and benefit from natural resources without consequence. Furthermore, human actions should not be evaluated according to ethical standards that are different from those used to evaluate other life forms. In design, human-centred value systems propose creating systems that address human needs rather than constraining machines or technologies or even human dignity and rights (Buchanan, 2001). Ecocentric design takes a step forward, proposing the urban environment as a network of dependencies between humans, nonhumans and the environment that does not focus solely on human well-being (Bergström et al., 2009).

Changes and clashes of values have generated new alternative ways of thinking about the dominance of humans, such as the shift in thinking from anthropocentricism to ecocentrism, which is based on the idea that all organisms are interrelated – not only with their environment but also through their environmental interrelationships (Eckersley, 2023); ecological feminism, or 'ecofeminism', which recognizes the connections between treating women, people of colour and the underclass, on the one hand, and the nonhuman natural environment on the other (Warren, 1997); and the idea of the pluriverse, which is understood 'as the ever-changing entanglements of humans and nonhumans that result from the Earth's ceaseless movement of vital forces and processes' (Escobar, 2015, p. 14).

In this chapter, the term more-than-human is used as a concept that, as described by Forlano (2017), expands our understanding of the multiple agencies, dependencies, entanglements and relations that make up our world, acknowledging the idea that the term does not 'negate human design; it only multiplies those designs in a larger field so that there are always many instead of only one' (Easterling, 2016, p. 16). This chapter presents two case studies, 'The Secret Life of a Material' and 'The Biggest Bread Soup in the World', exploring the collaborative creative process with nonhumans and nature as co-authors. This chapter aims to address the following question: 'How can we use a creative process to co-author and build empathy with the natural environment, materials and their affordances?'

Theoretical Framework

Utilizing a theoretical framework created within the concepts of more-than-human empathy, a material-based approach and a creative process, two case studies are examined. The case studies were carried out and evaluated by two of the authors of this chapter: Jenny Bergström and Heidi Pietarinen.

More-Than-Human Empathy

Lanzoni (2018, p. 17) states that the definition of empathy has changed radically through history but that repeated themes can be detected, stating that 'empathy depends on movement between the poles of similarity and difference, of distance and closeness, of immersion and alienation'. Lanzoni (2018) also points out that, although the popular definition of empathy describes our ability to understand the experience of other humans, the origin of the term lies in the arts and that it used to mean the ability to place ourselves in the world around us and its nature and objects. This notion of empathy to imagine the feelings and experiences of nonhumans is interesting to remember when considering more-than-human empathy in an arts-based process because it opens up possibilities for these agencies to play a part in creating conditions for empathy (Sinquefield-Kangas et al., 2022).

Kouprie and Sleeswijk Visser (2009, p. 440) refer to 'the empathic' in design as 'deep understanding of the user and his or her experience including the situation and feelings'. We are interested in exploring how this idea of empathic design can be understood when considering empathizing as not only understanding and grasping the feelings and experiences of other humans but also, as described by Pestana (2023), as essentially an act of stepping outside oneself and shifting focus. Nature is being increasingly recognized as an author in scientific publications, such as the Matuwarra River of Life (River of Life et al., 2021) and Wann Country (Foster et al., 2020). Co-authorship in the arts can create new ways of looking at both ethical and copyright questions while offering alternate views. In this context, artists become experimenters, collectors and natural agents. This necessitates a multifaceted exploration of contemporary material culture and requires a border between the artist's studio, laboratory or residency and between fields and disciplines at the intersection of art, science and society (Berger et al., 2020).

Material-Based Approach

In this chapter, we look at materiality in the context of authorship. Leonardi's (2012) definition of materiality addresses the topic of combining enduring arrangements of material and form across place and time. In the context of human interaction, materiality is referred to as socio-materiality, which assumes that it is mediated through the objects, tools or materials that relate to different social realities (Huifen and Qiuhong, 2012). Nimkulrat (2012,

p. 3) discusses 'the importance of craft practice and the actual making of work in contributing to knowing through immersive sensory experience'. In Nimkulrat (2012), there is a reference to the concept of material thinking by Carter (2004), which suggests that materials are active in the creative process and that there is an interaction between materials and the maker's hands, mind and eyes.

Creative Process

Lubart (2001) states that the creative process has been researched since the 1950s and Guilford's work on creativity and refers to Guilford (1950, 1967) when distinguishing between convergent and divergent thinking. Baer and Kaufmann (2012) refer to divergent thinking as the thinking that results in many, different and unusual ideas, and convergent thinking as the thinking focusing on finding a correct answer. Divergent thinking is still seen as an indicator of creativity, both in creativity research and in educational practice (Baer, 2014).

The creative process and divergent and convergent thinking are often connected with a design process, especially a double diamond model, where divergent and convergent thinking take turns (Saad et al., 2020). Art-based methods such as improvisation activities can enhance divergent thinking (Sowden et al., 2015). Many design concepts, such as design thinking (Henriksen et al., 2017), embrace the creative process. Designers are asked to create new solutions, which demands creativity when moving in problem and solution (Dorst & Cross, 2001) spaces.

Research Design

Methodologically, this chapter builds on arts-based research (ABR) approaches. At the core of this research case is cyclicality, which is derived from an action research approach (Kemmis & McTaggart, 2005), meaning that it is a long-term process in which each cycle operates independently but evolves based on the evaluation and results of previous cycles.

New research knowledge is generated through artistic and practice-based case studies that function as learning cycles. ABR is an overarching research approach that includes a wide variety of art forms, genres and practices (Leavy, 2020). ABR can also cultivate empathy and self-reflection through the disruption of dominant narratives by utilizing performance and storytelling (Leavy, 2017).

Leavy (2020) discusses the cultivation of empathy as one of the greatest strengths of ABR, describing how 'connecting people on emotional and visceral levels, artistic forms of representation facilitate empathy' (Leavy,

2020, p. 27). Although Leavy focuses mainly on the outcomes of artistic processes, we believe that this approach can also be applied during the creative process and that working collaboratively in a shared process can help build empathy both among participants and towards the material itself as we, through engagement, learn about the nature of the material. The case studies presented in this chapter could be described as empathy-building exercises between humans and more-than-human agents.

This chapter relies on the art-based creative process with nonhumans. Art-based processes have a strong potential for generating knowledge that may be used and applied by the makers and researchers of artistic practice, thus transitioning from tacit to explicit. This premise is the basis of practice-based (or practice-led) and artistic research (e.g. Koskinen et al., 2011; Mäkelä & Routarinne, 2006). Practice serves as knowledge generation in all of its stages, from ideation through prototyping and making to the final artefact, leaving the artist–researcher in the position of creating, documenting, contextualizing and interpreting artefacts. In the following case studies, field notes, observation, video recording, written process documentation and collected reflections of participants are used as a basis for analysis. In addition, placing materials at the centre of a process, as in the following case studies, can open up to co-authorship between the participants themselves as well as between participants and the material, as the process and outcomes are influenced by the participants, their bodies and their gaze (Dora, 2009).

Case Studies

'The Secret Life of a Material' studies a creative and experimental process with a natural material, renewable gluten plastic. 'The Biggest Bread Soup in the World' describes how Reykjavik's Lake Tjörnin behaves like a living thing. Operating at the intersection of art and design, the first case study discusses the human–environment relationship in a design-based context, and the second case study discusses the same relationship in an art-based context.

The Secret Life of a Material

For most designers, engaging with materials is viewed as a natural part of the design process. Designers, architects and other professionals in the field of crafts know that different materials have different characteristics and various aesthetic expressions. There is also a long history of material innovation and material science driven by the empirical experimentation of artists, artisans and designers (Mazé et al., 2013). Nimkulrat (2012) addresses learning and knowledge contribution as an immersive sensory experience, noting that materials and experimentation play an important role in design thinking (DT). As stated by Kelley and Kelley (2013, p. 130), 'The reason for prototyping is experimentation – the act of creating forces you to ask questions and make choices'.

The workshop 'The Secret Life of a Material' took place at Konstfack University of Arts, Crafts and Design in Stockholm, Sweden. Led by Jenny Bergström, the participating students were given the task of exploring novel material and new methods of application. Students were asked to document and reflect upon their processes, and parts of the working sessions were filmed. The material – renewable gluten plastic (Figure 12.1), which was provided and introduced by the material researcher Mikael Gällstedt from Innventia – is made from a side stream material from ethanol production and has the potential to become a sustainable substitute for oil-based plastic products (Patni et al., 2014).

One goal of the workshop was to challenge the students' established thinking patterns and, through unknown material, encourage them to employ a beginner's mindset. The workshop setup was based on the

FIGURE 12.1 *Mixing and getting to know the material renewable gluten plastic.*

belief that working with a material acting differently or even alien forces us to step away from our initial plans and allows the material to lead us through the process. In this sense, the material itself can be regarded as a more-than-human agent in the design process. Another goal was to broaden the possible outcomes and forms of applications for renewable gluten plastics through this mindset, making the assumption that the participants, with their specific skill sets and knowledge, would produce outcomes different from those developed by the researchers at Innventia. Another assumption was that working with renewable gluten plastic would lead the students in directions that could not be fully anticipated and that control over the design process would partly be in the hands of the material itself.

For the students, the process required to 'get started' was longer and more challenging than in a normal course setup. To get ideas and imagine the possibilities, the participants had to 'get to know' and build empathy with the material by working with it hands-on. The material and the process of handling it were challenging because they acted differently than the materials they were familiar with. This setup required the participants to explore the material freely to imagine what to do with it. Although most of the time we are not consciously aware of our cognitive activity when interacting with material forms, just as handling a material or tool may become highly automated over time (Overmann & Wynn, 2018, p. 463), this first encounter with a material behaving in unexpected ways required full focus and engagement that forced participants to put ideas, concepts and aesthetics to one side. This reversed way of working was at first challenging, but after some experiments (Figure 12.2), the students noticed that their lack of control over the material was helping them let go of their fear of failing, allowing the material to take the process in directions that they did not plan for.

Stepping away from the materials we are familiar with can add value to the creative process and lead to unexpected solutions. Encountering new material helps us enter the beginner's mindset and view the task and challenge with fresh eyes. Because it forces us to let go of control and embrace failure, it can help us reduce our fear of failure. Giving creative authorship to nonhuman agents, in this case natural material, can help us challenge settled thinking patterns. When reflecting upon the workshop, Mikael Gällstedt referred to gluten plastic as a 'living' material that changes over time and is influenced by its surroundings. This description implies that the material can be viewed as an active agent. By allowing material to influence the creative process, we can also broaden the range of outcomes (Figure 12.3) and increase our chances of generating unconventional ideas (Rizzo et al., 2020).

FIGURE 12.2 *Experimenting with light.*

FIGURE 12.3 *A wide range of material samples were created during the workshop.*

The Biggest Bread Soup in the World (Stærsta Brauðsúpa í Heimi)

'The Biggest Bread Soup in the World' is the nickname given to Lake Tjörnin, which is located in the city centre in Reykjavik, Iceland. Feeding birds along the lake shore is a popular pastime attraction for travellers and city dwellers, which has led to the lake being both metaphorically and poetically referred to as bread soup (Holownia, 2013).

In 2022–3, Heidi Pietarinen was in residence at Reykjavik, the former home of writer, illustrator and natural scientist Benedikt Gröndal (1826–1907). Lake Tjörnin was an integral part of the author's outdoor itinerary and relationship with birds during the entire period of residency. Both places, Gröndal's House and Lake Tjörnin, created the starting point for the direct experience of nature, spatial design knowledge and what one city attraction represents, in addition to different perceptions of material movements and bodily aspects. It created the centre for everyday gatherings where strangers were likely to meet.

Pietarinen became a moving observer and interpreter, documenting enchanting changes and being part of birds' life in Lake Tjörnin (see Siqueland, 2022). Field notes were produced throughout the processes through photography, video recordings and the writing of a travel diary. The birds' curiosity towards humans and vice versa gradually developed into a very direct, honest and empathic experience of nature – a mutual encounter, experience and learning opportunity. Daily walks to the lake were translated into field notes explaining what happened to the artist–researcher in certain places in nature's living laboratory, even though the city was her canvas (Pietarinen et al., 2022; Tuomaala, 2023). The field notes revealed empathic connections between artist–researchers (humans), birds and water (nature), allowing for a more general understanding to emerge based on exclusive private experiences in Lake Tjörnin (Figure 12.4).

The activities at Lake Tjörnin (Figure 12.5) also produced different perceptions of material movements: the temperature of the skin (feeling cold: [un]comfortable, enchanting, overwhelming, repellent), pain (extreme cold wind), watery eyes, runny nose and dry lips (dry air), cough (inhaling makes you understand the volume of the lungs) and slow, (un)steady and small steps on the enchanting ground (motion, a bodily aspect, presence). The changes in the weather were felt and affected what Pietarinen wrote in her notes. She was in nature and surrounded by animals and people. This conveys the idea that nature does not consist of objects but can be observed through different life forms and cocreators: What could the collaboration between an artist–researcher and birds be like? Direct experience of nature can represent significant features of empathy.

Today, birdwatching, like other observational studies of nature, is culturally accepted almost everywhere. Both the sky and sea are vast, life-filled environments, with birds, flying insects, wind-blown seeds, microbes

FIGURE 12.4 *The direct experience of nature can represent significant features of life.*

and drifting spores. All of this material is a biological mass in the atmosphere (Macdonald et al., 2021). Artistic co-authorship, with its communities, past and present artists and nonhumans, contributes to the design of plurality and the pluriverse by providing new views and ways to create authorship in a posthumanist society (Bacharach & Tollefsen, 2015).

During residency, Pietarinen explored the notion of biophilia, meaning 'love of life or living system' (Fromm, 1964, pp. 13, 37–45), a term established by E.O. Wilson in 1984 that refers to humans' positive responsiveness to nature. In many places in Iceland, such as a bursting geyser, a roaring

FIGURE 12.5 *View of sea from the city of Reykjavik.*

waterfall, a stormy sea and a fissure between two tectonic plates, the land behaves like a living thing: shifting, breathing and snorting. Biophilia is not only a concept to include nature in human lives, but also the initiation of a cycle for humans to improve their quality of life while living in an urbanized society (Amazing Geological Wonders Forces of Nature Exhibition, 2023). It is possible to learn nature literacy (i.e. the practice of Shinrin-Yoku, also known as Forest Bathing, has proven to improve positive health benefits for the human physiological and psychological systems). When you are connected to nature, you 'open doors' for everything you see and feel because nature mirrors us and we learn about ourselves (Hansen et al., 2017). At Lake Tjörnin, the loud honks, barks and cackles of Greylag geese build their own multilayered stories in the background. The birds are of various tempers; some are empathic, while others are greedy or indifferent. It felt gentle to be surrounded by the geese, surrendering to be touched or to almost touch a plumage (a tactile quality) – a smoothness and surface design that could also be visually evaluated (Karana, 2010; Pietarinen et al., 2022; Sonneveld & Schifferstein, 2008). Some of the geese approached leisurely, waddling and pausing to stand on one leg, acknowledging people's willingness to notice them as individuals and feed them with breadcrumbs. The birds' direct, open-minded and empathic approach to humans was impressive. Spatial design knowledge started to act as a mediator of empathy design, even though emotional values are difficult to pinpoint, predict and make tangible (Hasling & Ræbild, 2018). What can one city attraction represent? An artist–researcher can be considered a modern-day explorer, investigating, for example, bio-art's emotional impact and material narrative potential as a catalyst for change or opportunities (Siqueland, 2022).

'The Biggest Bread Soup in the World' field notes tell a story of the author's need to sort, catalogue and understand the Arctic environment; field notes become works of art (in Finnish, *taiteen teos*), not artwork.

FIGURE 12.6 *Author sorting and cataloguing to generate an understanding of the Arctic environment.*

The field notes pass on an understanding of the subjects (colours and patterns), narratives and challenges shared by artists as explorers in the Arctic region (Figure 12.6). On the other hand, this perspective reflects a split in our ways of relating to nature. According to one view, nature is something untouched that should only be observed or recorded; according to another, it can be brought indoors, or a laboratory or studio is temporarily removed, in which case you can interact closely with nature. The division is loaded with social meanings and raises the question of who has the right to define what a creature is like, who has the right to interact with it and how (Bolt, 2004; Kontturi, 2012; Macdonald et al., 2021).

Conclusion and Discussion

Engaging with materials that are unfamiliar to us can introduce new aspects to the creative process and lead to unanticipated outcomes. By letting the material guide us through the process and by viewing materials as something 'living', we grant authorship to the materials and allow them to play an active role in the creative process (Carter, 2004). Although two case studies offer limitations when drawing conclusions and more studies are needed, observations and first-hand experiences during the workshop 'The Secret Life of a Material' have confirmed that the interaction between materials and the maker's hands, mind and eyes can contribute to the feeling of building empathy towards the natural material.

When altering between divergent and convergent thinking in the design process, giving authorship to the material and relinquishing control can help us enter a beginner's mindset (Kelley & Kelley, 2013) and broaden the spectrum of possible outcomes. Through discussions and reflections with participants, we have learned that working with material that acts as a 'living thing' can help build empathy among participants through a shared experience as well as with natural material through hands-on experiments. This finding resonates with Nimkulrat's (2012) description of how the making of work and the sensory experience contributes to knowing and the concept that artistic forms can connect people on an emotional level as well as the cultivation of empathy as one of the greatest strengths of ABR (Leavy, 2020).

The field notes for 'The Biggest Bread Soup in the World' were designed to improve the relationship and build empathy between humans and (non) humans, showing that it is possible to learn nature literacy, much like how it is possible to document materials and bring them to life if we let the land behave like a living thing. The field notes became a living diary of birds, places and people, both reconnecting people with city life and acting as a mediator of empathy design, creating connections between humans and birds and allowing for a more general understanding to emerge based on exclusive private experiences in the city in Reykjavik. In line with contemporary trends in bio-art, a living diary of birds offers not only the possibility of sources of inspiration or Arctic raw materials but also shared experiences, collaborative practices, cooperation and shared authorship with different species – empathy towards nature.

Takeaway

Both case studies use the role of the designer and artist–researcher as a modern-day explorer as a starting point. Whether bringing nature inside to explore it in a studio or going out to live and experience nature, the creative process can be a collaborative process with nonhumans and nature as co-authors. This opens up new, unexpected directions in the making and in art and design outcomes, helping with empathy building between humans, nature, animals or natural materials.

References

Amazing Geological Wonders Forces of Nature Exhibition. (2023). *Perlan Museum. Reykjavik*. https://www.perlan.is/en-gb.
Bacharach, S., & Tollefsen, D. (2015). Co-authorship, multiple authorship, and posthumous authorship: A reply to Hick. *The Journal of Aesthetics and Art Criticism*, 73(3), 331–334. https://doi.org/10.1111/jaac.12187
Baer, J. (2014). *Creativity and divergent thinking: A task-specific approach*. Psychology Press. https://doi.org/10.4324/9781315806785
Baer, J., & Kaufman, J. C. (2012). *Being creative inside and outside the classroom: How to boost your students' creativity – and your own*. Sense Publishers. https://doi.org/10.1007/978-94-6091-840-7
Berger, E., Mäki-Reinikka, K., O'Reilly, K., Sederholm, H., & Schmidt, M. (2020). *Art as we don't know it*. Aalto University.
Bergström, J., Mazé, R., Redström, J., & Vallgårda, A.(2009). Symbiots: Conceptual interventions into urban energy systems. *Nordes 2009: Engaging Artifacts*, 29 August – 1 September, The Oslo School of Architecture and Design, Oslo, Norway. https://doi.org/10.21606/nordes.2009.008
Bolt, B. (2004). *Art beyond representation: The power of the image*. I. B. Tauris.
Buchanan, R. (2001). Human dignity and human rights: Thoughts on the principles of human-centered design. *Design Issues*, 17(3), 35–39. https://doi.org/10.1162/074793601750357178
Carter, P. (2004). *Material thinking: The theory and practice of creative research*. Melbourne University Publishing.
Cropley, A. (2006). In praise of convergent thinking. *Creativity Research Journal*, 18(3), 391–404. https://doi.org/10.1207/s15326934crj1803_13
Dora, V. D. (2009). Performative atlases: Memory, materiality, and (co-) authorship. *Cartographica: The International Journal for Geographic Information and Geovisualization*, 44(4), 240–255.
Dorst, K., & Cross, N. (2001). Creativity in the design process: Co-evolution of problem–solution. *Design Studies*, 22(5), 425–437. https://doi.org/10.1016/S0142-694X(01)00009-6
Easterling, K. (2016). *No, you're not*. e-flux.com. http://www.e-flux.com/architecture/superhumanity/66720/no-you-re-not/
Eckersley, R. (2023). *Environmentalism and political theory: Toward an ecocentric approach*. Routledge.
Escobar, A. (2015). Transiciones: A space for research and design for transitions to the pluriverse. *Design Philosophy Papers*, 13(1), 13–23.
Forlano, L. (2017). Posthumanism and design. *She Ji: The Journal of Design, Economics, and Innovation*, 3(1), 16–39.
Foster, S., Paterson Kinniburgh, J., & Wann Country. (2020). There's no place like (without) country. *Placemaking Fundamentals for the Built Environment*, 63–82. https://doi.org/10.1007/978-981-32-9624-4_4
Fromm, E. (1964). *The heart of man: Its genius for good and evil*. Harper & Row.
Guilford, J. P. (1950). Creativity. *American Psychologist*, 5, 444–454.
Guilford, J. P. (1967). *The nature of human intelligence*. McGraw-Hill.
Hansen, M. M., Jones, R., & Tocchini, K. (2017). Shinrin-Yoku (forest bathing) and nature therapy: A state-of-the-art review. *International Journal of*

Environmental Research and Public Health, 14(8), 851. https://doi.org/10.3390/ijerph14080851

Hasling, K., & Ræbild, U. (2018). *Sustainable design cards: A learning tool for supporting sustainable design strategies*. https://www.researchgate.net/publication/334041972_Sustainable_Design_Cards_A_Learning_Tool_for_Supporting_Sustainable_Design_Strategies

Henriksen, D., Richardson, C., & Mehta, R. (2017). Design thinking: A creative approach to educational problems of practice. *Thinking Skills and Creativity*, 26, 140–153.

Holownia, O. (2013). The world is confused pink sheep: Subversive uses of Icelandic themes in the poetry in Þorarinn Eldjan. In C. Kelen & B. Sundmark (Eds.), *The nation in children's literature: Nations of childhood* (pp. 69–80). Routledge.

Huifen, W., & Qiuhong, T. (2012, July). Study on socio-materiality convergence of IT architecture and social architecture: From critical realism view. In *2012 International Conference on Computing, Measurement, Control and Sensor Network* (pp. 421–424). IEEE.

Karana, E. (2010). *Meanings of materials*. LAP LAMBERT Academic Publishing.

Kelley, D., & Kelley, T. (2013). *Creative confidence: Unleashing the creative potential within us all*. William Collins Books.

Kemmis, S., & McTaggart, R. (2005). Participatory action research: Communicative action and the public sphere. In N. K. Denzin & Y. S. Lincoln (Eds.), *The Sage handbook of qualitative research* (pp. 559–603). Sage Publications.

Kontturi, K. (2012). *Following the flows of process: A new materialist account of contemporary art*. Turun yliopisto.

Koskinen, I., Zimmerman, J., Binder, T., Redstrom, J., & Wensveen, S. (2011). *Design research through practice: From the lab, field, and showroom*. Elsevier.

Kouprie, M., & Sleeswijk Visser, F. (2009). A framework for empathy in design: Stepping into and out of the user's life. *Journal of Engineering Design*, 20(5), 437–448. https://doi.org/10.1080/09544820902875033

Lanzoni, S. (2018). *Empathy: A history*. Yale University Press.

Leavy, P. (2017). *Research design: Quantitative, qualitative, mixed methods, arts-based, and community-based participatory research approaches*. Guilford Publications.

Leavy, P. (2020). *Method meets art: Arts-based research practice* (3rd ed.). Guilford Publications.

Leonardi, P. M. (2012). Materiality, sociomateriality, and socio-technical systems: What do these terms mean? How are they different? Do we need them? In P. M. Leonardi, A. B. Nardi, & J. Kallinikos (Eds.), *Materiality and organizing: Social interaction in a technological world* (pp. 25–48). Oxford University Press.

Lewis, S. L., & Maslin, M. A. (2015). Defining the Anthropocene. *Nature*, 519(7542), 171–180.

Lubart, T. I. (2001). Models of the creative process: Past, present and future. *Creativity Research Journal*, 13(3–4), 295–308. https://doi.org/10.1207/S15326934CRJ1334

Macdonald, H., Ruuska, I., & Ruuska, I. (2021). *Iltalentoja (Evening Flights)*. Gummerus.

Mazé, R. (Ed.). (2013). *SWITCH! Design and everyday energy ecologies*. Interactive Institute (RISE).

Mäkelä, M., & Routarinne, S. (Eds.). (2006). *The art of research: Research practices in art and design*. University of Art and Design Helsinki.

Nayar, P. K. (2018). *Posthumanism*. John Wiley & Sons.

Nimkulrat, N. (2012). Hands-on intellect: Integrating craft practice into design research. *International Journal of Design*, 6(3), 1–14.

Overmann, K. A., & Wynn, T. (2018). Materiality and human cognition. *Journal of Archaeological Method and Theory*, 26(2), 457–478. https://doi.org/10.1007/s10816-018-9378-y

Patni, N., Yadava, P., Agarwal, A., & Maroo, V. (2014). An overview on the role of wheat gluten as a viable substitute for biodegradable plastics. *Reviews in Chemical Engineering*, 30. https://doi.org/10.1515/revce-2013-0039

Pietarinen, H., Miettinen, S., & Sarantou, M. (2022). Biotaidetta tarkasteleva laboratorio [A laboratory examining bioart]. *Tahiti*, 12(3), 111–120.

Pestana, M. (2023). *More than human empathy: A caring approach to ecosystemic design*. Instituto Superior Técnico, Lisboa, Interactive Technologies Institute, Portugal. https://doi.org/10.21606/iasdr.2023.407

RiverOfLife, M., Taylor, K. S., & Poelina, A. (2021). Living waters, law first: Nyikina and Mangala water governance in the Kimberley, Western Australia. *Australasian Journal of Water Resources*, 25(1), 40–56.

Rizzo, A., Ekelund, B., Bergström, J., & Ek, K. (2020). Participatory design as a tool to create resourceful communities in Sweden. In *Co-creation of public open places: Practice – Reflection – Learning*. C3Places Project. Culture & Territory, volume 4 (pp. 95–107). Lisbon: Lusófona University Press. https://doi.org/10.24140/2020-sct-vol.4-1.6

Saad, E., Elekyaby, M. S., Ali, E. O., & Hassan, S. F. A. E. (2020). Double diamond strategy saves time of the design process. *International Design Journal*, 10(3), 211–222.

Sinquefield-Kangas, R., Ojala, A., & Niinistö, H. (2022). (Re)Thinking time: Materializing more-than-human empathy in student teachers' video artworks. *Studies in Art Education*, 63(1), 39–54. https://doi.org/10.1080/00393541.2021.2007714

Siqueland, A. (2022). Painting across the Atlantic. In *Down Иorth: North Atlantic Triennial, 2022*. Down Иorth: North Atlantic Triennial 13.10.2021–5.2.2022. Reykjavík Art Museum, Hafnarhús.

Sonneveld, M. H., & Schifferstein, H. N. J. (2008). The tactual experience of object. In H. N. J. Schifferstein & P. Hekkert (Eds.), *Product experience* (pp. 41–67). Elsevier.

Sowden, P. T., Clements, L., Redlich, C., & Lewis, C. (2015). Improvisation facilitates divergent thinking and creativity: Realizing a benefit of primary school arts education. *Psychology of Aesthetics, Creativity, and the Arts*, 9(2), 128–138.

Tuomaala, E.-K. (2023). *Empathy ecosystem*. https://ennikukka.com/new-page-1.

Warren, K. J. (1997). Taking empirical data seriously: An ecofeminist philosophical perspective. In K. J. Warren, K. Warren, & N. Erkal (Eds.), *Ecofeminism: Women, culture, nature* (pp. 641-649). Indiana University Press.

PART THREE

Approaches to Data Analysis for Developing Service Propositions

13

Empathic Approaches to Qualitative Data Analysis:

Calling for a Deeper Understanding of Research Problems

Ludovick Myumbo and Albert Tibaijuka

Abstract

Increasing doubts about the efficacy of the predominant qualitative epistemologies in knowledge production beg for alternative analytical approaches. In this context, an attempt to deploy empathetic approaches to qualitative data analysis is critical for providing a deeper understanding of research problems. With neutrality still being instrumental, this study identifies the inclusion of empathetic approaches as a tool for rigorous scrutiny of information and discursive power for articulating participants' life realities. Inspired by Wetter-Edman's (2013) conceptual reflection on empathy and employing semi-structured interviews and discourse analysis, this study interrogates researchers' experiences in data analysis to uproot the underlying opportunities for a better understanding of research problems, particularly the ability to recognize participants' inner emotional states. In doing so, the study calls for a qualitative epistemological revision to accommodate other pertinent methods of examining data.

Introduction

One of the fundamental aspects of being human is to feel, show and express emotions (Gilbert 2001a), that is, to be sincere and realistic to oneself and others. Despite being inconsistent with professional qualities, such as the insistence on detachment/objectivity to social science researchers (Lupton, 1994), this aspect plays a pivotal role in offering deeper understandings of human actions and inactions. Scholars interested in studying human beings as social actors, particularly those who want to know what people do and why they do things the way they do them, are not expected to ignore this aspect. Although qualitative researchers believe that professional detachment enables them to conserve a safe boundary between themselves and their research participants (Dickson-Swift et al., 2006), it is erroneous to think that such an approach offers strong analytical power for people's life realities and/or social worlds. In a real sense, as Kurtz (2014, pp. 22–23) observes, the employment of empathic approaches to data analysis offers more opportunities for a rich understanding of the lived experiences of research participants. The logic is that objectivity (i.e. creation of boundaries and neutrality) helps keep the brain operating during data analysis, but without emotions it is difficult to gain an in-depth understanding of people's life realities.

The proponents of detachment tendencies in qualitative studies reject the deployment of empathy in data analysis on the grounds that the moment researchers are emotionally involved in the exercise of collecting and analysing data, they become overwhelmed by the encountered social situations, thus turning a blind eye to critical analytical issues. This is precisely why the advocates of boundaries or distances in qualitative studies consider the detachment approach as being more pertinent and professional. The opposite is argued to raise concern about the integrity of the data and the true motives and agenda of the researcher (Drake, 2010). In other words, professional detachment or distancing is said to prevent qualitative researchers from becoming 'emotionally overwhelmed' by the often-distressing situations that confront their research participants, particularly in sensitive research (Lonne, 2003, p. 293).

The argument that has been widely shared by supporters of professional detachment is that objectivity enables social research to be free from biases, beliefs and value judgement, thus presupposing an independent reality that can be grasped and making it 'universally' acceptable. In other words, detachment or distancing in the course of collecting and analysing data means making judgements based on observable phenomena that are uninfluenced by emotions or personal prejudices. Arguably, the main point here is that, with the detachment approach, evidence is dispassionately examined, in the sense that the obtained results are independent of the beliefs, hopes or fears of any individual researcher – all obtained not by intuition and speculation but by actual observations. It is similarly argued that the conclusions reached as the result of detachment approaches to inquiry and investigation are independent of extraneous variables, such as race, culture, religious political affiliation and so forth (Pandey, 2014).

Nevertheless, the professional detachment norm is also seen as neither a possible nor desirable standard for social scientific scholarship. Those subscribing to it are considered insensitive, and the approach itself is viewed as one that has a narrow focus on quantitative measures and/or experimental methods that can be mechanically replicated. The detachment approaches are well argued to be attempts towards an accurate depiction of the social world, with little attention given to the inhabitants' perspectives and with the endeavour to arrive at universal and invariant truths (Bird, 2020, pp. 83–4).

Smith (1976), for example, objects to the norm of distancing, arguing that it encourages social scientists to treat people as objects, stringently constricting the understanding of others and fostering narrowly disciplinary inquiries. Feyerabend (1987) extends this objection by rejecting the so-called 'objective truths' in research endeavours, arguing that its application is only for the purpose of furthering our understanding of the problem. Intriguingly, Trawick (1992, p. 2) confesses that they had never been 'objective' or pretending to be disinterested or uninvolved in the lives of informants by setting feelings aside. A more or less similar confession is made by Harding (1993), who finds the practice of objectivity challenging, particularly because 'it permits scientists and science institutions to be unconcerned with the origins or consequences of their problematics and practices or with the social values and interests that these problematics and practices support' (p. 71). Harding insists that it is only when there is empathy that researchers can know the 'real' world of their participants (Thurairajah, 2019, p. 135). Perhaps in agreement with the above confessions, Dickson-Swift et al. (2008) see the craving for objectivity in science as an anomaly because it minimizes scholars' understanding of the emotional distress of research participants, which also hinders a candid understanding of their real worlds.

In general terms, the above-discussed opposition to professional detachment cements the seminal works of prominent scholars dating back to the 1960s. These include Kuhn's (1962) outlook, which saw a weakness in the choice of conceptual models in research, arguing that it is typically influenced by scientists' judgements regarding the problem-posing and problem-solving abilities of these models. Popper (1963), who questions how the object of knowledge is empirically and neutrally given, contended that it is associated with the empiricist tradition and tied to key issues in epistemology and the philosophy of science. Parallel to Popper's view is Simmel's (1977) criticism of the ability of the norm of objectivity to provide mirror images of events as they truly happen. Foucault (1977), on the other hand, considers epistemologies problematic in terms of their implications for concepts of rationality, knowledge, certainty and truth. Although not final among the influential studies but in this context offering a conclusive disapproval, Rorty (1980) sees the application of conceptual frames in social research as not being primarily or exclusively reflecting nature but rather because, in pragmatic ways, they enable scientists to offer credible explanations, conduct experiments, make predictions and carry on intelligible discussions. This tendency leads researchers to use terms and concepts as well as knowledge not belonging to the research participants and not focusing on more complex realities experienced by the subjects.

As Green et al. (2006, pp. 449–51) posit, professional distancing can be important, but it can also be flexible and elastic while still ensuring competent and appropriate practices. In this sense, an adamant insistence on professional detachment is largely misleading because a claim that a researcher has the ability to 'feel something different, or feel less' (Malcolm, 2012, p. 11) can lead to ethical uneasiness and distorted data analysis (Jackson et al., 2013). In this sense, as Gilbert (2001b) observes, 'boundaries must be negotiated and renegotiated, [...] as a balance is sought between the dangers and benefits of being too far in or too far out of the lives of the researched' (p. 12). Thus, the aim of the present study is to draw on very particular scenarios of qualitative data analysis in which, without 'wearing the shoes' of the research participants, the researchers are unlikely to capture the underlying meanings of the participants' actions and end up holding superficial interpretations of events. In this context, the question is whether qualitative researchers can be empathetic without compromising objectivity and authenticity in the course of data analysis.

Empathy: A Conceptual Reflection

Wetter-Edman (2013) considers empathy to be a skill that humans employ to understand each other's emotions. Researchers can also apply it to obtain the innermost feelings of their participants and, therefore, be able to understand their actions and inactions. Wetter-Edman's constitution of empathy has been shared by other scholars, such as Rughinis and Huma (2014, as cited in Miettinen et al., 2016), who refer to empathy as the capacity to understand the situation of another person. This means that empathy is a person's definition of the situation and symbolic universe in which the elements of the situation become meaningful and shape actions. By being empathetic, researchers have a chance to understand the needs of their participants and be informed of them (De Lille et al., 2012). In addition to neutrality requirements in the social sciences, the application of empathy helps researchers understand the elements that underpin the realities and situations experienced by their participants (Rumble et al., 2010) and eventually respond to their inner and emotional states (Patnaik, 2009).

Researchers who are empathetic are arguably wiser, particularly by demonstrating their ability to understand and share participants' inner lives (Brink et al., 2011, p. 1). This implies that they are revealing levels of understanding that are beyond conventional practices in the social sciences and are thoughtful of participants' emotions and feelings, which informs them about the associated meanings connected to what participants do and do not do.

These empathetic actions and decisions from researchers are considered to be part of the well-known approaches towards empathy, namely, cognitive, affective and compassionate. By being able to recognize what participants feel, researchers are said to apply cognitive empathy because they understand the intentions of their participants (Bloom, 2017). When they share emotions with them, that is, are able to imagine and experience a certain situation, from the participants' point of view, researchers are applying affective empathy (Brink et al., 2011, p. 2). Moreover, when dealing with their emotions and circumstances as a way of addressing their suffering, compassionate empathy is applied (Bloom, 2017; Riess & Neporent, 2018).

The above reflection on empathy has a suitable connection to van der Walt and Oosthuizen's (2021) view of *ubuntu* (an African ancient word meaning humanity to others). Much as *ubuntu* theorization is not part of the present study, the authors' reflections on the same topic offer pertinent explanations for the present reflection on empathy. van der Walt and Oosthuizen view *ubuntu* as humanism, empathy, respect, collective responsibility, altruism, understanding, respect for individual differences, knowledge of self and others, goodness and benevolence. This perspective informs the central argument of this book chapter; that is, qualitative data analysis will be more meaningful by deploying empathic actions.

Methodology

Our own experiences in qualitative research necessitated and helped us conduct individual semi-structured interviews with researchers who were purposively selected based on their experiences in conducting qualitative studies on sensitive topics. The research topics investigated by the participant researchers included the experiences of young women who sell sex for their livelihoods in urban Tanzania and the experiences of marginalized communities attempting to negotiate their stakes with foreign investors who interrupted their livelihood activities. Data collection consisted of individual one-on-one interviews at a location that was convenient and appropriate for each participant researcher. The interviews took approximately one hour. With the permission of the participant researchers, all interviews were audio-recorded to preserve spoken words.

Data analysis was carried out concurrently using the constant comparative method of a thematic approach based on the work of Corbin and Strauss (2008) and Charmaz (2002). The reason for choosing this method of data analysis was its applicability to a study design whose aim was to describe and make sense of a phenomenon (Thomas, 2006). We particularly liked this method of data analysis because it helped us relate the facts provided by participant researchers on incorporating empathy into data analysis, with models or theories emphasizing data-driven analysis (Braun & Clarke, 2006). In addition, thematic analysis as part of the general inductive method immersed us in the raw material to allow us to look for more pertinent patterns of relationships (Braun & Clarke, 2006, p. 79), which were likely to inform us about the necessities of applying empathy in the very same exercise. Hence, our analysis of the data involved six steps or stages of sorting and coding the data into themes in relation to our research questions and objectives (Braun & Clarke, 2006; Creswell, 2002; Thomas, 2006). We chose not to use any preconceived categories, as suggested by Kondracki et al. (2002). Instead, the categories in our study arose from the data and became the initial codes for the themes of our research (Patton, 2002). In the course of doing this, we were able to code data from the field diary as the initial step of data analysis and theme development. The above-mentioned data collection and analytical strategies helped us 'remain open in the study and explore a myriad of possibilities' (Hsieh & Shannon, 2005, pp. 12–78) of using the information from our informants to respond to the research question. This approach was particularly feasible because, since its inception, it was informed by the empathy debate as its frame of thinking.

Credibility and Ethical Considerations

Ethical approval for the present study was granted by the Research Ethics Committee of St. Augustine University of Tanzania (SAUT). We also made a deliberate effort to communicate clearly to the participant researchers the risks involved in our research to obtain informed consent. We (the participants and us) then agreed together on how the data would be stored and who would have access to it, and we decided to use pseudonyms as a way of protecting the participants' anonymity (e.g. Karnieli-Miller et al., 2009, p. 183).

Our positions as experienced qualitative researchers also helped to clarify some ethical questions. We became mindful of the cultural and social contexts, especially the right and wrong ways of acting, which protected us and the participants from being biased and less mindful of sociocultural differences. Moreover, our research was aware of reflexivity, that is, accounting for any power imbalances while also considering the validity and reliability of the data (Maxey, 1999). With reflexivity, we were made aware of emotional labour in qualitative research, which provided us with a better tool to conceptualize how to manage our emotional states. Rosenberg (1990) refers to this as an invocation of altered internal emotional states to carry out the cognitive work required for data presentation and analysis (see also Grandey, 2000). Therefore, it is safe to say that we were mindful of the possibility of affecting the data analysis process by inflicting our own feelings and emotions on what was raised by participant researchers and, thus, remained alert to the same.

Findings and Discussion

In this section, we present the voices of the participant researchers and place them at the centre of the conversation (data analysis). The aim is to demonstrate examples of very particular situations in which, without understanding and sharing participants' inner lives, researchers are unlikely to understand the underlying meanings of their participants' actions and end up holding superficial interpretations of events. The purpose behind this depiction of particular circumstances encountered by researchers is to unravel moments that require 'empathy' to arrive at understanding why something happens the way it does. It is about 'being human' and learning from humans to understand the 'social worlds'. Put differently, the intention is to illustrate specific events that challenge the 'detachment' constitution. That is, without entering social actors' inner emotional states, researchers run the risk of being far removed from people's life realities.

In doing so, we present the findings by first describing two aggrieving moments that drew emotions from the participant researchers and then exemplifying the relevance of being empathetic during the collection and analysis of the data. The following is a description of the two painful encounters.

The first is a qualitative study conducted in 2018 to explore the lived experiences of young rural women who trade sex for their livelihoods in Mwanza town. This was important, given that there was a lack of qualitative research allowing young women who traded sex for their livelihoods to recount their situations in their own voices and from their own perspectives. The data were collected through participatory narrative inquiry, which afforded a space for a group of six young rural women to tell stories of their lived experiences.

During the study, the researcher listened to the participants' stories containing details about different forms of abuse they had suffered at the hands of their 'bosses', clients and society at large. Some of the stories represented, in the participants' own words, how their lives had been shattered by incomprehensible acts of cruelty, violence and neglect. The details given were graphic, containing explicit and vivid descriptions of the dire circumstances to which they were subjected in their everyday lives.

Many of the participants' stories described situations that were still ongoing in their lives, reflecting the freshness of their memories. The impact of these stories on the researcher was something not to understate. Listening to and writing them down gave the researcher a heightened sense of trauma – as if it were happening now. The researcher explained, 'It was like reading accounts of child sexual abuse [...], they made me feel so much pain. Managing the sadness, horror and anger arising from these stories was incredibly difficult'.

Although the researcher and his participants were able to enter into a dialogue about the emotional demands of these stories in their various meetings, the former observed that, for the most part, the emotional experiences were deeply active and privately felt by both parties.

The second is from a study conducted to understand the perceived inactions of community members whose livelihood strategies were shattered after a foreign mining company invested in their area, evicting hundreds of families from their ancestral land. Among many stories, after the mining company installed itself in this area, it restricted all community members from entering into its concession. This was a piece of land used by the locals for farming, cattle grazing and artisanal mining. One compelling incident, among many, was a story from a cattle keeper. As community members were forbidden from entering the mining company's leasing area, cattle keepers lost grazing land, so they resorted to sneaking into the controlled area to feed their cattle. Most of the time, the cattle keepers were apprehended, the cattle impounded and the owners faced penalties, trespassing charges and sentences from the primary court.

Nevertheless, these cattle keepers did not refrain from intruding on the mining concession to graze their cattle. One cattle keeper shared his experience with our participant researcher:

> I was once chased by the mining security guards at night. They had rubber bullets and long sticks. I ran and fell into a watery ditch and broke my leg. They caught me and took me to the police station and then to the hospital and court. My cows were left unattended, but luckily, they found the way home with the exception of two. The court told me to pay 300,000 (approximately USD 120) as a penalty. I had to sell a cow to afford the penalty. Since the accident, I cannot walk properly.
>
> My leg is aching, and I can't go distances to graze my cows. It's the children and neighbours who help me. I don't know how I will survive this life. I have children studying in primary and secondary school. I depend on those cows to pay for their education. What business can I do? I did not go to school. There is an official from the mine who tells me every day to sell the cows. I once asked him, 'If I do that, how will I survive?' He then told me to sell the cows and build a house. Therefore, I asked him, 'Will I eat the house? From a cow, I sell milk and get something for my family. If there is money to pay for my child's tuition, I sell a cow. If you sell one cow, it can help you fulfil different needs'.

Drawing from the two encounters stated above, how would a researcher remain detached? From talking to the participant researchers, they confessed being touched by these gruelling moments. They admitted that, if it was not about getting into their innermost emotional states, showing concerns, understanding and sympathizing with them, the participants would probably not share these dreadful experiences. By being compassionate and thoughtful about their emotions and feelings, the researchers made their participants speak about things they would never have had a chance to speak about. This is particularly true because some of the stories shared by the participants reflect on issues considered taboo, such as sex or bribes, which are subjects that people would not be ready or able to speak about directly, especially to individuals not so close to them.

For example, many sociocultural settings in Africa and Tanzania in particular consider sexuality taboo. In such settings, people are not expected to talk about and/or reflect on their sexual engagement, let alone their lived experiences of selling sex. In this sense, promiscuous tendencies are even worse because they are considered illegal and socially reprehensible. Thus, by not reproaching his participants, the researcher made them open up and share information that would otherwise remain concealed and, therefore, be able to understand why things happen the way they do in that social setting. This revelation helps to answer the research question of this study that from what has been observed, researchers are able to be empathetic but also remain objective.

Of course, such abilities do not occur without limitations. By being empathetic, researchers are likely to experience emotional upheaval (Hochschild, 1979). The present study acknowledges that by going deep into participants' innermost feelings and/or appalling experiences, there is a risk of being carried away and forgetting the core issues to be captured and analysed, even in the worst-case scenario of being personally affected (e.g. being emotionally and mentally disturbed). Although acknowledging such risks, the current study contends the need to deploy empathy in qualitative data analysis; otherwise, it will be difficult to understand people's real-life situations. To address the possible risks, researchers are advised to consider 'reflexivity', that is, to account for any emotional labour, as well as to consider the validity and reliability of the analysed data (Maxey, 1999).

Conclusion

The present study intended to stretch the limits of detachment tendencies in qualitative data analysis because, as discussed in the chapter, they seem to deny researchers the opportunity to gain a deeper understanding of why their research participants do certain things and why they do them in certain ways. The analytical effort from this work illustrates that decisions to approach qualitative data collection and analysis with empathy offer researchers unique opportunities to understand people's real-life situations and obtain more contextual, logical, discursive and analytical power. In other words, an empathetic approach to qualitative data analysis enhances understanding of features that necessitate certain realities and situations experienced by the participants and, therefore, respond to their inner and emotional states.

To put it in more particular terms, the present study aimed to explain that qualitative data collection and analysis is not only an intellectual exercise, but also 'a process of exploration and discovery that is felt deeply' (Gilbert, 2001a, p. 9). Researchers can enter into participants' life realities and understand what they do from their real social settings, thus being in a more comprehensible position to offer valid and reliable information/ explanations. Therefore, the present study shows that it is more meaningful and possible for the researcher to be empathetic during qualitative data collection and analysis and to offer authentic findings without being emotionally taken by their subjects.

Takeaway

The best way to produce pertinent knowledge from social actors should start by reversing qualitative inquiry approaches to incorporate means that will enable inquirers to listen to, share and understand the innermost feelings of their participants. Contrary to the prevailing epistemic insistence on researchers' detachment, the proposed approach can lead inquirers into intersubjectivity. This is particularly true because the inclusion of empathetic approaches to data collection and analysis seems to offer appropriate instruments for in-depth interpretations of the findings, most often taken for granted by the detachment thesis.

References

Bird, F. (2020). A defense of objectivity in the social sciences, rightly understood. *Sustainability: Science, Practice and Policy*, 16(1), 83–98. https://doi.org/10.1080/15487733.2020.1785679

Bloom, P. (2017). Empathy and its discontents. *Trends in Cognitive Sciences*, 21(1), 24–31. https://doi.org/10.1016/j.tics.2016.11.004

Braun, V., & Clarke, V. (2006). Using thematic analysis in psychology. *Qualitative Research in Psychology*, 3(2), 77–101. https://doi.org/10.1191/1478088706qp063oa

Brink, T. T., Urton, K., Held, D., Kirilina, E., Hofmann, M., Klann-Delius, G., Jacobs, M. A., & Kuchinke, L. (2011). The role of orbitofrontal cortex in processing empathy stories in 4- to 8-year-old children. *Frontiers in Psychology*, 2, 80. https://doi.org/10.3389/fpsyg.2011.00080

Charmaz, K. (2002). Qualitative interviewing and grounded theory analysis. In J. F. Gubrium & J. A. Holstein (Eds.), *Handbook of interview research: Context and method* (pp. 675–694). Sage.

Corbin, J., & Strauss, A. (2008). *Basics of qualitative research: Techniques and procedures for developing grounded theory* (3rd ed.). Sage. https://doi.org/10.4135/9781452230153

Creswell, J. (2002). *Educational research: Planning, conducting, and evaluating quantitative and qualitative research*. Pearson Education.
De Lille, C., Roscam Abbing, E., & Kleinsmann, M. (2012). *A designerly approach to enable organizations to deliver product-service systems*. DMI.
Dickson-Swift, V., James, E., Kippen, S., & Liamputtong, P. (2006). Blurring boundaries in qualitative health research on sensitive topics. *Qualitative Health Research*, 16(6), 853–871. https://doi.org/10.1177/104973230628752
Dickson-Swift, V., James, E., Kippen, S., & Liamputtong, P. (2008). Risk to researchers in qualitative research on sensitive topics: Issues and strategies. *Qualitative Health Research*, 18(1), 133–144. https://doi.org/10.1177/104973230730900
Drake, P. (2010). Grasping at methodological understanding: A cautionary tale from insider research. *International Journal of Research and Method in Education*, 33(1), 85–99. https://doi.org/10.1080/17437271003597592
Feyerabend, P. (1987). *Farewell to reason*. Verso Books.
Foucault, M. (1977). *The order of things: An archaeology of the human sciences*. Random House.
Gilbert, K. (2001a). *The emotional nature of qualitative research*. CRC Press Books.
Gilbert, K. (2001b). Introduction: Why are we interested in emotions? In K. Gilbert (Ed.), *The emotional nature of qualitative research* (pp. 3–15). CRC Press Books.
Grandey, A. (2000). Emotion regulation in the workplace: A new way to conceptualize emotional labour. *Journal of Occupational Health Psychology*, 5(1), 95–110. https://doi.org/10.1037/1076-8998.5.1.95
Green, R., Raeleene, G., & Robyn, M. (2006). Professional distance and social work: Stretching the elastic? *Australian Social Work*, 59(4), 449–461. https://doi.org/10.1080/03124070600986010
Harding, S. (1993). Rethinking standpoint epistemology: What is 'strong objectivity'? In L. Alcoff & E. Potter (Eds.), *Feminist epistemologies* (pp. 49–82). Routledge.
Hochschild, A. (1979). Emotion work, feeling rules, and social structure. *American Journal of Sociology*, 85(3), 551–575. https://www.jstor.org/stable/2778583
Hsieh, H., & Shannon, S. (2005). Three approaches to qualitative content analysis. *Qualitative Health Research*, 15(9), 1277–1288. https://doi.org/10.1177/10497323052766
Jackson, S., Backett-Milburn, K., & Newall, E. (2013). Researching distressing topics: Emotional reflexivity and emotional labour in the secondary analysis of children and young people's narratives of abuse. *Sage Open*, 3(2), 1–12. https://doi.org/10.1177/2158244013490705
Karnieli-Miller, O., Strier, R., & Pessach, L. (2009). Power relations in qualitative research. *Qualitative Health Research*, 19(2), 279–289. https://doi.org/10.1177/1049732308329306
Kondracki, N. L., Wellman, N. S., & Amundson, D. R. (2002). Content analysis: Review of methods and their applications in nutrition education. *Journal of Nutrition Education and Behaviour*, 34(4), 224–230. https://doi.org/10.1016/S1499-4046(06)60097-3
Kuhn, T. S. (1962). *Structure of scientific revolutions*. University of Chicago Press.
Kurtz, C. (2014). *Working with stories in your community or organization: Participatory narrative inquiry*. Kurtz-Fernhout Publishing.

Lonne, R. (2003). Social workers and human service practitioners. In M. Dollard, H. Winefield, & A. Winefield (Eds.), *Occupational stress in the service professions* (pp. 281–309). Taylor & Francis. https://doi.org/10.4324/9780203422809
Lupton, D. (1994). *Medicine as culture: Illness, disease and the body in Western societies*. Sage.
Malcolm, I. (2012). 'It's for us to change that': Emotional labour in researching adults' learning: Between feminist criticality and complicity in temporary gendered employment. *Adult Education Quarterly*, 62(3), 252–271. https://doi.org/10.1177/0741713611402050
Maxey, I. (1999). Beyond boundaries? Activism, academia, reflexivity and research. *Area*, 31(3), 199–208. https://doi.org/10.1111/j.1475-4762.1999.tb00084.x
Miettinen, S. A., Sarantou, M., & Akimenko, D. (2016). Art and storytelling as an empowering tool for service design: South Australian case study. In P. Rytilahti & S. Miettinen (Eds.), *For profit, for good: Development organizations through service design* (pp. 74–80). Lapland University Press.
Pandey, K. (2014). Objectivity in social sciences. Social Science Research Network. http://dx.doi.org/10.2139/ssrn.2399299.
Patnaik, D. (2009). *Wired to care: How companies prosper when they create widespread empathy*. FT Press.
Patton, M. (2002). *Qualitative research and evaluation methods*. Sage.
Popper, K. R. (1963). *Conjectures and refutations: The growth of scientific knowledge*. Routledge.
Riess, H., & Neporent, L. (2018). *The empathy effect: Seven neuroscience-based keys for transforming the way we live, love, work, and connect across differences*. Sounds True.
Rorty, R. (1980). *Philosophy and the mirror of nature*. Blackwell.
Rosenberg, M. (1990). Reflexivity and emotions. *Social Psychology Quarterly*, 53(1), 3–12. https://doi.org/10.2307/2786865
Rumble, A. C., Van Lange, P. A., & Parks, C. D. (2010). The benefits of empathy: When empathy may sustain co-operation in social dilemmas. *European Journal of Social Psychology*, 40(5), 856–866. https://doi.org/10.1002/ejsp.659
Simmel, G. (1977). *The problem of the philosophy of history: An epistemological essay*. Free Press.
Smith, W. (1976). Objectivity and the humane sciences: A new proposal. In W. C. Smith & W. G. Oxtoby (Eds.), *Religious diversity: Essays* (pp. 158–180). Harper & Row.
Thomas, D. (2006). A general inductive approach for analysing qualitative evaluation data. *American Journal of Evaluation*, 27(2), 237–246. https://doi.org/10.1177/1098214005283748
Thurairajah, K. (2019). Uncloaking the researcher: Boundaries in qualitative research. *Qualitative Sociology Review*, 15(1), 132–147. https://doi.org/10.18778/1733-8077.15.1.06
Trawick, M. (1992). *Notes on love in a Tamil family*. University of California Press.
van der Walt, J., & Oosthuizen, I. J. (2021). Ubuntu in South Africa: Hopes and disappointments – A pedagogical perspective. *Perspectives in Education*, 39(4), 89–103. https://doi.org/10.18820/2519593X/pie.v39.i4.7
Wetter-Edman, K. (2013). Relations and rationales of user's involvement in service design and service management. In S. Miettinen & A. Valtonen (Eds.), *Service design with theory: Discussions on change, value and methods* (pp. 107–116). Lapland University Press.

14

Effective Analytical Approaches as Precursors for Proof-of-Concept Generation

Ella Björn, Satu Miettinen, Titta Jylkäs and Melanie Sarantou

Abstract

In addressing the preliminary step in developing new empathic services, researchers in the Empathy Business project, funded by Business Finland, aimed to ideate viable experiments or proofs of concept (PoCs). These PoCs serve as testing grounds for new and more empathic services and tools, establishing feasible and viable service concepts to provide users with more empathic outcomes (e.g. service experiences). Ideas for viable PoCs were generated using validated data analysis processes to create practical and workable approaches for real-world business needs in Finland. The process enabled researchers to establish effective experiments for prototyping, feasibility testing, evaluation and decision-making purposes – determining the most effective PoCs comprised of data collection through workshops, interviews and notetaking. The forthcoming data were analysed using affinity analysis to uncover hidden process patterns for supporting data-driven decisions that enhance customer experiences and drive profits. Data from the workshop observations were captured in the Miro environment and summarized using digital sticky notes, enabling more effective analyses because of their visibility and efficiency. The interviews were transcribed through an artificial intelligence transcription tool and similarly clustered in the Miro environment. Both datasets were merged seamlessly in this digital environment, enabling clear patterns to emerge, such as business compatibility, sustainability, new tools and methods, centralizing people, online workshops and *in situ* workshops. These effective analytical methods and tools allowed the researchers to generate viable concepts for PoCs, including an emotion reflection tool.

Introduction

The 'Empathy Business' project aims to identify, experiment with and evaluate how to digitalize service and business prototyping through creativity. This research identifies opportunities and challenges for utilizing empathy and creative prototyping while focusing on business needs. The project used design-driven action research to develop proofs of concept (PoCs) for digitized solutions for service design. A multidisciplinary team of researchers conducted pilot studies to test the design methods with creative companies (Content Union, Zone ii Oy, and Zone 2 Pictures Oy) to produce PoCs. The PoCs created in the project were as follows:

1. An AI-based concept for understanding emotions in service prototyping
2. An emotion reflection tool for collecting emotional data during a service experience
3. A guidebook for building empathy in hybrid workshop settings

PoC refers to an initial test of an idea, method or product, and it has become a standard term for the early stage of the design process (Cooper & Sommer, 2016; Jobin et al., 2021). PoC helps concretize early-state ideas of the service design process (Kuure et al., 2019) by showing its potential, feasibility and functionality in real-life settings for different audiences, such as clients, partners or team members. An independent *Les Sismo* design studio founded in 1997 adopted the term PoC for service and experience design (Jobin et al., 2021). *Les Sismo* reported that PoC is valuable not only for validation but also for exploration (Jobin et al., 2021). Jobin et al. (2021) argue that the success of PoC is in combining the proof of the known and proof of the unknown, which enriches PoC as a tool for testing design decisions by validating knowledge and experimenting.

In the Empathy Business Project, the PoC generation started by interviewing service design industry experts, performing affinity diagramming to analyse the data and collaboratively generating PoC ideas based on affinity diagram analysis. The generated PoCs were tested with multiple stakeholders, and the concepts were developed through testing and iteration. Feedback is collected through paper surveys to collect data relating to PoCs and analyse further ways to build ideas. One of the PoCs created in the Empathy Business Project was the emotion reflection tool, which aimed to better understand users' emotions during the service experience. The tool offers new ways for companies to collect user data and develop their services based on users' emotions. This PoC collects data and encourages users to focus on their emotions and feelings.

The emotion reflection tool collects an anonymized dataset through testing sessions during service experiences. The anonymized dataset is utilized for research purposes to create a business concept that could be utilized in organizations in the future. The target audience for the emotion reflection tool would be organizations such as museums, art galleries, tourism businesses or other service providers who are potentially interested in their customers' emotions throughout the service. The tool is a web-based application in which the user provides information about their emotional state while participating in the service experience. It offers creative ways to focus on emotions felt in the body by using colours to visualize emotions.

In this chapter, we discuss how different analytical approaches have been used in the Empathy Business project to support service design and research, both through expert interview analysis and as part of the PoC generation of the emotion reflection tool.

Theoretical Background

A growing need exists for empathy to be embedded in all spheres of business (New & Kimbell, 2013). For example, empathy is the main ingredient of successful service delivery (Goleman, 2004). This means that more empathic services, dialogues and interactions should exist between organizations and their employees and users. The need for empathy is widely discussed in business leadership, but the notion has been criticized for its superficial approach (New & Kimbell, 2013; Ohren, 2014). Empathy does not always promote inclusiveness, and the other person is viewed as another 'I', leaving the actual person behind (Cipolla & Bartholo, 2014). Empathy can also be over-reliant. Over-relying on empathy can promote single-mindedness and even reinforce otherness (Vink & Oertzen, 2018). Additionally, the relationship between empathy and creativity has remained unexplored (Sosa, 2019). The mechanisms and methods of empathy-building through artistic practices in organizations will be explored in this project.

The complex psychological construct of empathy is described as a 'tripartite classification' (Ekman, 2007; Powell & Roberts, 2017, p. 137; Riess & Neporent, 2018). This classification includes cognitive, affective and compassionate empathy and is expanding (Sarantou & Miettinen, 2022). In addition, kinaesthetic and digital empathy have recently been included in emerging research on empathy (Cuykendall et al., 2015; Reynolds & Reason, 2012). Digital empathy is a technology-related skill involving ethical communication and digital interaction with others (Sreejun & Chatwattana, 2023). Digital empathy assists users in using technology ethically by relating to users who are different from them, refraining from misconduct and looking the other way when others are being bullied online (Sreejun & Chatwattana, 2023). Empathy can be both the outcome of a process and feedback into iterative and cyclical processes.

A model for creating empathy, based on Kouprie and Sleeswijk-Visser's model (2009), was proposed by Miettinen et al. (2016). This model is based on the discovery, immersion, connection, detachment and planning cycles. Several design tools have been developed to create empathy and user insights. Gaver et al. (1999) employ cultural probes that build empathy and help contextualize the user experience. Mattelmäki et al. (2014) revisit the concept of empathy deployed in design research during the 1990s. Empathic design is an interpretative approach to creating new solutions in users' lives focusing on experiences, emotions and meaningful everyday practices to stimulate innovation and human-centred solutions.

Emotions are closely linked to empathy and consider both affective and cognitive aspects. People make meanings and judgements about the world through inner cognitive processes (Desmet, 2002) and by sharing affective feelings (Norman, 2004). Sharing emotions fosters empathy and understanding among the actors involved in the process (Davis, 1996). In addition to cognitive aspects, emotions are embodied (Colombetti & Thompson, 2008). Merleau-Ponty (1962) states that we are connected to the world through our bodies and bodily sensations from experience. Empathy can be viewed through embodiment because our body reacts to emotional experiences through bodily feedback (Desmet, 2002). Emotional reactions can be visceral, behavioural or reflective and can be formed consciously, unconsciously or automatically (Norman, 2004). Emotions can also be evoked by copying other people's emotions (Desmet, 2002) and are formed through our cultural and experiential bodies (Ståhl et al., 2009). Our bodies are cultural because the existing cultural norms guide our behaviour, while previous experiences also affect emotions. Understanding emotions and the body's role in creating sense-awakening experiences helps in the design of services that facilitate creativity and self-expression. We believe that emotions do not appear only in the mind but are sensed and felt in different parts of the body. Embodied emotions can be studied using colours, among others, for emotion recognition and self-reflection (Umair et al., 2018) as well as self-awareness and regulation (Huang et al., 2015). Colours may more effectively convey emotions than verbal expressions (see Etcoff & Magee, 1992; Sutton et al., 2016; Young et al., 2013). Emotional visualizations can help individuals understand one another, build empathy and overcome language barriers (Şemsioğlu et al., 2022).

Because of digitalization and rapid technological advancement, empathic solutions in service design and services have become even more relevant. Although technologies bring ethical concerns of misuse and control, new digital solutions can offer ways to remind and refresh thoughts to address basic human needs to better understand different perspectives, cultures and ways of perceiving the world. The possibilities for new technologies are enormous when empathy is involved in the early stages of design processes. This requires a deep focus on how we perceive empathy towards others and ourselves and how to design solutions and experiences that bring different minds, attitudes, emotions and viewpoints together seamlessly and meaningfully.

Research Design

The methodology of the 'Empathy Business' research project strongly emphasized business research (Zikmund et al., 2013) and consisted of action research using ethnographic methods and service design. These methods supported a mixed methods approach or triangulation that was used to capture all the data (Creswell, 2013; Muratovski, 2021). Ethnographic research was conducted through observations, interviews and focus group discussions with the design organizations utilizing digital solutions to determine the state-of-the-art experience of using digital solutions in service design and business development. Twenty in-depth interviews were conducted to understand user needs and the state-of-the-art technologies employed for this digital solution. By analysing and understanding users' needs for service offerings, the insights were clustered according to various topics. This research developed digital tools for service designers and design and business development processes.

The interviews were conducted with senior service designers and managers during the spring of 2023. The interviewees represented a global group of designers in Europe, Latin America and Asia who were working (a) as a senior in-house designer in a major corporation, (b) as a senior or partner in a medium-sized design consultancy or (c) as a senior designer in a small design consultancy (Miettinen et al., 2024). The interviews focused on the future of service design as a practice, the skill sets needed by the service designers and the topics that the service designers are dealing with. These requirements outlined what kinds of tools would be required in the future. The interviews had a preset of questions that the interviews followed, but the interviewer left much room for the research participant's intuition and flow of talk. This set the tone of the interviews (Miettinen et al., 2024).

The interviews were recorded and transcribed using an AI-powered tool to make automated notes in the Stream application. After this, the interviewers made notes, including the critical outputs of the interviews, and exported them to the Miro board. The interviews were analysed using the affinity diagram method to make sense of the large amounts of data (Lucero, 2015). The Miro AI function was first used to create initial clusters. Then, the research group worked collectively to categorize and reorganize the clusters to create an affinity diagram. This led to identifying the possible research questions and PoC ideas based on each cluster (see Figure 14.1). The collective analysis was performed through digital workshops to reach a consensus on the analysis results. This analysis process included support from AI in several phases: creating transcripts, transferring interviewer notes to Miro and creating initial clusters for the affinity diagram.

The extensive dataset and thorough collaborative analysis process gave the project and PoC generation a solid basis for data-driven design. Data-driven design refers to the process of designing where data inform

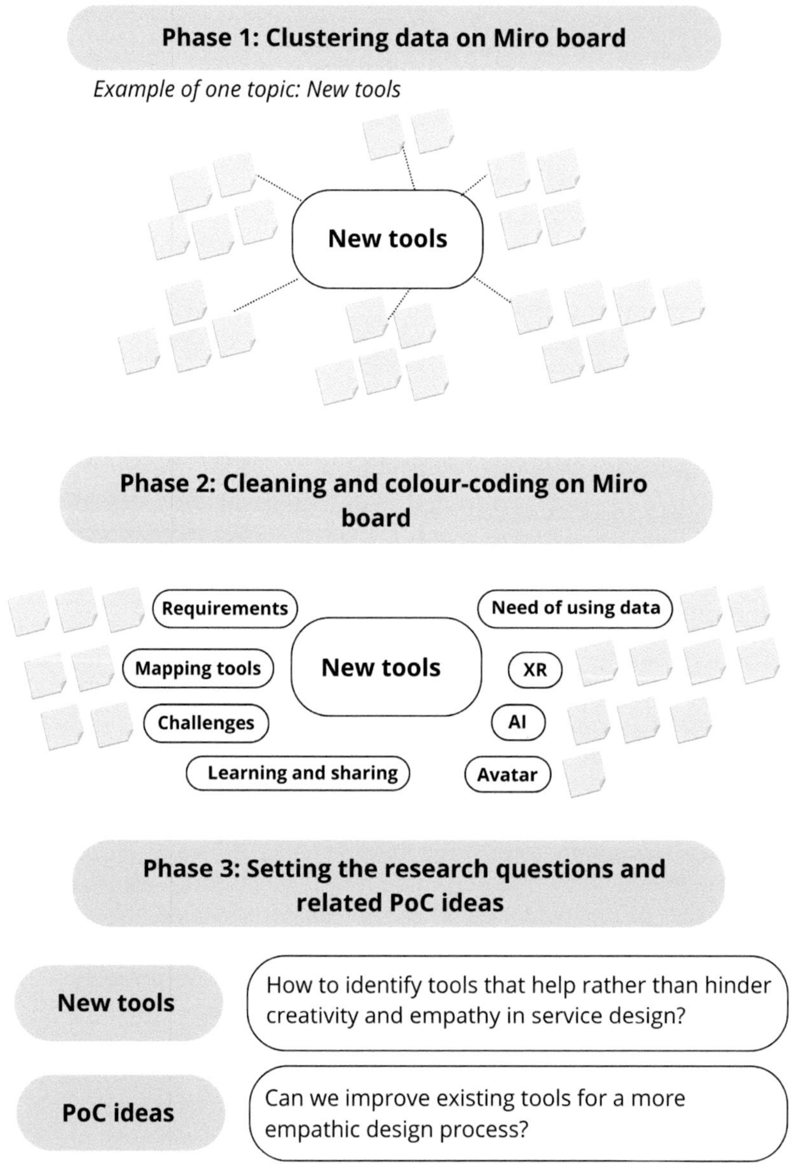

FIGURE 14.1 *Affinity diagram of the Miro board.*

creative decisions (Cantamessa et al., 2020). Based on the analysis of the interview data, eight cluster topics were identified: *sustainability*, *business compatibility*, *new tools*, *designers' skills*, *art-based methods*, *people in the centre*, *online workshops* and *physical workshops*. Two to five research questions were identified for each cluster. Eleven PoC ideas were created, three of which were prioritized for the Empathy Business project, one of which was the emotion reflection tool.

Based on affinity diagramming, new tools to enhance empathy and creativity are needed in service design, and one way to research empathy is through embodied emotions. The research group proposed the hypothesis that the creative use of colours may help individuals express the embodied emotions evoked during different service experiences. The hypothesis was tested by prototyping and testing the emotion reflection tool, which utilizes creativity and intuition for emotional self-reflection through colours. The tool prototype was tested during three different service experiences with different target groups, including students in the service design workshop, visitors to the art exhibition and tourism guides participating in tourism service prototype testing.

The PoC investigates the proof of the known and proof of the unknown by making the concept 'double proofed', which provides a greater understanding and discovery of new dimensions of the concept and how they should be tested (Jobin, 2021). Unknown peripheral variables do not directly impact hypothesis validation (Jobin, 2021). We both validated and explored whether using colours may offer new ways of understanding a user's embodied emotion reflection. We based our exploration on previous studies showing that visualized emotions may convey emotions more effectively than verbal expressions (see Etcoff & Magee, 1992; Sutton et al., 2016; Young et al., 2013).

The emotion reflection tool was tested by service design students, tourism guides and local people visiting an art exhibition. The target groups were chosen to represent different possible users of the services, service design students and tourism staff to explore whether the tool could be used in the creative or tourism industries in the future. The first test was conducted for the service design students to test the tool in an agile manner and identify any significant weaknesses before testing with companies and actual services. All the testing sessions of the tool were conducted during a sense-awakening part of the experience. In the service design workshop, the tool was tested during a meditative moment outdoors, while the participants were asked to pay attention to their senses evoked in nature. In the second experiment, the tool was used while the participants were watching a performative art show at the opening of the art exhibition. The third experiment collected data during two sense-awakening activities during the tourism service prototype, which included making a campfire outdoors and listening to local mythological stories and drum playing in a covered hut. All the experiments aimed to understand the embodied emotional data that can be collected creatively during an experience and the possible challenges and benefits of using colours.

The emotion reflection tool was tested by prototyping an online and paper survey form of the prototype (see Figure 14.2). The online version of the prototype was used in the first testing session of the service design workshop, and the paper form was used for opening the art exhibition and testing the tourism service prototype. The first session was conducted for the service design students, who tested the prototype with their mobile phones in the service design workshop related to emotions. The last two testing sessions used a paper form instead of the online prototype because the service organizers required participation to be as simple as possible.

The users were asked to reflect on their emotions by answering which body part their current feelings or emotions occur in and what colour best describes that specific emotion or feeling. The actual meaning of the colours was not studied deeply by acknowledging that the colour meanings are culture dependent and are already well studied. Instead, the purpose was to determine whether the tool could offer ways to creatively reflect embodied feelings. This was studied by asking about the meanings of the colour and body part selections and how they perceived demonstrating and reflecting emotions through colours.

The PoC generation process was conducted iteratively, and different stakeholders were involved throughout the process. The expert interview analysis offered valuable data on business needs within service design and directed the generation of the PoC. The process also involved the researchers and company partners of the Empathy Business project in ideating the PoC ideas and collecting feedback on the PoC's possibilities, weaknesses and risks. The emotion reflection tool was developed iteratively based on the feedback and observations of the users testing the tool.

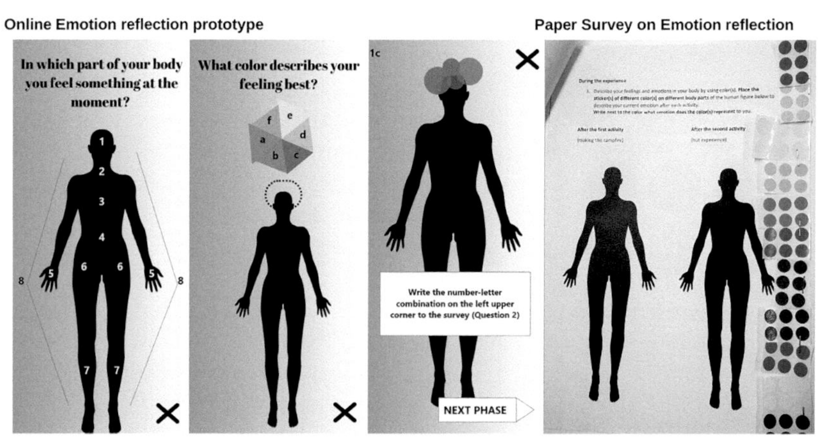

FIGURE 14.2 *A screenshot of the emotion reflection online prototype (left).*

Findings

Based on the affinity diagram, the need for new digital tools in service design was identified. The need to overcome language barriers and the ability to evaluate service interactions were identified in the data. People want to feel recognized and empowered, which requires emotional intelligence from the service designer. Combining the elements and needs raised in the clusters of new tools, designer skills and art-based methods (ABMs) led to the creation of an emotion reflection tool to support intuition and improvisation in expressing emotions in the body (see Figure 14.3).

The results of the PoC testing supported the interview data by showing that colours helped to express emotions that may sometimes be harder to express verbally. The PoC testing sessions showed that the participants felt good that their aspirations and feelings were being asked about, supporting the feeling of being heard. However, some participants mentioned that they were worried about how the colours would be interpreted and whether their answers would be misunderstood. According to this result, the tool was developed for another testing session by adding a question about the emotion the specific colour represented for them. A peripheral variable (Jobin, 2021) of the findings was that the colours themselves raised the interest of some of the participants, even before they knew what they were about. It was also found to be more creative than traditional surveys. Although it was found to be creative, 11 out of 38 participants answered that describing emotions through colours in the body was perceived as quite challenging. Two participants mentioned that it was difficult. Some participants initially found it confusing, but it made sense to self-reflect their feelings and emotions. These aspects were taken into consideration when developing the tool further.

The emotion reflection PoC test showed that using stickers in the paper tool allowed the participants to use colours more intuitively and allowed

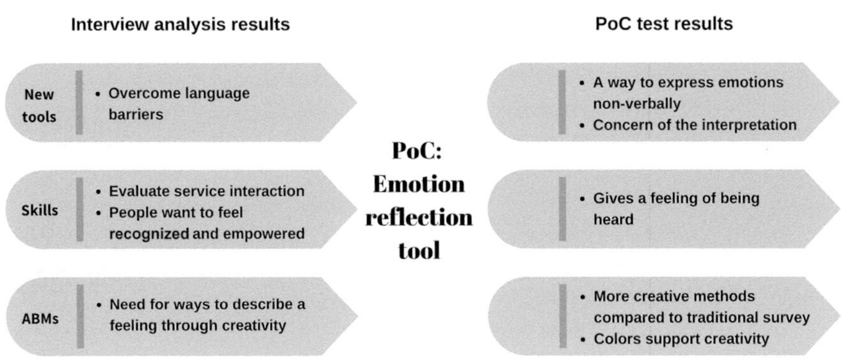

FIGURE 14.3 *Results of affinity diagramming and PoC testing.*

more freedom of choice compared with the online prototype. According to this finding, the tool was developed iteratively to enhance creativity and intuition in emotional expression. The creative and free selection of different colours was an essential element of the tool because it offered ways for the user to use creativity to express oneself without overthinking the meaning of different colours. The users trusted their intuition, which was used for the creative expression and reflection of one's emotions. This feature was missing from the first version of the tool's online prototype and will be added when the online tool is further developed.

Using analytical approaches, such as collective affinity diagram analysis for PoC generation, showed that collective mapping, coding, decoding and clustering through reflective discussions provided valuable insights for the data analysis and revealed various ideas that may not appear or be identified when working solo. Empathy was practised and enacted in coworking processes in which researchers from different backgrounds worked collaboratively. The research team included people with arts, service design, engineering, social, economics and business sciences backgrounds. This meant that an empathic approach between analysts with different theoretical backgrounds could be used because a mix of different backgrounds offered a platform for fruitful discussion and interpretation of the data from various perspectives.

Additionally, the combination of collaborative analysis and the utilization of AI tools gave the research team trust in the data-driven process. AI made the process faster because of its tedious work, such as transcribing and translating, but the team collaboratively formed the core of the analysis. The collaborative approach provided validation and rigour for the analysis. Additionally, novel views of the data emerged because of the collation of different experiences and theoretical backgrounds.

Discussion

The analytical approach in the Empathy Business project combined advanced AI-powered tools with collective analysis by the project team. The concern of using AI in data analysis for information loss, the nuances of data being misunderstood or interpretations being misleading was overcome by the project team taking over the sense-making of the analysis. The combination of machine and human intelligence allowed the project team to focus on the meaning of the data rather than on tedious tasks.

The analytical and creative process, from expert interviews to PoC testing, has provided new information about how human emotions are reflected and represented during a service experience. As the findings show, the main benefit of the emotion recognition tool was that it offered a creative method of providing a feeling of being heard and self-reflecting the current embodied and emotional self. Giving room for self-reflection may offer more memorable experiences and emotional attachment through personal recollection (Ståhl et al., 2009).

Following Miettinen et al.'s (2016) empathy creation model, the participants can be seen as creating empathy towards themselves by going through the first three steps of the model. The tool supports participants in discovering their own emotions, immersing themselves in the experience through self-reflection, and connecting to the service experience on a different level. According to the PoC testing findings, some participants reflected on their emotional experiences through their own personal backgrounds and roots, giving the exercise a deeper reflective meaning.

Boehner et al. (2005, p. 201) state that the focus in design should be moved from 'helping computers to better understand human emotion to helping people to understand and experience their own emotions'. This can be achieved when designing an interaction that leaves surfaces open for users to appropriately recognize themselves socially, emotionally or bodily through the interface and avoid reductionism (Höök, 2008). The emotion reflection tool utilized this idea to support intuitive self-expression and improvisation. In addition to finding the method creative, it was also mentioned that feeling different emotions simultaneously was experienced positively.

Conclusion

This chapter illustrates the integration of empathy into service design through PoC design for an emotion reflection tool generated after a qualitative data analysis process supported by AI. AI tools were utilized to support the data analysis, and digital tools were explored as part of the PoC prototyping and testing practice. The discussion of emotions is growing, and different methods of researching and discussing emotions have become essential for creating empathy and understanding one another. These methods could be further developed to address groups' emotions, focusing on one's own emotions and understanding more about others and how different feelings affect us and our ways of working with others.

Takeaway

Nonverbal communication is important when expressing one's feelings during the service experience. This gives users the feeling of being heard and understood. This research advances service design discourse by demonstrating how empathy can deepen user engagement and satisfaction through the use of an emotion recognition tool to support self-expression.

Collaborative data analysis, combined with AI tools, keeps the research team in control of the sense-making of research results. When a team with various backgrounds analyses data collaboratively, it creates empathy for the research topic and for the team alike. Likewise, the combination of AI tools and collaborative analysis increases trust in the analysis results.

Acknowledgements

We wish to acknowledge the participants who took part in the project. We truly appreciate the valuable time, insights and feedback they contributed to the research.

Empathy Business project (2023–4) has received funding from Business Finland under grant agreement No. 7784/31/2022.

References

Boehner, K., DePaula, R., Dourish, P., & Sengers, P. (2005). Affect: From information to interaction. In *Proceedings of 4th decennial conference on critical computing: Between sense and sensibility* (pp. 59–68). Århus.

Cantamessa, M., Montagna, F., Altavilla, S., & Casagrande-Seretti, A. (2020). Data-driven design: The new challenges of digitalization on product design and development. *Design Science*, 6, e27. https://doi.org/10.1017/dsj.2020.25

Cipolla, C., & Bartholo, R. (2014). Empathy or inclusion: A dialogical approach to socially responsible design. *International Journal of Design*, 8(2), 87–100.

Colombetti, G., & Thompson, E. (2008). The feeling body: Toward an enactive approach to emotion. In W. Overton, U. Muller, & J. Newman (Eds.), *Developmental perspectives on embodiment and consciousness* (pp. 61–84). Psychology Press.

Cooper, R. G., & Sommer, A. F. (2016). The agile–stage-gate hybrid model: A promising new approach and a new research opportunity. *Journal of Product Innovation Management*, 33(5), 513–526. https://doi.org/10.1111/jpi

Creswell, J. W. (2013). *Research design: Qualitative, quantitative and mixed methods approaches*. Sage.

Cuykendall, S., Soutar-Rau, E., Cochrane, K., Freiberg, J., & Schiphorst, T. (2015). Simply spinning: Extending current design frameworks for kinesthetic empathy. In *Proceedings of the Ninth International Conference on Tangible, Embedded, and Embodied Interaction* (pp. 305–312). ACM.

Davis, M.H. (1996). *Empathy: A social psychology approach*. Routledge.

Desmet, P. (2002). *Designing emotions*. Delft University of Technology.

Ekman, P. (2007). *Emotions revealed. Recognising faces and feelings to improve communication and emotional life* (2nd ed.). Holt Paperback.

Etcoff, N. L., & Magee, J. J. (1992). Categorical perception of facial expressions. *Cognition*, 44(3), 227–240.
Gaver, B., Dunne, T., & Pacenti, E. (1999). Design: Cultural probes. *Interactions*, 6(1), 21–29.
Goleman, D. (2004). What makes a leader? *Harvard Business Review*, 82(1), 82–91.
Höök, K. (2008). Knowing, communication and experiencing through body and emotion. *IEEE Transactions on Learning Technologies*, 1(4), 248–259.
Huang, Y., Tang, Y., & Wang, Y. (2015). Emotion map: A location-based mobile social system for improving emotion awareness and regulation. In *Proceedings of the 18th ACM Conference on Computer Supported Cooperative Work & Social Computing* (pp. 130–142). CSCW '15.
Jobin, C., Hooge, S., & Le Masson, P. (2021). The logics of double proof in proof of concept: A design theory-based model of experimentation in the unknown. In *Proceedings of the Design Society* (Vol. 1, pp. 3051–3060). Cambridge University Press.
Kouprie, M., & Sleesvijk-Visser, F. S. (2009). A framework for empathy in design: Stepping into and out of the user's life. *Journal of Engineering Design*, 20(5), 437–448.
Kuure, E., Jylkäs, T., & Miettinen, S. (2019). Service design creating value for industrial corporations through AI proofs of concept. In E. Bohemia, G. Gemser, N. Fain, C. D. Bont, & R. A. Almendra (Eds.), *Proceedings of the Academy for Design Innovation Management Conference* (pp. 620–628). Academy for Design Innovation Management. https://doi.org/10.33114/adim.2019.08.322
Lucero, A. (2015). Using affinity diagrams to evaluate interactive prototypes. In J. Abascal, S. Barbosa, M. Fetter, T. Gross, P. Palanque, & M. Winckler (Eds.), *Human–computer interaction – INTERACT 2015. Lecture notes in computer science* (Vol. 9297). Springer.
Mattelmäki, T., Vaajakallio, K., & Koskinen, I. (2014). What happened to empathic design? *Design Issues*, 30(1), 67–77.
Merleau-Ponty, M. (1962). *Phenomenology of perception* (Vol. 26). Routledge.
Miettinen, S., Nam, K., Björn, E., Jylkäs, T., Pääkkönen, T., Chen, K. H., & Kontio, T. (2024). *Open dataset of 20 interviews with senior service designers and managers for the Empathy Business research project* [Dataset]. Zenodo. https://doi.org/10.5281/zenodo.10810666
Miettinen, S., Sarantou, M., & Akimenko, D. (2016). Art and storytelling as an empowering tool for service design: South Australian case study. In P. Rytilahti & S. Miettinen (Eds.), *For profit, for good: Development organizations through service design* (pp. 74–80). Lapland University Press.
Muratovski, G. (2021). *Research for designers: A guide to methods and practice*. Sage.
New, S., & Kimbell, L. (2013). Chimps, designers, consultants and empathy: A 'theory of mind' for service design. In *Proceedings of Cambridge Academic Design Management Conference* (pp. 139–152). CADMC 13.
Norman, D. (2004). *Emotional design: Why we love (or hate) everyday things*. Basic Books.
Ohren, D. (2014). The limits of empathy in business ethics education. *Business Ethics Journal Review*, 1(18), 113–119.

Powell, P. A., & Roberts, J. (2017). Situational determinants of cognitive, affective, and compassionate empathy in naturalistic digital interactions. *Computers in Human Behavior*, 68, 137–148.

Reynolds, D., & Reason, M. (Eds.). (2012). *Kinesthetic empathy in creative and cultural practices*. Intellect Books.

Riess, H., & Neporent, L. (2018). *The empathy effect: Seven neuroscience-based keys for transforming the way we live, love, work, and connect across differences*. Sounds True.

Sarantou, M., & Miettinen, S. (2022). Introduction. In M. Sarantou & S. Miettinen (Eds.), *Empathy and business transformation* (pp. 1–12). Taylor & Francis.

Şemsioğlu, S., Karaturhan, P., & Yantaç, A. E. (2022). EMOTE: An interactive online tool for designing real-time emotional AR visualizations. In *Proceedings of the 13th Augmented Human International Conference* (pp. 1–8). ACM.

Sosa, R. (2019). Teaching (with) empathy and creativity in design. In *Proceedings of the DRS Conference LearnXDesign 2019* (pp. 1533–1600). Design Research Society.

Sreejun, S., & Chatwattana, P. (2023). The imagining learning model with enquiry-based learning via augmented reality to enhance creative products and digital empathy. *Journal of Education and Learning*, 12(2), 52–59.

Ståhl, A., Höök, K., Svensson, M., Taylor, A. S., & Combetto, M. (2009). Experiencing the affective diary. *Personal and Ubiquitous Computing*, 13, 365–378. https://doi.org/10.1007/s00779-008-0202-7

Sutton, T. M., & Altarriba, J. (2016). Color associations to emotion and emotion-laden words: A collection of norms for stimulus construction and selection. *Behaviour Research Methods*, 48, 686–728.

Umair, M., Latif, M. H., & Sas, C. (2018). Dynamic displays at wrist for real time visualization of affective data. In *Proceedings of the 2018 ACM Conference Companion Publication on Designing Interactive Systems* (pp. 201–205). ACM.

Vink, J., & Oertzen, A. S. (2018). Integrating empathy and lived experience through co-creation in service design. In *Proceedings of the ServDes2018 Conference. Service design proof of concept* (pp. 18–20). Linköping University Electronic Press.

Young, S. G., Elliot, A. J., Feltman, R., & Ambady, N. (2013). Red enhances the processing of facial expressions of anger. *Emotion*, 13(3), 380–384.

Zikmund, W. G., Babin, B. J., Carr, J. C., & Griffin, M. (2013). *Business research methods*. Cengage Learning.

15

Extending Visual Analysis Techniques for Empathy Building and Cross-Cultural Design

Tori Simpson, Henry Hosker, Tokushu Inamura, Yasuyuki Hirai and Melanie Sarantou

Abstract

Although empathy building is common within workshop practices, empathy among facilitators towards participants during post-workshop data analysis is underexplored. This chapter provides a preliminary justification that addressing this oversight will improve the quality of insights and further researchers' understanding of participant output. This analysis was driven by the question of how best to extend visual analysis techniques to empathically understand the diverse meanings of visual and textual artefacts left behind by workshop participants while understanding the cultural nuances of the collected data. Two workshops were run in two different zoological gardens with a total of twenty-three participants – comprising zoo visitors, animal-keeping staff, administrators and members of the general public – using prompts and sketching as instruments to collect information about the general public's attitudes towards avian species in Fukuoka, Japan. The results were analysed by a cross-cultural, multidisciplinary team using an autoethnographic focus group.

Introduction

The work discussed in this chapter examines how more-than-human species might be incorporated as equal stakeholders within a service design process, here with the aim of working towards and imagining a future with greater cohesion and a more positive relationship between humans and more-than-human species within built environments and modern society. The chapter focuses particularly on acknowledging other species as equal entities that share the spaces in question, and therefore necessitate consideration in the design of spaces and the interactions that exist within them. Apart from a few studies focusing on animal adoption services (Shore, 2005; Tu & Chou, 2014; Wang et al., 2020; Xu & Pan, 2021), a limited number of studies have examined the relationship between service design and animal–human perspectives. To develop these methods, we first focused on an obviously designed interaction between humans and avian species in a zoo context. This chapter details the analysis methods used to generate insights using data collected in workshops conducted in zoos (Figure 15.1) that aimed to understand the relationship between zoo visitors and avian species. Figure 15.2 illustrates the overarching project of which this analysis method is part. Workshops were conducted in Kyushu, Japan, through collaboration between three institutions – Kyushu University, The Royal College of Art and Imperial College London – and industry partners – Fukuoka Zoo, Omuta Zoo and the Tomodachiya Picture Book Museum.

The purpose of the described workshops was to generate deep insights that could later be used to inform the design of zoo services that can equally consider the perspectives of the visitors and animals. When incorporating diverse stakeholders, it is critical to employ empathic approaches that develop a deep and diverse understanding of stakeholders and their perspectives, including their needs and desired outcomes (Gallego & Hernández, 2022). Methods must be developed for researchers to foster cognitive and emotional empathy in the design process to gain deeper user insight (Gasparini, 2015). To gain this insight, it is particularly important that analysis methods are developed to bolster generative participatory research.

Elements that foster empathy towards animals may be fruitful for future service development within underserved animal care and well-being settings (Webber et al., 2022). The focus of the chapter is on the analytical processes used by the designer–researchers to understand the empathy expressed via illustrations developed by the visitor–participants during the workshop. Throughout the stages of data collection and analysis resulting from the workshops, the designer–researchers explored the visitor–participants' empathy towards other species and visitor–participants' ability to shift their perspectives towards understanding other species, as expressed in their illustrations.

In service design contexts, the literature has established workshops as avenues for codesigning services that are empathic towards stakeholders (Kuure, 2020). Sanders et al. (2010) categorize the types of activities that are undertaken during workshops into three categories: (a) making

FIGURE 15.1 *Group discussion of workshop outputs with the participants.*

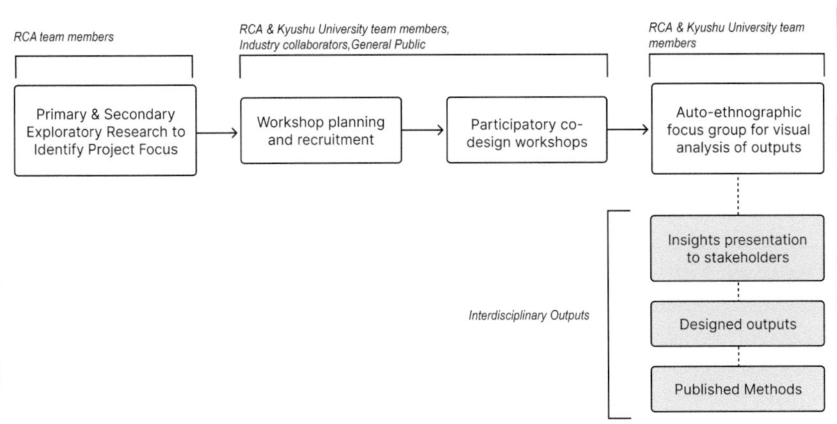

FIGURE 15.2 *Flowchart illustrating the overarching context that gave rise to this visual analysis methodology.*

tangible things; (b) talking, telling and explaining; and (c) acting, enacting and playing. These activities can generate a multitude of data, including visual and textual outputs from participants, which can lead to insightful information (Ozkaynak et al., 2021). In this context, the animals, zoological garden experts, managers, staff and visitors, in addition to designers and researchers from the Faculty of Design, Kyushu University, the Royal College of Art and Imperial College London, were the immediate stakeholders in the workshops and wider project.

The produced visual output from this study can contain many qualities, but their semiotic implications are subjective and culturally informed, given the cross-cultural context in which the workshop occurred. Furthermore, the team dynamics between such cultural readings influence the insights drawn. As collaborative design work is increasingly being conducted in cross-cultural groups, important methods have been developed to address this issue (Lee, 2004). Visual analysis methods that consider such cultural aspects provide room for exploration, especially in relation to empathic and collaborative modes. Through ongoing collaboration between the designer–researchers and the stakeholders of the two zoological gardens, novel visual analysis methods that deeply consider cultural dynamics within the stakeholder group have been developed and tested.

Given the opportunities presented by conducting workshops in a cross-cultural setting and exploring a unique topic, the discussion of the chapter highlights the value of the demonstrated visual analysis techniques in expanding empathic modes of work within service design contexts. The findings illuminate opportunities for fostering empathic capacity building among cross-cultural designer–researcher teams and for fostering a more empathic understanding of nonhumans in service design settings, such as zoological gardens; however, the value communicated in the developed processes expands beyond this particular context. Meaning these methodologies and the principles that inform them are transferrable to any design project led by a cross-cultural team, conducted within a complex or sensitive environment, or including a large variety of stakeholders.

Therefore, the below sections present a brief overview of the role of arts-based methods (ABMs) in creating empathy and the value of visual analytical approaches in service design settings, particularly considering designers working in cross-cultural teams. In the following section, the terms are defined, and the theoretical scope of the study is clarified. Later sections will present the methodological approach, concluding by explaining the value of the novel visual analysis approach and its broader application in service design as well as the specific potential to add value within cross-cultural teams in service design.

Theoretical Framework

As mentioned, the current study's methodology was developed by a cross-cultural multidisciplinary team from three institutions – Kyushu University, Royal College of Art and Imperial College London. The individuals within the team came from diverse cultural backgrounds, with many team members not originating from Japan, the country in which the present work was conducted. Furthermore, the team was composed of individuals specializing in a variety of fields, including design engineering, architecture, strategic design, service design, social sciences and art. Therefore, this work brings together theory and practice from a range of disciplines.

Participatory Research in Service Design

Service design can be categorized as the exploration of an open problem space involving different actors, including users. Researchers have defined the roles, knowledge and cultures that can enhance the shaping of services as a value creation process. This type of research enables the design of processes and organizational aspects that promote value cocreation within a service context (Morelli et al., 2021). It has been well established that participatory design methodologies such as workshops can create value within a service design research context by helping designer–researchers understand users' perspectives, wants and needs to design inclusivity more effectively (Gallego & Hernández, 2022). This in-depth understanding of the landscape is essential when designing a service, particularly one that addresses a complex issue with varied stakeholders, as the current work does. To apply participatory design research methods within this context, an analysis method must be developed to understand the diverse meanings communicated within participatory, generative sessions for these to be translated into insights that can be used to inform the design of services.

In the present study, empathy is defined as the ability to enter another person's inner world, to gain understanding of what they might be feeling, and to imply meaning (Kalisch, 1973). Empathy in design can be fostered through storytelling, using participatory methods to connect individuals and stakeholder groups by sharing experiences, emotions and values, hence leading to innovative and valuable design solutions (Strickfaden et al., 2023). This is of particular importance when designing for diverse stakeholders or stakeholders who are unconventional in their inclusion, with whom participants may not naturally empathize – in this case, more-than-human species (Bouche, 2021). Here, the more-than-human world refers to the concept coined by Abram (1996) and included in the title of his seminal work. Abram relates the human to the more-than-human, as is 'the human community to the wider ecology' (Abram, 1996, p. 4). To maintain the multiplicity of meaning present in the workshop outcomes and uncover generative insights during the analysis process, it was also important to build empathy between the researchers and workshop participants, as was done in the autoethnographic focus group. Diverse generative approaches can stimulate empathy in service contexts. Gallo (1994) discusses the different cognitive and emotional responses that can generate empathy, for example, by considering and imagining how others feel and understanding their emotions, in contrast to existing paradigms that prioritize understanding people from an objective and rationally guided perspective. Gallo (1994) explains that empathy helps people imagine how others feel, expanding their emotional experience rather than deepening it, encouraging the use of speculative 'as if' scenarios to explore affective and cognitive dimensions when dealing with people, which is a useful aspect for service designers. Within the context of the present work, 'as if' scenarios were used as prompts for drawing.

Moreover, to help visitor–participants explore current human–animal relationships and their potential futures, these 'as if' scenarios were carried forward to the analysis prompts used by the designer–researchers to interpret the drawings produced during the workshops.

Art-Based Methods for Empathy Building in Service Design

When collected in participatory research and context-specific settings, a wide variety of data from ABMs can foster multiple understandings of a research phenomenon (Seppälä et al., 2021). Formal qualitative inquiry that employs creative approaches to comprehend and express the subjectivity of human experience while generating new knowledge through research is known as art-based research (ABR) (Baden & Wimpenny, 2014; Leavy, 2017). As the bricolage of ABR, ABMs can be applied in all phases of enquiry by employing an end-to-end arts-based approach, from initiating and defining the research problem to data collection, analysis and dissemination (Leavy, 2017). A rich array of ABMs can be used in various contexts, art forms, genres and materials, including creative writing, literature, visual and performing arts and drama (Leavy, 2017). According to Baden and Wimpenny (2014), arts-informed research is an engaging and complex research area that calls for a closer examination of the relationships between art-based inquiry, researchers and participants, including both human and more-than-human stakeholders.

To build empathic relationships between researchers, workshop participants and other stakeholders, one first needs to challenge methods of analysis and interpretation and how they can support collaborative research (Baden & Wimpenny, 2014). Sarantou and Miettinen (2022) establish the value of arts-based approaches and ABR in participatory service design contexts. ABMs enable the collection of diverse data types, such as textual, notational, visual, voice, observational and performative data, that can emerge from a research context. Arts-based approaches have been applied in evaluation methodology (Simons & McCormack, 2007), and visualization techniques, such as mapping and service journeys, are widely applied in service design practice (Segelström, 2010; Trischler & Zehrer,

2012). However, arts-based analysis is not widely used in service design, despite the use of ABMs in service design for defining research problems and data collection (Miettinen et al., 2016). This method applies ABM to a service design process – both within primary research workshops and visual analysis of produced data – to create and analyse stories and holistic pictures of relationships, interactions and the environments within which they exist. This rich qualitative data collected at all stages allows for empathy building at multiple levels – between visitor–participants and avian species during workshops as well as between designer–researchers and visitor–participants during the autoethnographic focus group used to analyse the workshop output. This unique use of end–end ABM within service design extends the potential applications of ABM in service design and opens up the potential for further research within service design analysis.

Given the circumstances in which this project took place, this theoretical framework cannot be completed without discussing the implications of conducting design research in cross-cultural settings with a multicultural and interdisciplinary team. In a collaborative design setting, work is increasingly being conducted in culturally heterogeneous teams, which can help in uncovering new perspectives (Halskov & Christensen, 2018). Cross-cultural settings stimulate awareness of multiple ontologies and the ways of creating understanding, enabling different backgrounds and life experiences to come together, having space to follow cultural cues (Lloyd & Oak, 2018) and giving all participants a voice and space to express that can lead to robustness through the transferability of knowledge across contexts (Aaker et al., 2001). However, it is also important to tailor the development of these methods in this way, given that working in cross-cultural teams can present challenges. Within cross-cultural research, designer–researchers need to question their understanding of and position in relation to power and how power affects ethical research (Boivin & CohenMiller, 2024; Halskov & Christensen, 2018). The negative effects of research need to be mitigated through an awareness of power stemming from researchers' own cultural contexts (Boivin & CohenMiller, 2024). For this reason, this methodology employs empathy not only at the aforementioned levels but also within the research team during the analysis process, allowing for the inclusion and understanding of the variety of cultural and disciplinary perspectives brought to the table. The aim of this methodology is to foster frame-switching, with the analysis generating a final set of multiple perspectives and insights understood by all stakeholders (Clemmensen et al., 2018).

Research Design

The visual analysis described in this chapter examines the data produced during two drawing workshops conducted in Japan in July 2023 (Figure

15.3, Figure 15.4). The workshops were conducted in collaboration with industry partners, Fukuoka Zoo, Omuta Zoo and the Tomodachiya Picture Book Museum. The purpose of these workshops was to understand the relationships between humans and birds from a human perspective.

To fully investigate the results of these workshops, the designer–researcher team devised a visual analysis process in the form of an autoethnographic focus group that applies ABMs, fostering empathy towards examining multiple meanings understood from visitor–participant drawings. Specifically, the present study focused on three aims: understanding the empathy built between participants and the drawings of avian species, building empathy between the research team and workshop participants and creating an effective workflow for a cross-cultural multidisciplinary team, effectively including each member's perspective.

The research and analysis team consisted of seven members, each with a different cultural background, education and professional experience. The team also did not share a common first language, and only some team members shared a common cultural background and language with the workshop participants. Therefore, it was important to design a visual analysis method that included all perspectives and facilitated a workflow that valued different interpretations and multiple meanings while giving way to frame-switching, which means understanding the problem space with a new perspective shared by all members. Because the workshop participants

FIGURE 15.3 *Participants in the workshop, creating the drawings analysed.*

FIGURE 15.4 *One participant's drawings depicting their most memorable interaction with a bird (left), the same interaction from the bird's perspective (middle) and how they imagine that interaction may look in the future (right). Each drawing is accompanied by explanatory notes.*

were mainly Japanese, the analysis also relied heavily on Japanese team members to translate the annotations of each drawing and explain various metaphors and semiotics embedded within the drawings, which enabled the sharing of cultural knowledge (Clemmensen et al., 2018).

Rationale for a Methodological Approach

Exploring the question, 'How can design research evolve to include more-than-human stakeholders?' presented an opportunity to incorporate an end-to-end arts-based approach at every stage of the research process (Lee et al., 2023). The workshops asked the participants to create drawings, and the analysis methodology prompted creative notetaking by the designer–researchers and visual representations of key themes. Using an art-based approach in the analysis also allowed for the multiplicity of meaning communicated by the participants and further expanded by the diverse group within the research team to remain present in the research findings. Particularly relevant in the analysis portion of the work, these ABMs allowed for an approach that built empathy at multiple levels, both between participants and more-than-human actors, as well as between the researchers and participants. The inclusion of multiple experience-based perspectives and use of nonverbal methods of expression to elicit this empathy are characteristic of design thinking and highly valued in a discipline where designing to include varied perspectives and user needs is imperative (Zaff et al., 1993).

These utilized methods were qualitative and provide subjective knowledge to the research environment, and this exploratory knowledge was important when trying to empathize and engage with participants' inner worlds. This novel analysis methodology resulted in generative, multiple research insights rather than a reductive synthesis of user perspectives, which was a particularly important approach when trying to understand and disseminate conveyed meanings stemming from different cultural backgrounds.

This methodology was also explicitly designed to be used within a cross-cultural, multidisciplinary team, where each member brought their own skill set, cultural perspectives and inherent biases. Collaborative approaches to design more frequently occur in culturally heterogeneous teams, which may lead to new team perspectives and new challenges with respect to collaboration and mutual understanding (Halskov & Christensen, 2018). The analysis method prioritized included different interpretations of the visual outputs based on the researchers' own cultural knowledge, while making this cultural knowledge accessible to all team members (Clemmensen et al., 2018).

Step-by-Step Application of Methodology

The application of the methodology is discussed by presenting the primary review phase, which entailed personal reflection on the workshop by the researchers, followed by analysis and consolidation of insights.

Primary Review

The visual analysis process began with a personal reflection on the workshops. Overall, the team discussed their general experiences in facilitating and observing the workshop participants. Once reminded of the workshop context, each team member was given pens, pencils, and A3 sheets of paper. Team members were asked to take notes in any form, creating a personal reflective document throughout the process; an example of these notes is shown in Figure 15.6. They were asked to give particular emphasis to identifying key recurrent themes. Each set of drawings was then individually examined. The data included the three drawings produced by each workshop participant – two examples of which are shown in Figures 15.4 and 15.5 – based on the prompts 'Imagine your most memorable interaction with a bird', followed by 'Imagine the same interaction but from the perspective of the bird' and 'Imagine that interaction in 50 years' time'. The participants then analysed these drawings as a set of three drawings reflecting an individual storytelling experience.

Team members took turns based on a rota to comment on specific aspects of each drawing set. This rota was predefined and implemented to encourage each team member to share their thoughts on all facets of the drawings. Discussion ensued until there was a general consensus regarding completeness.

These analysis points are as follows: The 'Dictionary' asked the designer–researchers to consider specifically what they saw in the images and nothing else. The intent was to have the team members reflect on the raw image and focus on the lines, colours, scale of objects and so forth. The second prompt was used to draw attention to the relationships present in the drawings. This was done by examining factors such as scale, emotional expressions, line of sight of figures and body language. Third, attention was given to the meaning of attempting to understand the interpretations of the relationship between the workshop participant and birds, leaning on general themes of the drawings, such as the inclusion (or lack thereof) of the natural environment and physical, human or societal structures that were examinable. Finally, the final prompt-focused perspective was presented in the set of drawings. Multiple elements were considered in this process, such as from a conservational point of view, as well as societal and sentimental changes regarding avian species. The third, explicit drawing of how the participant imagined the interaction changing in the future, providing a particular touchpoint from which this

FIGURE 15.5 *One participant's drawings depicting their most memorable interaction with a bird (left), the same interaction from the bird's perspective (middle) and how they imagined that interaction may look in the future (right). Each drawing is accompanied by explanatory notes.*

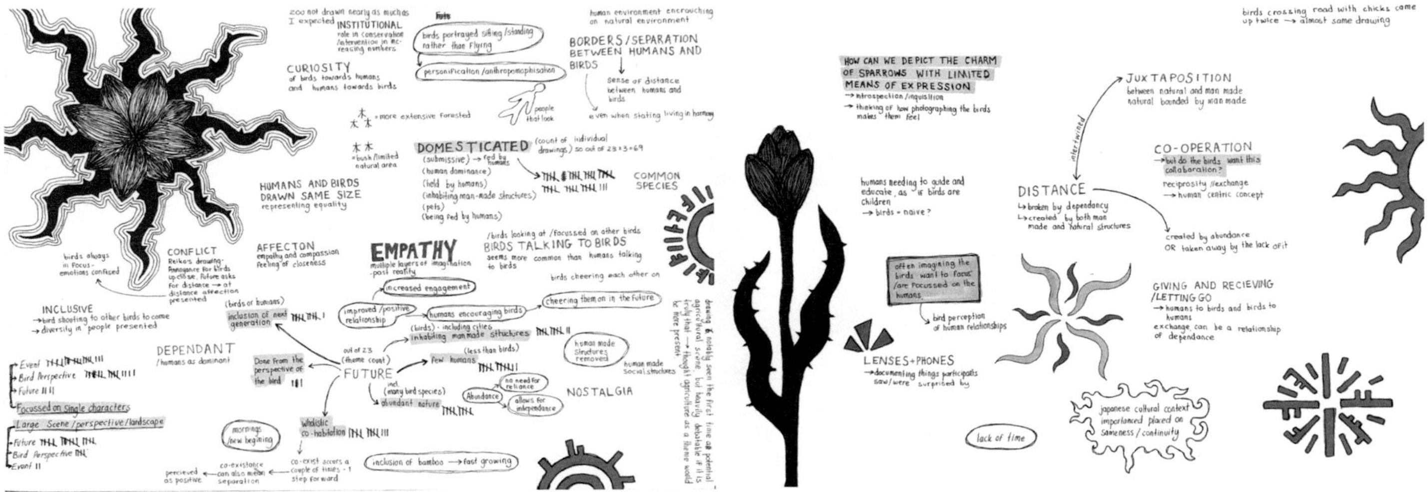

FIGURE 15.6 *Notetaking of analysis team members during the autoethnographic focus group.*

future perspective could be drawn; however, using the progression of the drawings also proved to demonstrate trends each participant focused on and the participant's outlook on the future and transitory points required to move towards it.

The intention of including these prompts and structures within the analysis process was to discursively identify every component of drawing, regardless of whether it was viewed as contextually or culturally relevant. Moving deeper with each prompt, symbolism and intention were derived from each set of drawings. Finally, the divergent point of the analysis – learning from and working with particular challenges – was determined by introducing and attempting to understand the emotional elements of each drawing set, as intended by the workshop participant. This pyramidal top-down approach enabled discussions to organically progress and diverge, which was intended to significantly increase the number of insights and viewpoints generated about each drawing set.

Insight Consolidation and Analysis

Once all the drawings had been considered and deeply discussed by the group, each team member had a complete illustration of the recurring themes. Next was a guided consolidation of those thoughts into concise phrases or small sketches. Each team member was asked to write three phrases relating to each of the following prompts: birds' role in society, mindsets and attributes, and imagined futures. These prompts were highly contextual to the research question of the workshop and essentially coalesced the key research questions with which the team entered the workshop. For other visual analyses, these questions should reflect the desired thematic insights and research questions of the research. At this stage, as a collective, the designer–researcher team had a concise collection of reoccurring themes expressed using phrases and accompanying visuals. After completing the consolidation phase, the team revisited the three prompts, taking turns presenting one phrase at a time. These themes were clustered into insights. This clustering, as shown in Figure 15.7, was a progressive process in which groupings were redefined during the first few rounds.

This was the conclusion of the process used during the visual analysis. Each step was purposefully included to introduce creative and analytical thinking alongside arts-based practices to deepen the processes and outcomes. The final body of work (Figure 15.8) was reflective and considered, and themes and thoughts were successfully incorporated from each team member.

FIGURE 15.7 *Clustering team member illustrations of recurring themes identified during the autoethnographic focus group.*

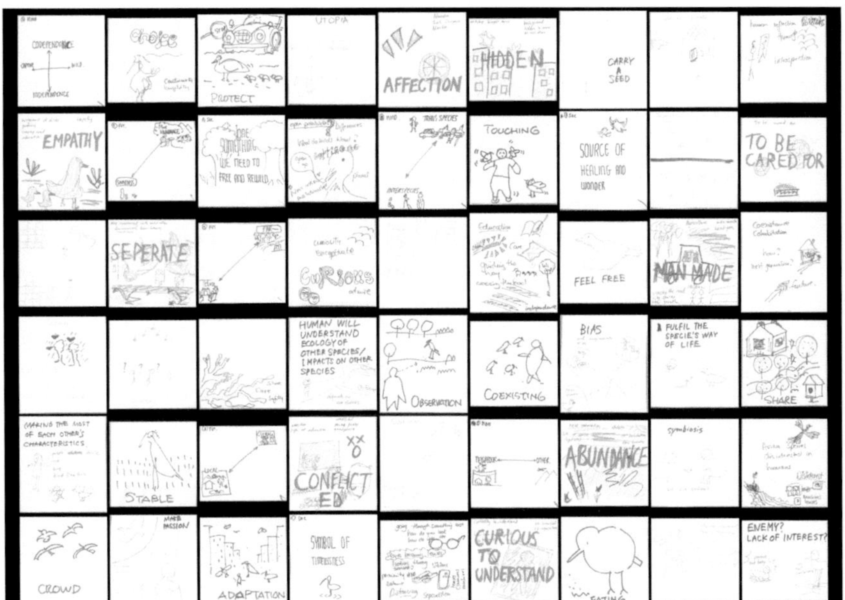

FIGURE 15.8 *Team members' illustrations of reoccurring themes, identified during the autoethnographic focus group.*

Findings

Regardless of whether the user is designed to be human or nonhuman, it is irrefutable that the involved stakeholders (including the designers themselves) benefit when the research process and design itself are informed by empathy building and perspective shifts towards end-users. This case becomes stronger as the difference in stakeholder experience widens and the number of included perspectives increases (Smeenk et al., 2016).

The premise that empathy is a necessary element in all present and future designs has been well established (Kouprie & Sleeswijk-Visser, 2009). This chapter exemplifies that, given the inclusion of shifted perspectives towards understanding more-than-human species and designer–researcher perspectives through autoethnographic methods, empathy building increases in importance as experiences diverge. Furthermore, given the lack of inclusion of more-than-human species in design up until this point, there are very few methodological approaches for this empathy building, hence its inclusion through and across all levels of workshop involvement (Poikolainen Rosén et al., 2022). This empathy building during the workshops and further additions within the analysis was a means for shifting participant and team member perspectives towards better considering other species, later enabling considered, targeted service designs that better reflected the needs of all stakeholders for whom they were designed. Large, complex problems can only be addressed when the design and research process is tailored towards gaining a holistic view of all stakeholder needs. The use of emotional and cognitive empathy has the potential to expand viewpoints but needs greater attention from the research community within design to develop transferable methodologies by which to employ empathy when analysing complex issues, often in complex circumstances (Gasparini, 2015). This chapter provides a discussion of one of the most pressing issues and prompts empathy by developing methodologies for analysis. The use of multiple outlets, and the reinforcement of empathy across stages meet this call, resulting in higher-quality insights.

Value of Arts-based Methods in Cross-cultural Research and Multidisciplinary, Cross-cultural Research Teams

Overcoming cultural barriers in understanding, disseminating and explaining empathic research is a challenge for any team. The methods employed here helped overcome those barriers by introducing the structures and freedoms to purposefully enable clear communication regarding the discussed topics. Not only were creative activities utilized in the initial workshop, but visual analysis and further visual media during said analysis were also leveraged to effectively communicate this empathy. The injected arts-based activities during the analysis and usage of affinity mapping generated an autoethnographic analysis that extended empathy building to being between the designer–researchers and visitor–participants. This enabled greater discussion, which, under cross-cultural conditions, was necessary for inclusion and consensus. This collective consensus permitted a nonreductionist, multiple, cocreation of meaning of participant and peer sentiment, regardless of cultural background, which is essential to the value of the output of the analysis.

Often, cultural backgrounds can stand in the way of complete team understanding and, therefore, limit the depth of the outputs reached. This chapter reflected on a method that overcame these barriers sufficiently to produce conclusions and insights verifiable through multidisciplinary, cross-cultural agreement. Through the use of additional ABMs and subsequent documentation, these conclusions became cemented and demonstrable.

This method utilized and demonstrated ABMs as a powerful methodology for documenting and measuring the empathy of workshop participants, in this particular context, towards avian species. The experimental methods developed in the analysis and described in the chapter were used as a means to develop deep empathy among all involved stakeholders. What happened as a supplementary result of the process was that this developed empathy was documented throughout the process for both workshop participants and team members. Empathy has historically been difficult to document (Surma-aho & Hölttä-Otto, 2022). The particular method demonstrated in this chapter was practical, engaging and effective and, therefore, is proposed as a versatile, cross-cultural methodology for building, analysing and documenting multiple levels of empathy within a workshop setting.

Conclusion and Transferability

This novel analysis method could be applied in a variety of different contexts, particularly within projects that focus on innovating towards solutions to complex problems, where a holistic and diverse perspective of the problem space is needed for researchers to accurately understand the design issue and wider context within which it exists. This chapter concerns a particular scenario in which visually complex workshop outputs were disseminated and insights generated, indicating a distinct usage of the methodology. With this being said, it is believed that the methodology is meaningful and useful in multiple contexts beyond those previously described. A step-by-step breakdown was given to allow researchers to translate and expand upon this methodology in a variety of contexts.

Takeaway

By reviewing the experimental methods employed in the complex visual analysis of empathy-related drawn artefacts, this chapter has presented cross-culturally applicable advancements in understanding and employing empathy in a service design context. Through an innovative end-to-end arts-based approach, the team demonstrated an effective analysis tool for the subject of empathy towards diverse stakeholders. The methodology provides enriched insights by enabling diverse interpretations informed by variable cultural nuances, contributing to service design by suggesting the application of ABMs within the analysis portion of design research.

References

Aaker, J., Kumar, V., & Day, G. S. (2001). *Marketing research* (Vol. 12). John Wiley & Sons.

Abram, D. (1996). *The spell of the sensuous: Perception and language in a more-than-human world*. Pantheon.

Baden, D., & Wimpenny, K. (2014). Art-based research. In N. K. Denzin & Y. S. Lincoln (Eds.), *The Sage handbook of qualitative research* (pp. 519–534). Sage.

Boivin, N., & CohenMiller, A. (Eds.), (2024). *How to understand and unpack power in qualitative research in multicultural contexts* [How-to Guide]. Sage. https://doi.org/10.4135/9781529691115

Bouche, B. (2021). At the limits of species: D. H. Lawrence's critical anthropomorphism in 'Fish'. *Études Lawrenciennes*. https://doi.org/10.4000/lawrence.2805

Clemmensen, T., Ranjan, A., & Bødker, M. (2018). How cultural knowledge shapes core design thinking – A situation specific analysis. *CoDesign*, 14(2), 115–132. https://doi.org/10.1080/15710882.2017.1399146

Gallego, C., & Hernández, G. C. (2022, May). The roles of empathetic competency and participative design in organisational stakeholder relationships. *Strategic Design Research Journal*, 15(2), 121–134. https://doi.org/10.4013/sdrj.2022.152.04

Gallo, D. (1994). Educating for empathy, reason, and imagination. In G. D. Rawlings (Ed.), *Proceedings of the Annual Conference of the American Association for the Advancement of Curriculum Studies* (pp. 43–60). State University of New York Press.

Gasparini, A. (2015, February). Perspective and use of empathy in design thinking. In *ACHI, The Eighth International Conference on Advances in Computer-Human Interactions* (pp. 49–54).

Halskov, K., & Christensen, B. T. (2018). Designing across cultures. *CoDesign*, 14(2), 75–78. https://doi.org/10.1080/15710882.2018.1459101

Kalisch, B. J. (1973). What is empathy? *The American Journal of Nursing*, 73(9), 1548–1552. https://doi.org/10.2307/3422614

Kouprie, M., & Sleeswijk-Visser, F. S. (2009). A framework for empathy in design: Stepping into and out of the user's life. *Journal of Engineering Design*, 20(5), 437–448. https://doi.org/10.1080/09544820902875033

Kuure, E. (2020). *Service design workshops in design practice* [Doctoral Dissertation]. University of Lapland. https://urn.fi/URN:ISBN:978-952-337-235-1

Leavy, P. (2017). *Research design: Quantitative, qualitative, mixed methods, arts-based, and community-based participatory research approaches*. Guilford Press.

Lee, K. (2004). Design methods for cross-cultural collaborative design project. In J. Redmond, D. Durling, & A. de Bono (Eds.), *Futureground – DRS International Conference 2004, Melbourne, Australia*, 17–21 November. https://dl.designresearchsociety.org/drs-conference-chapters/drs2004/researchchapters/128

Lee, S.-H., Yang, M., de Weck, O. L., Lee, C., Coughlin, J. F., Klopfer, E., & Ochsendorf, J. (2023). Service design in action: Transformation, consideration, and system thinking. *Proceedings of the Design Society*, 3, 3145–3154. https://doi.org/10.1017/pds.2023.315

Lloyd, P., & Oak, A. (2018). Cracking open co-creation: Categories, stories, and value tension in a collaborative design process. *Design Studies*, 57, 93–111. https://doi.org/10.1016/j.destud.2018.02.003

Miettinen, S. A., Akimenko, D., & Sarantou, M. (2016). Narrative-based art as means of dialogue and empowerment. In S. Golchehr, R. Ainley, A. Friend, C. Johns, & K. Raczynska (Eds.), *Mediations: Art & design agency and participation in public space* (pp. 137–149). Royal College of Art.

Morelli, N., de Götzen, A., Simeone, L. (2021). An Approach to Service Design. In: Morelli, N., de Götzen, A., Simeone, L (Eds.), *Service Design Capabilities*. Springer Series in Design and Innovation, vol 10, pp.9–26. Springer, Cham. https://doi.org/10.1007/978-3-030-56282-3_2

Ozkaynak, M., Sircar, C. M., Frye, O., & Valdez, R. S. (2021, May). A systematic review of design workshops for health information technologies. *Informatics*, 8(2), 34.

Poikolainen Rosén, A., Normark, M., Wiberg, M. (2022). Towards more-than-human-centred design: Learning from gardening. International Journal of Design, 16(3): 21–36. https://doi.org/10.57698/v16i3.02

Sanders, E. B. N., Brandt, E., & Binder, T. (2010, November). A framework for organising the tools and techniques of participatory design. In *Proceedings of the 11th Biennial Participatory Design Conference* (pp. 195–198). Association for Computing Machinery.

Sarantou, M., & Miettinen, S. (Eds.). (2022). *Empathy and business transformation*. Taylor & Francis. https://doi.org/10.4324/9781003227557

Segelström, F. (2010). *Visualisations in service design* [Doctoral dissertation, Linköping University Electronic Press].

Seppälä, T., Sarantou, M., & Miettinen, S. (2021). *Arts-based methods for decolonising participatory research*. Taylor & Francis. https://doi.org/10.4324/9781003053408

Shore, E. R. (2005). Returning a recently adopted companion animal: Adopters' reasons for and reactions to the failed adoption experience. *Journal of Applied Animal Welfare Science*, 8(3), 187–198.

Simons, H. W., & McCormack, C. (2007). *Participatory evaluation: Using visual images to explore program quality*. Teachers College Press.

Smeenk, W., Tomico Plasencia, I., & van Turnhout, K. (2016). Exploring the potential of wearables in the context of animal–human interaction. In *Adjunct Proceedings of the 2016 ACM International Joint Conference on Pervasive and Ubiquitous Computing and Proceedings of the 2016 ACM International Symposium on Wearable Computers* (pp. 43–46). Association for Computing Machinery. https://doi.org/10.1145/2968219.2968251

Strickfaden, M., Ruiz, A., & Thomas, J. (2023). (Re)storying empathy in design thinking. In Y.-G. Ghim & C. (S.) Shin (Eds.), *Interdisciplinary practice in industrial design. AHFE 2023 International Conference. AHFE Open Access* (Vol. 100, pp. 57–67). AHFE International. https://doi.org/10.54941/ahfe1002971.

Surma-Aho, A., & Hölttä-Otto, K. (2022). Conceptualisation and operationalisation of empathy in design research. *Design Studies*, 78, 101075. https://doi.org/10.1016/j.destud.2021.101075

Trischler, J., & Zehrer, A. (2012). Service design: Suggesting a qualitative multistep approach for analysing and examining theme park experiences. *Journal of Vacation Marketing*, 18(1), 57–71.

Tu, Y. H., & Chou, W. H. (2014). Service design blueprinting for stray animals: Concept of establishing joint adoption platform for stray animals. In *Computer, Intelligent Computing and Education Technology* (pp. 302–305). CRC Press.

Wang, D., Chen, C. W., You, Y. Y., Tsai, S. C., Hong, S. M., Chen, S. Y., & Tang, H. H. (2020). Optimising the adoption process in public animal shelters through service design thinking. In S. Boess, M. Cheung, & R. Cain (Eds.), *Synergy – DRS International Conference 2020, 11–14 August*. https://doi.org/10.21606/drs.2020.173

Webber, S., Cobb, M. L., & Coe, J. (2022, May 9). Welfare through competence: A framework for animal-centric technology design. *Frontiers*, 9. https://doi.org/10.3389/fvets.2022.885973

Xu, L., & Pan, Y. (2021). Research on the design of service process for adoption of stray animals. *The Japanese Journal of Ergonomics*, 57(Supplement-2), K9–K9.

Zaff, B. S., McNeese, M. D., & Snyder, D. E. (1993). Capturing multiple perspectives: A user-centered approach to knowledge and design acquisition. *Knowledge Acquisition*, 5(1), 79–116. https://doi.org/https://doi.org/10.1006/knac.1993.1003

PART FOUR

Design for Empathic Digital Approaches in Service Improvements

16

Empathic Development for Service Prototyping:

Challenges and Opportunities that Emerged During Design Thinking Workshops with Analogue and Digital Tools

Kuan Hui Chen and Kiwoong Nam

Abstract

Covid-19 has accelerated the integration of digital collaborative platforms, which are now essential tools for service development in the workplace. However, these platforms often introduce technical and human challenges that hinder effective communication and collaboration. Methods: This chapter employs ethnographic observation to examine the use of four different tools – virtual reality (VR), Lego®, craft and digital tools – in service development workshops. It assesses their effectiveness in facilitating service prototyping and addresses the challenges posed by a lack of physical interactions. Results: The findings indicate that, although VR and artificial intelligence (AI) can simulate human behaviours and potentially improve digital collaboration, they fail to facilitate emotional interactions and empathy among designers. This gap significantly affects the service design process and necessitates the effective amalgamation of digital and analogue tools. Additionally, given the variety of digital tools enabled by AI-embedded platforms, digital tools can diverge to generate creative ideas. Discussion: The disparities between digital and physical interactions in service prototyping present significant challenges in empathizing with users' emotions but also delineating opportunities through the effective use of both tools. There is a crucial need for rigorous research to understand the impact of these technologies on human behaviour in the context of service development.

Introduction

Service prototyping is an essential step in empathizing with users. Through a simulated service environment, prototyping enables designers to identify unnoticed impediments to development processes. Moreover, since the outbreak of Covid-19, digital prototyping tools have been equipped with seamless collaborative technologies. They facilitate borderless service development beyond physical limitations. However, the plethora of technologies can also lead designers to become lost in the digital jungle. The speed of developing digital tools surpasses the pace of designers' adaptation to new tools. Meanwhile, designer empathy is cultivated through both direct and indirect experience (Bennett & Rosner, 2019; Kouprie & Visser, 2009). Because direct experiences often entail significant costs or are sometimes impracticable, indirect user experiences through innovative tools become crucial.

This leads to the main research question of this chapter: 'How can designers effectively choose prototyping tools for empathic services?' In other words, it is necessary to identify when and how different types of digital and analogue tools can work together to create meaningful solutions. Selecting the right tools for specific goals is crucial for designing cost-effective and empathetic services (Thomas & McDonagh, 2013). Conventional digital technologies alone often fail to provide an intuitive environment for designers (Grässler & Taplick, 2019) and require inspiration from physical elements (Shugrina, 2023). Analogue tools also necessitate digital iterations for complex design tasks (Isa & Liem, 2023). Therefore, this research is particularly important now as numerous challenges and possibilities are emerging and a clear trajectory for the empathetic contribution of design tools needs to be established.

Emerging Service Prototyping Tools

Artificial intelligence (AI) technologies pervade every aspect of our lives. Embedded within existing services, the AI tools for innovation such as high-quality text-to-image apps, and accurate cancer detecting technology, are more significant than ever before. VR technology has undergone dramatic developments in recent years. Coupled with VR hardware, such as Apple's Vision Pro, public and business interest in AI and VR has reached its zenith (Ahmed et al., 2023). Given that services involve users' interactions with either people or systems, the efficacy of technologies in service delivery is heavily contingent upon users' physical and emotional responses to these technologies (Bakari et al., 2021; Blut et al., 2016). The burgeoning trend of employing AI and VR is pivotal for fostering empathy among users (Zwoliński et al., 2023) and is gaining popularity in service design (Bu et al., 2021; Colley et al., 2020). Nonetheless, the critical perspectives on technology use highlight concerns not only regarding the ethical implications of emerging technologies (Dhirani et al., 2023; Klein, 2022; Longo et al., 2020) but also about their practical efficacy (Bakari et al., 2021; Lee et al., 2020). Because of its inherently interpersonal nature, the design of services is greatly dependent on the designer's empathy and empathic qualities of the service system. As a result, it is imperative to explore how service prototyping tools relate to the cultivation of designers' empathy. The present study aims to identify the challenges and opportunities for empathic service development through the use of emerging (technology) and established (analogue) service prototyping tools.

Empathy in Service Prototyping

Analogous to mock-ups in product design, service design necessitates a prototyping process that enables designers to test solutions and identify potential user-perspective bottlenecks. This step is crucial for ensuring the successful delivery of services. Experts recommend an innovative lab setting to fully grasp the holistic user experience (Miettinen et al., 2012). Defining service prototyping, Razek et al. elucidate the following:

> A process for service development that employs service representations or components thereof to explore an idea, communicate a concept, and evaluate a design while encompassing various intangible and tangible aspects, diverse activities, and multiple stakeholders.
>
> (2018, p. 1)

Accordingly, designers' empathy is paramount in developing meaningful services for users (Almendra & Silva, 2022). However, cultivating empathy can be challenging, often necessitating intensive and direct experiences. This challenge is exacerbated when designers aim for prototype services with limited information or for lesser-known markets. Fortunately, advancements in technology and service prototyping methodologies have been mitigating this information barrier. Nevertheless, it is vital to ascertain whether these tools and methods are effective in fostering genuine empathy towards users (Soto et al., 2023). Moreover, designers' reluctance to employ digital tools for creative work persists, stemming from the perception of digital tools as impersonal and individual proficiency with these tools (Frich et al., 2021). Therefore, it is essential to understand the challenges encountered during the prototyping process from the designer's perspective to deliver empathic services.

Design Thinking and Prototyping Tools

Since its conceptualization by the IDEO company (Brown, 2008), the five stages of design thinking (empathize, define, ideal, prototype and test) have been widely adopted in numerous design practices. The nonlinear nature of the design thinking process enables designers and practitioners to continuously refine their ideas prior to the launch of products and services. In doing so, design thinking methodology facilitates the generation of creative ideas underpinned by empathy (Ho et al., 2023; Kim, 2020; Sakama et al., 2018). To conduct the creative workshop, the present study utilized a combination of digital and analogue tools. Although the experiment aimed to investigate the role of digital technologies in promoting empathy, integrating analogue tools revealed the challenges inherent in digital tools. Indeed, case studies incorporating LEGO® with digital tools (Borum et al., 2014) and pens with interactive monitors (Geyer et al., 2010) underscore the significance of utilizing both media to enhance creativity and collaboration in workshops. In essence, analogue tools facilitate tactile experiences and bolster designers' emotional connection to users, thereby leading to the development of empathic services.

Methods

Both analogue and digital tools can be used to carry out service prototyping, depending on the required fidelity of the design stages (Razek et al., 2017). Analogue tools are equally meaningful in developing digital services (Sezen, 2024). Nevertheless, as addressed in the main research question – 'How can designers effectively choose service prototyping tools?' – designers' interactions with both tools are the primary focus of the current study. In a structured situation, observation facilitates the identification of scenarios in which design can contribute most significantly (Cash et al., 2015). Consequently, this chapter employs ethnographic methods – namely observation and discussion – to identify challenges that arise during the design process. To achieve this goal, the researchers organized a structured workshop featuring design thinking steps with two Finnish researchers following introductory lectures on service design given one day before the workshop.

The participants were master's degree students studying business administration at a higher education institution in Lausanne, Switzerland. The cohort consisted of nine female and twenty-three male students (32 students in total), with their educational backgrounds being distributed across economics (7), design (5) and engineering (20). Despite their diverse backgrounds, the participants are proficient in design thinking and service design because the master's course includes basic service design and design management modules. The students were organized into groups of four, resulting in eight groups in total, with consideration given to achieving

Persona Profile: Jukka

Jukka

Age: 45
Family: Wife and two children (8 & 10 years old), aging parents
Profession: Engineer

Insights

An aging population in Finland has forced many adult children to care for their parents. One of the ways for Jukka to take care of his parent is to provide them with groceries in addition to his own family as going to grocery store is difficult for them.

"When I am caring for a parent I need a way to know which food to buy so I can ensure a good diet for my parents."

"As I have a busy schedule, I need an easy and fast way of getting the grocery store"

"I have a family and have to buy larger amounts of groceries. Therefore, I need an easy way to collect all the food and to make sure that I have everything I need."

Persona Profile: Liisa

Liisa **Fluffy**

Age: 23
Family: Boyfriend, dog
Profession: Student

Insights

Having a dog in Rovaniemi is quite popular. Liisa often walks to a local supermarket with her dog, Fluffy. In Finland it is not allowed to take dogs inside a supermarket and so Liisa needs to leave Fluffy outside. Liisa is also on a tight budget since she doesn't have other income than the public student allowance.

"Because I worry about Fluffy too much, I hurry with my shopping and often forget to buy something on my list."

"To save money, I look closely for any sales when I go to buy groceries and may plan my cooking based on what I find at a reduced price."

"I want to combine the walk with Fluffy and doing grocery shopping so that I save time."

FIGURE 16.1 *Two Finnish personas for the workshop.*

a homogeneous mix of educational backgrounds. A basic introduction to design thinking, service design and relevant tools (personas, service blueprints and customer journey maps) was provided prior to the workshop to ensure that the students were familiar with the design practices employed during the session.

Pilot Cases

Initially, the participants were presented with a design brief. The workshop participants were tasked with proposing systemic and/or physical design solutions that could consider the constrained space available both outside and within the premises. Subsequently, the researchers introduced two unfamiliar Finnish people (refer to Figure 16.1) to evaluate how the participants would derive design solutions empathically. The participants had the freedom to select their target persona at their discretion.

The workshop participants included Swiss-based master's programme students who had no prior experience with shopping at Finnish grocery stores. The introduction of the two Finnish personas aimed to assess the students' ability to develop empathy towards an unfamiliar market using both analogue and digital tools. A Miro board was prepared in advance to document the discussions and progress made during the workshop.

Workshop Design

Given the workshop's limited duration, the experiment omitted the 'test' phase and focused on the first four design thinking processes: empathize,

	Empathise (30 mins)	Define (30 mins)	Ideate (30 mins)	Prototype (30 mins)
VR	Team 1 & 2	Team 7 & 8	Team 5 & 6	Team 3 & 4
Lego	Team 3 & 4	Team 1 & 2	Team 7 & 8	Team 5 & 6
Craft	Team 5 & 6	Team 3 & 4	Team 1 & 2	Team 7 & 8
Digital	Team 7 & 8	Team 5 & 6	Team 3 & 4	Team 1 & 2

FIGURE 16.2 *The workshop design.*

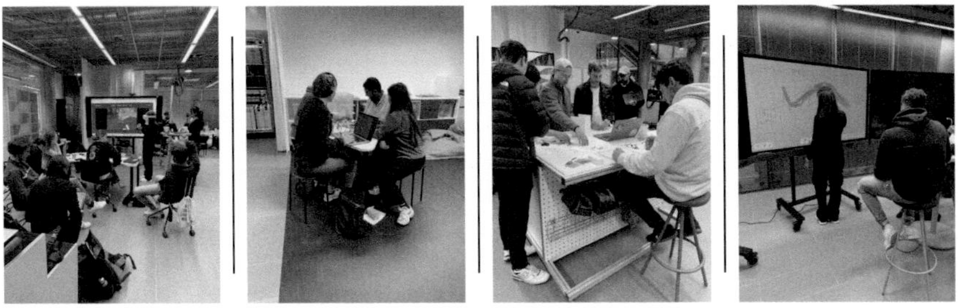

FIGURE 16.3 *The four service prototyping workstations. Photograph by Nam (2023).*

define, ideate and prototype. Each team employed different tools for the various design thinking processes to explore the challenges and opportunities presented by these tools. The design of the workshop process is depicted in Figure 16.2.

Combining digital and analogue tools for design thinking activities enables designers to flexibly brainstorm ideas (Frich et al., 2021). Each team allocated 30 minutes to each step, utilizing different tools. These tools served as workstations. At the VR station, the participants immersed themselves in a virtual replica of a grocery store located in Lapinrinne, Rovaniemi, using VR headsets in the customizable platform Spatial.io (www.spatial.io). LEGO® and craft materials were used in different workshop stages. For example, team 3&4 used LEGO® for the Empathize step while team 1&2 played with LEGO® for the Define step. The participants had access to generative AIs, web search facilities and an interactive screen within the lab. The colour-coded groups in Figure 16.2 represent the sequence of the workshop activities. The four workstations are depicted in Figure 16.3.

Researchers allowed students to choose their personas in Figure 16.1 freely. Four teams selected Jukka's persona, which involved elderly parents, while the other four teams opted for Liisa's persona, which included a dog. All teams recorded their progress on a structured Miro board. Teams were invited to present their design outcomes to their classmates within a five-minute time frame. As a result, the participants had the opportunity to share their experiences of utilizing various tools across the different design thinking processes with others during the discussion session.

Results and Findings

All the teams successfully created draft design solutions for each case. They presented their prototypes using a variety of tools, including paper houses, short videos, AI-generated images, LEGO® and digital whiteboards. Feedback on the use of four different tools for the initial four stages of design thinking was recorded on the Miro board.

Combining Digital and Analogue Tools for Creating Effective Design Options

Digital tools have traditionally been more effective in convergent thinking processes (Frich et al., 2021). However, when combined with analogue tools, such as LEGO® and paper crafts, as in the present study, designers' creativity and the efficacy of the tools can be maximized (Zboinska, 2019). The participants' feedback is summarized in Table 16.1.

Those participants utilizing digital tools for the initial stages of the design thinking process (empathize, define and ideate) provided positive feedback on the use of digital tools for service design. However, some groups using digital tools in the prototype stage noted that the tools from previous stages were sufficient for prototyping solutions, and they expressed a preference for analogue tools (LEGO® and craft) for visualizing service design outcomes. In addition, a few groups employed generative AIs to gain a deeper understanding of the persona and to visualize how their ideas might

TABLE 16.1 *Feedback on digital tools for design thinking steps*

Design thinking steps with digital tools	Group feedback
Empathize	*'Miro was great (for making progress notes). Everyone worked at the same time. Fast responses and reflection (with team members) are great. Interactive'* *'We used Google to have inspiration about what does exist'* *'We also used ChatGPT and recreated the behaviour of the persona and the family environment with Dall-E'*
Define	*'We could visualize directly to the interactive whiteboard. It reduces time to brainstorm ideas … We took different gadgets (for brainstorming ideas) and it was fun and easy to use'* *'Drawing on a whiteboard allowed us to have a better visualization of the problem as a whole'*
Ideate	*'We like digital tools for the Ideate step because it was quick'* *'Since we discussed (various ideas) in the team, we only used the interactive whiteboard'*
Prototype	*'We were a little confused because using digital tools seems like duplicating the same things from previous steps'* *'LEGO® may work better in this step'*

FIGURE 16.4 *AI-generated image from Dall-E. Image generated by OpenAI's DALL-E model based on researchers' prompt.*

be actualized. The group's AI-generated image is depicted in Figure 16.4. The AI-generated image assisted participants in identifying design elements pertinent to the challenges. The combination of digital tools (interactive whiteboards, web searches and generative AIs) with analogue experiences stimulated participants' creativity and led to the proposition of design alternatives.

As detailed by the participants in Table 16.1, the use of multiple tools facilitated the prompt and efficient generation of design work. The interplay between analogue and digital platforms is especially crucial for eliciting innovative solutions (Pershina et al., 2019) and for integrating knowledge (Caccamo et al., 2023). As a result, the experiment substantiates the efficacy of utilizing a variety of tools in service prototyping.

VR Space Has Limitations

The participants enjoyed the VR space but provided critical feedback on using VR for service prototyping. The feedback for using VR in the workshop is summarized in Table 16.2.

Despite recent advancements in VR technology, the current platforms do not yet possess the capability to provide an authentic replica of a physical space. The absence of sensory experiences beyond the visual field has led to difficulties comprehending real-world challenges. Current VR technologies fall short when it comes to delivering a holistic sensory experience to

TABLE 16.2 *Feedback on the VR experience in the workshop*

Design thinking steps with VR experience	**Group feedback**
Empathize	*'VR was useful for identifying the (design) focus'* *'In a real situation, I sometimes bumped into the other shoppers. However, the VR cannot provide this experience'*
Define	*'VR was not realistic. I cannot touch products or feel fellow shoppers'*
Ideate	*'VR was not useful because ideas have been generated in previous steps'* *'VR may be useful for the first step (empathize) not Ideate step'*
Prototype	*'VR was not necessary (because we completed the task with the other tools)'* *'VR (only) supported our sketch for the idea'*

users (Harley et al., 2018). Beyond technological constraints, the use of VR headsets is associated with health issues such as physical discomfort, spatial mismatch and psychological issues (Davey & Hancock, 2019) as well as equality concerns relating to the accessibility of expensive VR devices (Trahan et al., 2019). Therefore, these findings underscore the need to investigate the philosophical and methodological implications of VR for service design.

Empathizing with Various Tools

The two personas presented were unfamiliar to the workshop participants. The researchers were interested in observing how the participants would empathize with these people using various tools. The participants were instructed to make notes on design considerations to comprehend the targets' needs and motivations. The participants' notes were categorized into the following list, as shown in Table 16.3.

For example, the note 'The dog is outside!' was categorized as 'Frustration' in Emotional notes, and 'Shopping should not be physically challenging' was categorized as 'Store layout' in the Other notes category. In doing so, the researchers summarize the data from the empathize step, as shown in Figure 16.5.

Although the quantity of notes may not fully reflect participants' empathy towards the targets, the analysis of notes from the empathize stage underscores the significance of informative tools in understanding users, particularly when designers are unfamiliar with them. At the craft station during the empathize step, the teams restricted to using only craft materials

TABLE 16.3 *The categorized participants' notes*

Emotional notes	**Other notes**
Care	Cost efficiency
Frustration	Time efficiency
Fun	Informative
Relax	Legal issue
Worry	Store layout
	System improvement

Workstation	Emotional notes	Other notes
VR	2	10
Craft	1	13
Lego	5	13
Digital	8	21

FIGURE 16.5 *Analysis of participants' notes.*

found it especially challenging to understand the situation and targets. One participant at the craft station remarked the following:

> 'There was the limitation (of brainstorming persona's needs and motivation) because we were only allowed to use our own imagination'.

A lack of stimulating materials can impede the designer's exploration of the user's world, making empathy more difficult to achieve (Kouprie & Visser, 2009). In essence, tools that simulate the environment or provide pertinent information are valuable for designers in developing empathy with their targets.

In summary, designers' authentic experiences are instrumental in fostering empathy towards users (Kouprie & Visser, 2009; Woodcock et al., 2018). As a result, the primary challenge lies in the technical limitations of VR, which fail to deliver tactile and lifelike experiences. This necessitates a re-evaluation of VR's role in the design process. Furthermore, the outputs from generative AIs are not yet sufficiently detailed. As seen in the participants' Dall-E output in Figure 16.4, while the objects may resemble dogs, there is scope for improvement in detail. However, the experiment

FIGURE 16.6 *The combined use of AI tools (Dall-E & Adobe Photoshop Generative Fill).*

also demonstrated that such technical challenges can be surmounted by using a combination of tools. For instance, participants' original AI creations with Dall-E could be refined using Adobe Photoshop's generative fill, as illustrated in Figure 16.6.

Furthermore, analogue tools, such as LEGO® and craft materials, continue to play a vital role in evoking emotions that contribute to a designer's empathy. Given that the visualization of ideas stimulates creativity and team dialogue (Eppler et al., 2013), engaging with LEGO® and sketching ideas as a team assists participants in generating innovative concepts. Any tools that visualize a designer's creativity can facilitate the exchange of emotional intelligence among team members (Callahan, 2018). Hence, visualization through different tools contributes to collective empathy with users and the brainstorming of diverse ideas.

Discussion

The present study has utilized a creative workshop with assigned tools for design thinking steps to examine the challenges and opportunities for utilizing analogue and digital prototyping tools. This novel method fosters designers' empathy towards users with a holistic experience of design challenges. As described in the results and findings section, digital tools enhance the intuitive environment for designers by stimulating their creativity. For example, the role of VR in service design is to foster a desired environment to boost creativity, not to replicate the real world (Gong & Georgiev, 2020; Graessler & Taplick, 2019; Lee et al., 2021). Furthermore, this research is based on a single case study. In light of this, further investigation into VR's role in connecting multiple design platforms and augmenting designers' creativity is warranted. This aligns with the need to establish philosophical grounds for service prototyping tools. Moreover, the present study confirms the synergistic impact of AI when combined with other design tools, with AI-embedded tools elevating designers' creativity (Shugrina, 2023; Zhang et al., 2022). However, selecting the most suitable AI tools is increasingly challenging because of the plethora of available tools. Therefore, it is vital to experiment with various combinations of design tools and AI technologies to effectively empathize with users.

However, empathy and inclusivity must be carefully balanced when selecting the most effective combination of prototyping tools. Designers should adopt a holistic view of the targets and their environment to deliver equitable and meaningful solutions from both business and societal perspectives. In this context, the concept of a pluriverse in design (Escobar, 2018), which refers to a 'world where many worlds fit', is significant for appreciating multiple realities. Therefore, investigating the optimal combination of digital and analogue tools for establishing unique yet discrete worlds is worth considering in future studies.

Conclusion

One might simultaneously be a university researcher, a dad, a music enthusiast and a fan of Haruki Murakami. These multifaceted identities interact with and influence the value of a service offering. Emerging technologies have the potential to provide a comprehensive analysis that is essential for meaningful service provision. This chapter has investigated the application of four distinct tools for service prototyping. The amalgamation of analogue and digital tools aids in the spontaneous generation of innovative ideas. Despite the existing challenges, the expansion of AI and VR tools extends the scope of digital tools within both convergent and divergent design thinking processes. Now, we can better reflect on the philosophical and methodological paths for employing AI and VR in service prototyping.

Takeaway

It is important to test the potential of emerging technologies for empathetic service design development, and using the familiar short workshop format is a useful method of service design research. The key insights gleaned from this chapter encompass the architecture of the service prototyping workshop, which utilizes both analogue and digital tools. Initially, a variety of tools employed at each stage of the design thinking process offer practical guidance for the use of distinct tools at specified stages of design, thereby enabling researchers and practitioners to construct workshops with greater efficacy. This chapter's conclusions advocate for the expanded development of pluriverse design in terms of both the tools and outcomes of design.

Acknowledgements

This research was supported by Business Finland (Empathy Business project, grant number: 7784/31/2022).

References

Ahmed, N., Wahed, M., & Thompson, N. C. (2023). The growing influence of industry in AI research. *Science*, 379(6635), 884–886.

Almendra, R., & Silva, F. (2022). The central role of empathy in service design. In D. Raposo, N. Martins, & D. Brandão (Eds.), *Human dynamics and design for the development of contemporary societies. AHFE 2022 International Conference. AHFE Open Access* 25, 163–168, AHFE International. https://doi.org/10.54941/ahfe1001385

Bakari, A. R., Christina, C. H., Ninditya, D. P., & Iffan, M. (2021). Digital customer service: The effectiveness of communication in technology revolution 4.0. *International Journal of Research and Applied Technology (INJURATECH)*, 1(2), 403–409.

Bennett, C. L., & Rosner, D. K. (2019, May). The promise of empathy: Design, disability, and knowing the 'other'. In *Proceedings of the 2019 CHI Conference on Human Factors in Computing Systems* (pp. 1–13). Association for Computing Machinery.

Blut, M., Wang, C., & Schoefer, K. (2016). Factors influencing the acceptance of self-service technologies: A meta-analysis. *Journal of Service Research*, 19(4), 396–416.

Borum, N., Brooks, E. P., & Frimodt-Møller, S. R. (2014). The resilience of analogue tools in creative work practices: A case study of LEGO Future Lab's team in Billund. In M. Kurosu (Ed.), *Human–computer interaction. Theories, methods and tools. HCI 2014. Lecture notes in computer science*, 8510, 23–34. Springer. https://doi.org/10.1007/978-3-319-07233-3_3

Brown, T. (2008). Design thinking. *Harvard Business Review*, 86(6), 84.

Bu, L., Chen, C. H., Ng, K. K., Zheng, P., Dong, G., & Liu, H. (2021). A user-centric design approach for smart product-service systems using virtual reality: A case study. *Journal of Cleaner Production*, 280, 124413.

Caccamo, M., Pittino, D., & Tell, F. (2023). Boundary objects, knowledge integration, and innovation management: A systematic review of the literature. *Technovation*, 122, 102645.

Callahan, E. (2018). Cross-cultural empathy: Learning about diverse users in design thinking process. In C. Stephanidis (Ed.), *HCI International 2018 –*

Posters' extended abstracts. HCI 2018. Communications in computer and information science, 850, 236–240. Springer. https://doi.org/10.1007/978-3-319-92270-6_32

Cash, P., Hicks, B., Culley, S., & Adlam, T. (2015). A foundational observation method for studying design situations. *Journal of Engineering Design*, 26(7–9), 187–219. https://doi.org/10.1080/09544828.2015.1020418

Colley, A., Suoheimo, M., & Häkkilä, J. (2020, November). Exploring VR and AR tools for service design. In *Proceedings of the 19th International Conference on Mobile and Ubiquitous Multimedia* (pp. 309–311). Association for Computing Machinery. https://doi.org/10.1145/3428361.3432074

Davey, S., & Hancock, P. (2019). Virtual reality technologies. In B. Akhgar (Ed.), *Serious games for enhancing law enforcement agencies: Security informatics and law enforcement* (pp. 43–64). Springer. https://doi.org/10.1007/978-3-030-29926-2_3

Dhirani, L. L., Mukhtiar, N., Chowdhry, B. S., & Newe, T. (2023). Ethical dilemmas and privacy issues in emerging technologies: A review. *Sensors*, 23(3), 1151.

Eppler, M. J., Öste, H. F., & Bresciani, S. (2013, July). An experimental evaluation on the impact of visual facilitation modes on idea generation in teams. In *2013 17th International Conference on Information Visualisation* (pp. 339–344). IEEE.

Escobar, A. (2018). *Designs for the pluriverse: Radical interdependence, autonomy, and the making of worlds*. Duke University Press.

Frich, J., Nouwens, M., Halskov, K., & Dalsgaard, P. (2021, May). How digital tools impact convergent and divergent thinking in design ideation. In *Proceedings of the 2021 CHI Conference on Human Factors in Computing Systems* (pp. 1–11). Association for Computing Machinery.

Geyer, F., Klinkhammer, D., & Reiterer, H. (2010, November). Supporting creativity workshops with interactive tabletops and digital pen and paper. In *ACM International Conference on Interactive Tabletops and Surfaces* (pp. 261–262). Association for Computing Machinery. https://doi.org/10.1145/1936652.1936706

Gong, Z., & Georgiev, G. V. (2020). Literature review: Existing methods using VR to enhance creativity. In *Proceedings of the Sixth International Conference on Design Creativity (ICDC 2020)* (pp. 117–124). Cambridge University Press. https://doi.org/10.35199/ICDC.2020.15

Graessler, I., & Taplick, P. (2019, July). Supporting creativity with virtual reality technology. In *Proceedings of the Design Society: International Conference on Engineering Design* (Vol. 1, No. 1, pp. 2011–2020). Cambridge University Press.

Harley, D., Verni, A., Willis, M., Ng, A., Bozzo, L., & Mazalek, A. (2018, March). Sensory VR: Smelling, touching, and eating virtual reality. In *Proceedings of the Twelfth International Conference on Tangible, Embedded, and Embodied Interaction*. (pp. 386–397). Association for Computing Machinery. https://doi.org/10.1145/3173225.3173241

Ho, C., Sun, J., & Wu, K. (2023, April). Exploring creativity of designers in children's library navigation APP design with design thinking workshop. In *Proceedings of the 2023 8th International Conference on Information and Education Innovations*. (pp. 180–186). Association for Computing Machinery. https://doi.org/10.1145/3594441.3594470

Isa, S. S., & Liem, A. (2023). The impact of LEGO and analog objects in co-creating and prototyping ideas. *International Journal of Design Creativity and Innovation*, 11(1), 1–25.

Kim, T. (2020). The effect of creative potential on innovation behavior: Focusing on design thinking. *Journal of Distribution Science*, 18(8), 65–74.

Klein, A. Z. (2022). Ethical issues of digital transformation. *Organizações & Sociedade*, 29, 443–448.

Kouprie, M., & Visser, F. S. (2009). A framework for empathy in design: Stepping into and out of the user's life. *Journal of Engineering Design*, 20(5), 437–448.

Lee, S., Oh, H. Y., & Choi, J. (2020). Service design management and organizational innovation performance. *Sustainability*, 13(1), 4.

Lee, J. H., Yang, E., & Sun, Z. Y. (2021). Using an immersive virtual reality design tool to support cognitive action and creativity: Educational insights from fashion designers. *The Design Journal*, 24(4), 503–524.

Longo, F., Padovano, A., & Umbrello, S. (2020). Value-oriented and ethical technology engineering in industry 5.0: A human-centric perspective for the design of the factory of the future. *Applied Sciences*, 10(12), 4182.

Miettinen, S., Rontti, S., Kuure, E., & Lindström, A. (2012). Realizing design thinking through a service design process and an innovative prototyping laboratory: Introducing service innovation corner (SINCO). In Design Research Society 2012, Bangkok: Conference Proceedings: *Re: search: Uncertainty, Contradiction, Value* (pp. 1202–1214). Chulalongkorn University.

Pershina, R., Soppe, B., & Thune, T. M. (2019). Bridging analog and digital expertise: Cross-domain collaboration and boundary-spanning tools in the creation of digital innovation. *Research Policy*, 48(9), 103819.

Razek, A. R. A., Van Husen, C., & Imran, S. (2017). Introduction to a service prototyping tool box. In *Serviceology for services: 5th International Conference, ICServ 2017, Vienna, Austria, July 12–14 2017, Proceedings 5* (pp. 135–143). Springer International Publishing.

Razek, A. R. A., Van Husen, C., Pallot, M., & Richir, S. (2018, June). A proposed research framework and model for service prototyping. In *2018 IEEE International Conference on Engineering, Technology and Innovation (ICE/ITMC)* (pp. 1–8). IEEE.

Sakama, N., Mori, H., & Iba, T. (2018). Creative systems analysis of design thinking process. In B. Akhgar (Ed.), *Collaborative innovation networks: Building adaptive and resilient organizations. Studies on Entrepreneurship, Structural Change and Industrial Dynamics* (pp. 103–113). Springer. https://doi.org/10.1007/978-3-319-74295-3_9

Sezen, T. I. (2024). Analog prototyping for digital game design. In *Encyclopedia of computer graphics and games* (pp. 102–104). Springer International.

Shugrina, M. (2023). Designing intelligent tools for creative people. *IEEE Computer Graphics and Applications*, 43(3), 102–110. https://doi.org/10.1109/MCG.2023.3260964

Soto, M., Miettinen, S., & Alhonsuo, M. (2023, November). The prototyping service experience as a tool of empathy and research: From Finland to Chile. In *Entanglements and flows: Service encounters and meanings* (pp. 440–457). Linköping University Electronic Press.

Thomas, J., & McDonagh, D. (2013). Empathic design: Research strategies. *The Australasian Medical Journal*, 6(1), 1–6. https://doi.org/10.4066/AMJ.2013.1575

Trahan, M. H., Smith, K. S., Traylor, A. C., Washburn, M., Moore, N., & Mancillas, A. (2019). Three-dimensional virtual reality: Applications to the 12 grand challenges of social work. *Journal of Technology in Human Services*, 37(1), 13–31.

Woodcock, A., McDonagh, D., Osmond, J., & Scott, W. (2018). Empathy, design and human factors. In *Advances in usability and user experience: Proceedings of the AHFE 2017 International Conference on Usability and User Experience, July 17–21, 2017, The Westin Bonaventure Hotel, Los Angeles, California, USA* (Vol. 8, pp. 569–579). Springer International Publishing.

Zboinska, M. A. (2019). Influence of a hybrid digital toolset on the creative behaviors of designers in early-stage design. *Journal of Computational Design and Engineering*, 6(4), 675–692.

Zhang, X., Mao, X., Yin, Y., Chai, C., & Zhang, T. (2022, December). Melting your models: An integrated AI-based creativity support tool for inspiration evolution. In *2022 15th International Symposium on Computational Intelligence and Design (ISCID)* (pp. 97–101). IEEE.

Zwoliński, G., Kamińska, D., Haamer, R. E., Pinto-Coelho, L., & Anbarjafari, G. (2023). Enhancing empathy through virtual reality: Developing a universal design training application for students. *Medycyna Pracy*, 74(1), 199–210.

17

Empathy and Sustainability at the Core:

Leveraging AI Tools in Service Design

Nicholas Stevens, Yangyang Zhao and Alma Leora Culén

Abstract

Integrating values, including empathy, into more-than-human service design is becoming increasingly critical in the rapidly evolving landscape of digitally augmented practices. This type of service design needs to integrate social, economic and environmental sustainability values, focusing on human and nonhuman actors and the planet. This chapter investigates the role of increasingly powerful artificial intelligence (AI) and, more specifically, generative artificial intelligence (GAI) tools, as exemplified by ChatGPT, that can support empathic and sustainable service design. By examining the role of empathy in service design, we investigate how design practices and processes are challenged or given new opportunities with the inclusion of GAI tools. More concretely, we explore opportunities for human–AI collaboration to expand the range of actors and dimensions of empathy. Our contribution to sustainable service design includes developing a framework in which the GAI enriches the understanding of empathy in complex design processes. Additionally, the breadth of this framework is illustrated via a variety of collaborations between industry and academia.

Introduction

Practising service design amid rapidly developing technologies and our increasingly connected lives, service and logistic platforms and knowledge infrastructures, along with accelerating environmental and societal issues, such as climate breakdown and polarization, is highly challenging. This raises the following question: Can emerging generative artificial intelligence (GAI) technologies empower service design practices to address these challenges, and if so, in what ways?

Since the 1990s, empathy has become a central concept in service design (Bearden et al., 1998; Behara & Chase, 1993; Sarantou & Miettinen, 2022). The current debate on adopting GAI for empathic problem solving has spurred recent research, from technology development to design-related applications (Schmetkamp, 2020). GAI generates responses to specific prompts based on broader observations, hence making each response bespoke. This unique capability is very appealing and eagerly anticipated. However, concerns about hasty adoption were also observed with other similar novel technologies, such as virtual reality (VR). Although research has shown VR's potential to increase empathy (Martingano et al., 2021), there has also been speculation and misrepresentation of VR's capabilities, such as being an 'empathy machine' (Martingano et al., 2021).

The concept of artificial empathy in AI, which is rooted in psychology and neuroscience, highlights its potential to emulate human-like empathic capacity in interactions (Asada, 2015). Even before the widespread adoption of GAI, Verganti et al. (2020) discuss AI's support for innovation processes, suggesting that it transforms designers' practices with minimal impact on core design principles. This prompted us to reconsider how adopting GAI might present opportunities to develop service design as a discipline while delivering more desirable societal outcomes. How might service designers employ GAI, such as ChatGPT, to make empathic contributions towards addressing crucial societal concerns, such as sustainability? Additionally, could this aid in evolving the practice and rethinking the current frameworks of service design?

This chapter outlines the research conducted within the context of a project-based course in which six teams of upcoming designers undertook more holistic design processes, collaborating with industry partners to address complex real-world challenges. ChatGPT was included in its design toolkit to assess its potential contribution to the process. Five of the six projects were considered cases because their processes and outcomes offered insights into the questions driving the present research.

Our contribution is twofold. First, we derive the GAISE (GAI supported empathy) framework, which demonstrates how the GAI might contribute to empathic service design. The existing frameworks, such as the one from

Gray et al. (2010) for using empathy in design, are incorporated into the GAISE. However, GAISE aims to extend beyond the typical user-centred approach to include crucial concerns such as nature and broader societal needs in service design. Second, we illustrate the application of GAI in concrete projects, helping to understand how the framework was derived and how GAISE can be used in more-than-human design.

Background

The following sections outline empathy, design, technology, complexity, and their relations.

Framing Empathy

The designer's observation, affinity and understanding of others are significant and represent differentiating characteristics from those of many other professions (Surma-aho & Hölttä-Otto, 2022). This empathic capacity has become a foundational tenet of service design, fostering valuable connections between users and designers (Brown, 2008; Koskinen et al., 2003; New & Kimbell, 2013). Through the designer's identification and understanding of users, better services and products are developed: people identify with products or services more easily, users' experiences are enhanced, and some of the normative aspects of designing for others are alleviated (DiSalvo, 2012, p. 16).

Among the most common mappings or frameworks in service design is the empathy map (Gray et al., 2010), as shown in Figure 17.1, which

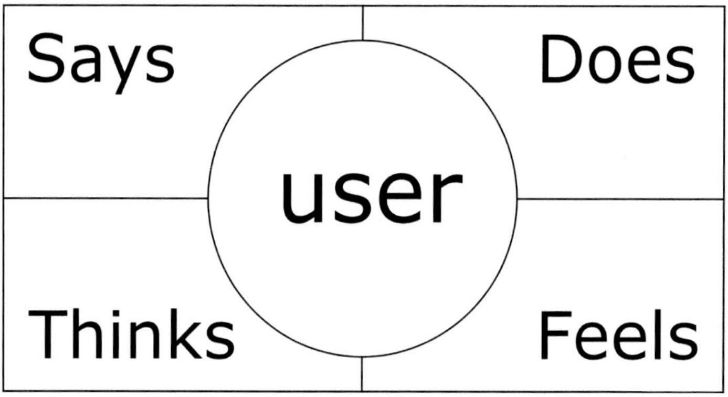

FIGURE 17.1 *Empathy mapping. Gray et al. (2010) (redrawn by authors).*

situates the user at the centre, informing the designer about what the user says, does, feels and thinks.

Empathic understanding is also integral to the user-centred design (UCD) process and 'design thinking' (Brown, 2008). However, these processes have faced criticism for various reasons (Kimbell, 2011; Kolko, 2018; Nussbaum, 2011; Sevaldson, 2018). Among the criticisms is the attempt to provide a simplified step-by-step process for what is inherently a messy, reflective and complex process. Critics argue that these approaches can be tokenistic in their use of empathy or can be constructed as simple toolkits to serve commercial interests. Additionally, there is concern that adopting a UCD, or 'design thinking', may overlook multiple important aspects of the issue at hand. Although designers focus on the user, what and who else may be deprioritized?

As a modification to the UCD process, the human-centred design (HCD) process aims to include people who are affected by a service or design process but who are not direct users. However, this may still leave out other entities, such as the environment or democratic values, such as inclusion and diversity. Another criticism is that user- or human-centred processes tend to make general assumptions about the design process and may limit the exploration of other design approaches (Suchman, 2006, p. 239) and the potential of other design processes from being undertaken. As a result, emerging designers, such as the students involved in our study, may perceive this as the standard approach to 'design'. This impression may also extend to those outside the field of design.

Design and Its Attempts to Navigate Technology and Complexity

A recent review of AI applications (Yüksel et al., 2023) reveals its utility for various design practices (e.g. initial design, design evaluation and optimization), highlighting the importance of choosing AI tools for a fit to design problems. For instance, prior service research has found that, as a social robot, AI offers more economic value to customers in terms of financial (Flavián et al., 2022) and marketing services (Liu-Thompkins et al., 2022). With the progression of AI technologies, especially large language models such as ChatGPT, design practices are experiencing a dual shift towards a data-driven approach and a broader landscape. Researchers have identified significant opportunities for AI applications to advance broader, socially sustainable values (Lee, 2021). Technology not only changes the processes used in design, but also transforms the environments in which design operates. Technological advances enable designers to undertake design processes more effectively and comprehensively, while also creating rapidly evolving complex situations in which these processes must occur.

Design typically integrates new technological advances into its workflow. However, the rapid uptake of recent AI offerings across all industries, the current lack of rigorous auditing of both AI systems and their applications, previous technology missteps such as facial recognition (Raji et al., 2020) and recognizable biases in ChatGPT responses regarding gender (Gross, 2023), among other aspects, imply that caution is needed. A more critical, reflective approach would consider the ontological cycle in which these technologies change the terrain in which design is practised, potentially necessitating new design approaches.

Consequently, the broader field of design faces calls to take on more responsibility for its outcomes, orient those outcomes socially, diversify its access and expand its areas of concern. Transition design (Irwin et al., 2015), as illustrated in Figure 17.2, is one attempt to address some of these calls; it aims to help designers pivot to address these broader concerns through a holistic approach that acknowledges the systemic, temporal and complex nature of issues. Transition design places specific attention on the designer's mindset and posture, contrasting many other design approaches in which a detached viewpoint is commonly assumed. This attention helps designers empathize in manners that are not solely focused on the user but also accommodate the larger context in which design occurs, such as society and the environment. This illustrates some of the complexities and self-reinforcing entanglements of empathy, service design, sustainability, society and AI.

Although some basic understanding of AI in and for design is starting to emerge, we contribute to these initial understandings by focusing on how the GAI supports empathy in complex design processes.

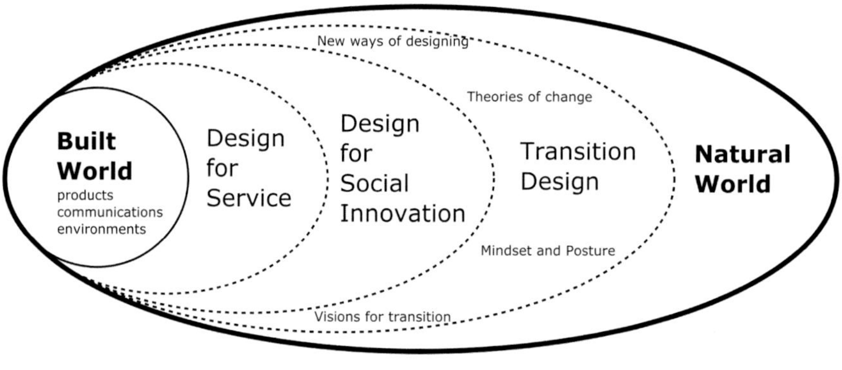

FIGURE 17.2 *This paper outlines the broader focus of transition design, Irwin et al. (2015), and is redrawn by the authors to include the main components of the framework: mindset and posture, visions, theories of change and new ways of designing.*

The Case

Our explorations concerning empathy were carried out in the context of research-based teaching at a leading European university, in which we explored how GAIs, such as ChatGPT (OpenAI, 2023), could be integrated into complex design situations. Our course is an advanced, elective bachelor's course in transition design (Irwin et al., 2015), the design for change from current situations to more desirable and sustainable versions, seeking long-term and systemic solutions. This implies complex design tasks. The course focused on democratization and inclusion, and semester-long teamwork on each project was carried out in collaboration with industrial partners. Rather than attempting to simplify the wicked problems facing design teams, we

TABLE 17.1 *Overview of project themes, briefs, outputs, collaborations and empathic qualities (the last column on empathic qualities will be subsequently described and explained)*

Theme	Project no.	Description	Output of project	In collaboration with	qualities of empathy			
					emotional		cognitive	
					proxy	delivery	accuracy	breadth
Social inclusion	1	Empowering older adults to decide on services in care-facilities	an app enabling the elderly to participate in decision-making	A design consultancy providing such services to the municipality	■	■		■
	2	Can activities at tourist destinations, like dog sledding, be more approachable to the visually challenged?	Auditory services, tangible devices supporting direct bodily experiences, e.g., the velocity changes	A research consortium, which included tourist destination developers	■	■		
Democratizing finance	3	Can trading be done responsibly? Find points of impact, ethical and sustainability challenges	Mobile interactions and parts of the system prototyped to support ethical and sustainable trading	A start-up developing a service for trading	■	■	■	■
	4	What do ordinary people need to know about crypto (AI-driven) trading? (learning, ethics)	Mobile interface interactions, an app including future learning and instruction videos	A start-up developing a service for trading	■	■	■	■
Democratizing production	5	How do we situate and develop DIY opportunities in existing manufacturing organizations?	DIY products like large digital buttons and touchless command of machines for knitting	An organization making interior design products and clothing, employing differently abled people		■	■	

considered whether they might be considered more holistically with the help of ChatGPT, allowing for acknowledging and embracing complexity more easily. Having our informatics students explore ways that GAI technology might be leveraged to encourage positive societal transitions is not only interesting and useful but also ethically responsible when large parts of their education are dedicated to technology development.

The twenty-three students who elected this course were divided into six design teams addressing six different topics through their projects. Five were relevant to this research (the sixth project resulted in a tangible serious game platform rather than a service and was therefore not further discussed in this chapter) and are summarized in Table 17.1, which briefly describes the design tasks, project outputs and industrial collaborators. The last column – the qualities of empathy – is placed in the table but will be explained later because it represents our findings concerning the kinds of empathy that emerged through the study.

As shown in Table 17.1, two teams engaged with two distinct briefs related to social inclusion, two teams engaged with briefs related to the democratization of trading, and the last team looked into the democratization of manufacturing processes.

The course acknowledged the multiple crises the world faces, aiming to build students' abilities to work through complex issues in a flexible, considerate, but generative manner. It introduces these future design professionals to new, broader, more-than-human (MTHD) (Wakkary, 2021) conceptions of design or designing-with that acknowledge and reprioritize other actors in addition to humans, such as digital assemblages (including GAI), other species or even environments. In addition, transition design (Figure 17.2), rather than using categorizations such as interaction, service or industrial design, focuses on four primary areas of interest: theories of change (how to work with long-term change and scale), mindset and posture, visions for transition (shaped by stakeholders in a participatory fashion, with a focus on clarity and social desirability) and new ways of designing (integrative, holistic). The GAI supported both learning and design.

Methods

The qualitative case study, which was based on the five design projects, is the second piece of research emerging from the same course setup. The first research article considered the ways in which ChatGPT supports creativity and teamwork (Culén et al., 2024). For the present study, audio recordings and ChatGPT conversation logs from four initial design sessions were collected. Additionally, an anonymized questionnaire was administered, and semi-structured interviews with each team were conducted. For our purpose in this chapter, the same dataset was revisited with a crucial difference – it was examined to identify and inform various aspects of empathy provided by the GAI. Furthermore, throughout the course, observations were made by the two authors, who were mentors of the projects. This involvement allowed them to follow the students' evolving processes and how they utilized ChatGPT's capabilities within their projects through weekly design critiques/reflections. In addition to these course-based data, background literature and mappings were actively considered sources of information and inspiration when discussing empathic qualities.

Thus, the main method was the reflexive discourse between the authors. Multiple sessions took place over the course of several months, during which the themes, design processes and reflections on empathic AI use were discussed. Similar to how Odom et al. (2016) teases out the qualities of research products, we used reflective, iterative and discursive processes to tease out and frame the observed qualities of empathy.

Through this work, the authors had the opportunity to observe first-hand multiple and varied instances of design performed differently. However, these first-hand observations and reflections give the study an ethnographic component that aligns well with more-than-human (Wakkary, 2021) design variants in which the situated nature of design is acknowledged and leveraged.

Qualities of Empathy and the GAISE Framework

Our thematic analysis of the available material, such as the final reports and transcripts of the recordings of students using ChatGPT, led to iterative reflections among the authors concerning the patterns of use of the GAI to support empathic exchange and design. Our material clearly showed that the established framework (Gray et al., 2010) was still used and could be supported by ChatGPT, with ChatGPT serving as a user proxy. However, we identified instances in which ChatGPT contributed to an empathic exchange in a novel manner. The contributions occurred in most of these instances, even though the GAI was not explicitly engaged to provide empathic output. Interestingly, in contrast to contributing to the empathic relations between designers and users, the novelty was the recognition of and contribution to the relations between the GAI and other actors, both human and nonhuman. We find user placement at the centre of the grey empathy map (Figure 17.1) to be limiting and counterproductive in complex design. These instances led us to situate GAI at the centre of our framework, reiterating our more-than-human intentions. However, the GAI also served as a proxy for user groups in projects where they were essential (e.g. inclusion). In what follows, we describe our GAISE framework (see Figure 17.3).

The GAISE framework comprises four qualities or aspects of empathy. The first quality, the empathic proxy, is similar yet broader than empathy with the user, which is central to the traditional framework. However, in complex domains and designing for sustainability transitions, a user is not the only entity towards which the designer is empathic – nature and the societal consequences of design also deserve an empathic review. Second, for empathic accuracy, the GAI is used to gain more information on the topic

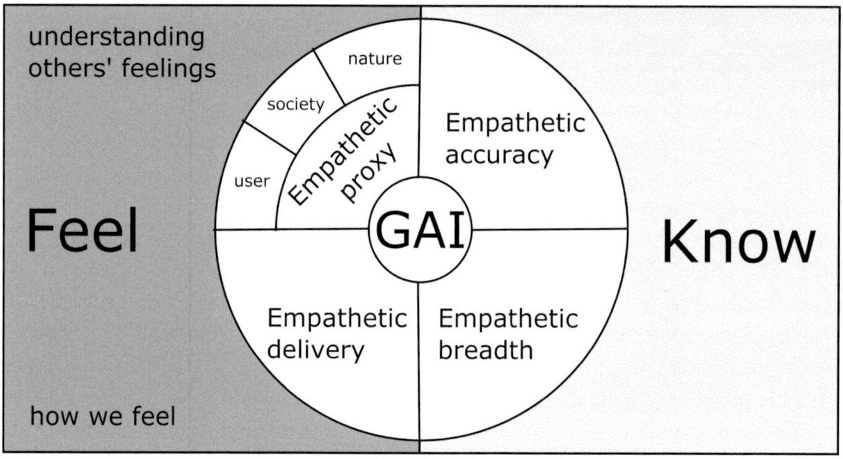

FIGURE 17.3 *GAISE (GAI Supported Empathy) framework for GAI-assisted empathy in complex design spaces.*

area to gain a cognitive understanding of the complex situation, which helps contextualize the problem better and improves empathic accuracy (Ickes, 1993). The third quality, empathic delivery, relates to the way in which a GAI might empathically divulge information, provide support and so forth without being judgemental or disrespectful when communicating with the design team. Finally, empathic breadth refers to the range of aspects and plurality of entangled issues typical of complex domains and the ways in which AI can turn designers' attention to this plurality.

Each of these qualities is discussed and exemplified in the discussion section.

Discussion

The extremely rapid uptake of GAI across various fields, including service design, indicates that the use of such tools might become commonplace in the near future. Thus, understanding how these tools might support empathy and design processes more generally is essential at the juncture of service design and the complex societal changes under way. The GAISE framework aims to provide a foothold for designers who may lack experience with the individuals for whom the designs might be relevant, the design domain or the design tools and techniques to use when designing services in a more-than-human manner. To clearly communicate our points, we divide the section into two parts: the first discusses GAISE and the examples from our case, while the second addresses some points of consideration, including the limitations of the study.

GAISE: Examples and Discussion

Regarding empathy, the initial work showcased here illustrates the benefits of the GAI, rapidly expanding designers' perceptions of the intricacies of designing in complex situations and mitigating the need for simplification and reduction. Here, we discuss in detail, with examples from our study, various aspects of how the GAISE framework functions in design.

Working directly with specific user groups in a project can often be challenging for various reasons, such as privacy concerns, ethical issues, ill health, cost, timing or logistics. In such instances, the use of conversational GAIs can be valuable in providing examples of concerns or issues relevant to the work of a team. In our study, two projects clearly utilized ChatGPT in this manner.

The team in Project 1 (see Table 17.1) could not directly contact elderly care home residents because of ethical restrictions and the short time frame needed to obtain necessary permission. None of the team members had elderly relatives whom they could approach. To mitigate this lack of access to their core user team, they consulted ChatGPT about various aspects of ageing and care to gain insight into the issues likely pertinent to the project.

Similarly, the team in Project 2 inquired about how it feels to be a tourist with impaired vision, especially when engaging in activities such as dog sledding. This use of the GAI provided an empathic connection to the users in their absence, effectively using the GAI as a proxy. This illustrates the concept of using the GAI as an empathic proxy or representation of the user.

In contrast, the teams in Projects 3 and 4 emphasized protecting the planet and the sustainability of the environment, resulting from the (future) widespread engagement in trading cryptocurrency. They consulted the GAI on various aspects of protecting nature, effectively using the GAI as a proxy for nature under the stress of massive computations and financial transactions.

The team engaged in Project 5 focused on social sustainability and the inclusion of as many people as possible in social structures, exemplifying the use of AI as a social proxy. In their case, they were looking for opportunities to make running production machines easier for a broader range of differently enabled employees in endeavours such as creating interior design articles, shoes or clothing.

These examples demonstrate the diverse applications of the GAI as a proxy for users, social structures or nature, showcasing its potential to provide empathic connections and insights in situations where direct interaction is challenging.

Empathic Accuracy

In service design, achieving empathic accuracy plays a pivotal role in crafting experiences that resonate with users on an emotional level, fostering heightened satisfaction and loyalty. As a result, service designers aim to comprehend not only the immediate context in which users interact with the service but also the broader societal and cultural factors that influence their emotions and behaviours. Our course scaffolds learning about this broader context and grasping the significance of doing so. One of the tools integrated into the course curriculum is the Winterhouse matrix (Cannon et al., 2013), which frames design across various social pathways, ranging from individual designers creating specific products or services to designing interventions capable of affecting substantial and lasting changes at the cultural level.

In Project 5, the team utilized ChatGPT to better understand the broader context pertinent to their manufacturing case. They asked ChatGPT to complete the Winterhouse matrix, the new tool with which they were becoming acquainted. They received a nine-step process in return. Although the process was represented differently than the matrix outlined on the Winterhouse site (Cannon et al., 2013), the team initially failed to recognize that they were essentially the same. Rejecting the nine-step process, the team continued to prompt ChatGPT to fill the matrix cell by cell. Finally, they understood the intent of the matrix and the essence of the process, rather than focusing on a particular representation. This iterative process of clarification may resemble other forms of learning. Nevertheless, the current study provided rapid results regarding a comprehensive understanding of the broader contextual nuances of their design case while enhancing their grasp of new tools (both the ChatGPT and the Winterhouse pathway matrix). This process did not directly forge an emotional connection with the task for the team. However, cognitive understanding provides a logical scaffold for the subsequent development of emotional connections and comprehension.

Empathic Delivery

The characteristics of a delivery method have a considerable influence on the reception of information (Hardee, 2003). For instance, valuable information might be dismissed solely because of the condescending manner in which it is conveyed. These characteristics often take on greater importance when there is a power imbalance between those providing and receiving information (Henderson, 2003).

The way in which the GAI delivers information is one of the main areas of focus in GAI development. Many variants are already available, and some have strong credentials (Concannon & Tomalin, 2023) concerning an empathic rapport with the user.

Examples of how ChatGPT can deliver information were provided by all project teams. When teams were asked in a questionnaire at the end of the project if they considered GPT in any way similar to a teammate, they unequivocally refused. However, although 'teammate' might not be an accurate description, the teams did work with the information provided by the GPT and probed it to provide further examples, similar to how one would relate to an experienced teammate or mentor. Moreover, unlike with human team members, knowledge from the ChatGPT could be solicited anytime, day or night without condescension, patronization or other inferences. It seemed to cater better to students' needs than a time-pressured expert in the field might.

Therefore, although the overall experience of collaborating with GPT might not be considered similar to that of working with a teammate, the mediation provided by GPT greatly aided the project work, partly because of the empathic way of delivering information.

Empathic Breadth

Navigating complex design domains with numerous actors requires methods capable of handling inherent intricacies. Furthermore, detailing the relationships between these actors, identifying sites of current tension and potential interventions necessitates thorough analysis without oversimplification. Giga mapping (Sevaldson, 2011) is useful in this regard. However, the use of GAI turned out to be very beneficial in the fleshing out of this mapping, providing relevant input that could otherwise have been overlooked. The use of the GAI for mapping was particularly beneficial in domains where designers lacked familiarity, knowledge and personal experience. Projects 1, 3 and 4 all leveraged ChatGPT to assist in compiling Giga maps for areas in which they possessed limited background knowledge, such as elderly care structures and financial trading systems. Partner organizations assessed these Giga maps, and teams collaborated with them to complete them accurately.

In contrast to emphatic accuracy, this quality explicitly focuses on knowledge about the domain, making the design team feel more confident in the process. This quality was central for understanding the design domain, providing a very different quality from the common paradigm where users are experts in what they say, do, think and feel. This quality helps designers feel they have more mastery of the domain and understanding of it without reducing complexity. Instead, the focus is on relational content and how things fit together.

Points of Consideration

First, we acknowledge that the outlined research considers a small number of cases within the academic setting. However, the research and design were performed in collaboration with organizations in the industry, and we expect that similar observations and conclusions could be drawn from additional studies executed in a similar manner in a purely industry setting. Nevertheless, more work along these lines is needed, and rapid and accurate technological changes must be considered continuously. A change in the ChatGPT version from 3.5 used in the present study to version 4 demonstrated increased accuracy.

Second, this is a qualitative and interpretative study in which our reflections, observations and other collected data were central to this reflective process, revealing an autoethnographic aspect of our work. Such first-person approaches are becoming increasingly common within more-than-human design (Wakkary, 2021).

Third, the accessibility of the GAI, the ease at which outputs can be obtained, the variety of ways in which they might be integrated and the initial promising indications point to the likelihood that there will be significant uptake of the use of the GAI within the design process as this new technology is exploited. As with the uptake of any new technology, these issues and concerns need to be actively and critically considered. The uptake should not replace the need to involve as wide a range of participants with lived experiences as possible. We must also endeavour that these practices are not undertaken in a superficial, tokenistic manner (Vink & Oertzen, 2018) but are used to address these complex intertwined issues in a considered manner.

Takeaway

Integrating values of social, economic and environmental sustainability into more-than-human service design that focuses on human and nonhuman actors and the planet is becoming increasingly critical in the rapidly evolving landscape of digitally augmented practices. This chapter has investigated how leveraging GAI in human–AI collaborations that enriches understandings of empathy to include more-than-human elements. Our contribution lies in developing a framework for the use of the GAI in complex design processes incorporating these four new dimensions of empathy, including empathic breadth, empathic delivery, empathic proxy and novel variants of conventional empathic accuracy.

References

Asada, M. (2015). Development of artificial empathy. *Neuroscience Research*, 90, 41–50. https://doi.org/10.1016/j.neures.2014.12.002

Bearden, W. O., Malhotra, M. K., & Uscátegui, K. H. (1998). Customer contact and the evaluation of service experiences: Propositions and implications for the design of services. *Psychology and Marketing*, 15(8), 793–809. https://doi.org/10.1002/(SICI)1520-6793(199812)15:8<793::AID-MAR5>3.0.CO;2-0

Behara, R. S., & Chase, R. B. (1993). Service quality deployment: Quality service by design. In *Perspectives in operations management* (pp. 87–99). Springer US. https://doi.org/10.1007/978-1-4615-3166-1_5

Brown, T. (2008). Design thinking. *Harvard Business Review*, 86(6), 84–92, 141.

Cannon, C., Chochinov, A., & Narezo, G. (2013). *Social design pathways*. http://www.winterhouseinstitute.org/pathways

Concannon, S., & Tomalin, M. (2023). Measuring perceived empathy in dialogue systems. *AI and Society*, 39, 2233–2247. https://doi.org/10.1007/s00146-023-01715-z

Culén, A. L., Karahasanovic, A., Stevens, N. S., Makokha, J., & Zhao, Y. (2024). ChatGPT: Mediating complex design processes. In C. Gray, E. Ciliotta Chehade, P. Hekkert, L. Forlano, P. Ciuccarelli, & P. Lloyd (Eds.), *DRS 2024: Resistance, Reflextion, Recovery, Reimagination, Boston*, 23–28 June. https://doi.org/10.21606/drs.2024.628

DiSalvo, C. (2012). *Adversarial design*. The MIT Press. https://doi.org/10.7551/mitpress/8732.001.0001

Flavián, C., Pérez-Rueda, A., Belanche, D., & Casaló, L. V. (2022). Intention to use analytical artificial intelligence (AI) in services: The effect of technology readiness and awareness. *Journal of Service Management*, 33(2), 293–320. https://doi.org/10.1108/JOSM-10-2020-0378

Gray, D., Brown, S., & Macanufo, J. (2010). *Gamestorming: A playbook for innovators, rulebreakers, and changemakers*. O'Reilly Media, Inc.

Gross, N. (2023). What ChatGPT tells us about gender: A cautionary tale about performativity and gender biases in AI. *Social Sciences*, 12(8), 435. https://doi.org/10.3390/socsci12080435

Hardee, J. (2003). An overview of empathy. *The Permanente Journal*, 7(4), 51–54. https://doi.org/10.7812/tpp/03-072

Henderson, S. (2003). Power imbalance between nurses and patients: A potential inhibitor of partnership in care. *Journal of Clinical Nursing*, 12(4), 501–508. https://doi.org/10.1046/j.1365-2702.2003.00757.x

Ickes, W. (1993). Empathic accuracy. *Journal of Personality*, 61(4), 587–610. https://doi.org/10.1111/j.1467-6494.1993.tb00783.x

Irwin, T., Kossoff, G., Scupelli, P., & Tonkinwise, C. (2015). *Transition Design 2015*.

Kimbell, L. (2011). Rethinking design thinking: Part I. *Design and Culture*, 3(3), 285–306. https://doi.org/10.2752/175470811X13071166525216

Kolko, J. (2018). The divisiveness of design thinking. *Interactions*, 25(3), 28–34. https://doi.org/10.1145/3194313

Koskinen, I., Battarbee, K., & Mattelmeaki, T. (2003). *Empathic design*. IT Press.

Lee, K. (2021). A systematic review on social sustainability of artificial intelligence in product design. *Sustainability (Switzerland)*, 13(5), 1–29. https://doi.org/10.3390/su13052668

Liu-Thompkins, Y., Okazaki, S., & Li, H. (2022). Artificial empathy in marketing interactions: Bridging the human-AI gap in affective and social customer experience. *Journal of the Academy of Marketing Science*, 50(6), 1198–1218. https://doi.org/10.1007/s11747-022-00892-5

Martingano, A. J., Hererra, F., & Konrath, S. (2021). Supplemental material for virtual reality improves emotional but not cognitive empathy: A meta-analysis. *Technology, Mind, and Behavior*, 2(1). https://doi.org/10.1037/tmb0000034.supp

New, S., & Kimbell, L. (2013). Chimps, designers, consultants and empathy: A 'theory of mind' for service design. *Proceedings of the 2nd Cambridge Academic*

Design Management Conference, September (pp. 139–152). http://www.lucykimbell.com/stuff/New_Kimbell_Empathy_2013.pdf

Nussbaum, B. (2011). Design thinking is a failed experiment. So what's next. *Fast Company*, 6, 97–105. https://www.fastcompany.com/1663558/design-thinking-is-a-failed-experiment-so-whats-next

Odom, W., Wakkary, R., Lim, Y. K., Desjardins, A., Hengeveld, B., & Banks, R. (2016). From research prototype to research product. *Conference on Human Factors in Computing Systems – Proceedings* (pp. 2549–2561). https://doi.org/10.1145/2858036.2858447

OpenAI. (2023). *ChatGPT 3.5* (3.5). https://chatgpt.com/

Raji, I. D., Gebru, T., Mitchell, M., Buolamwini, J., Lee, J., & Denton, E. (2020). Saving face: Investigating the ethical concerns of facial recognition auditing. *AIES 2020 – Proceedings of the AAAI/ACM Conference on AI, Ethics, and Society* (pp. 145–151). https://doi.org/10.1145/3375627.3375820

Sarantou, M., & Miettinen, S. (2022). *Empathy and business transformation*. Taylor & Francis.

Schmetkamp, S. (2020). Understanding A.I. – Can and should we empathize with robots? *Review of Philosophy and Psychology*, 11(4), 881–897. https://doi.org/10.1007/s13164-020-00473-x

Sevaldson, B. (2011). Nordic design research conference: GIGA-mapping: Visualization for complexity and system thinking in design. *Nordes*, 0(4). http://nordes.org/opj/index.php/n13/article/view/104/88

Sevaldson, B. (2018). Beyond user centric design. *OCAD University Open Research Repository*, 1, 11. http://openresearch.ocadu.ca/id/eprint/2755/1/Sevaldson_Beyond_2018.pdf

Suchman, L. (2006). *Human–machine reconfigurations*. Cambridge University Press. https://doi.org/10.1017/CBO9780511808418

Surma-aho, A., & Hölttä-Otto, K. (2022). Conceptualization and operationalization of empathy in design research. *Design Studies*, 78, 1–16. https://doi.org/10.1016/j.destud.2021.101075

Verganti, R., Vendraminelli, L., & Iansiti, M. (2020). Innovation and design in the age of artificial intelligence. *Journal of Product Innovation Management*, 37(3), 212–227. https://doi.org/10.1111/jpim.12523

Vink, J., & Oertzen, A.-S. (2018). Integrating empathy and lived experience through cocreation in service design. *ServDes2018 – Service Design Proof of Concept, June* (pp. 471–483). http://www.servdes.org/wp/wp-content/uploads/2018/07/37.pdf

Wakkary, R. (2021). *Things we could design*. The MIT Press. https://doi.org/10.7551/mitpress/13649.001.0001

Yüksel, N., Börklü, H. R., Sezer, H. K., & Canyurt, O. E. (2023). Review of artificial intelligence applications in engineering design perspective. *Engineering Applications of Artificial Intelligence*, 118, 105697. https://doi.org/10.1016/j.engappai.2022.105697

18

Empathetic AI:

Integrating Artificial Intelligence and Service Design for 3H Care in Singapore

Bo Gao, Di Wang, Robin Chan Chung Leung, Zhiwei Zeng, Benny Toh Hsiang Tan, Yang Qiu, Huiguo Zhang, Xinjia Yu, Jun Ji, Bo Huang, Eng Sing Lee, Hui Hwang Teong, Ray Tian Rui Lai, Wai Han Hoi and Chunyan Miao

Abstract

The integration of artificial intelligence (AI) and service design in healthcare presents opportunities for creating empathetic and user-centred digital services. This case study introduces the AI-assisted 3H care (A3C) system, which combines intelligent agents and service design methodologies to support personalized, predictive and empathetic 3H care in Singapore. 3H is collectively known as one of three chronic conditions (hyperglycaemia, hypertension and hyperlipidaemia). The A3C system framework, comprising four layers (user needs, data, intelligent agents and 'HAI touchpoints' (human-centred AI touchpoints)), enables dynamic adjustments to user states and incorporates AI within the information architecture. Pilot studies have demonstrated the system's usability and positive impact on users' health awareness and behaviours. This paper also discusses the evolving role of service designers in the era of AI, highlighting the need for collaboration with diverse stakeholders and the adoption of data-driven, user-centred and ecosystem-oriented approaches.

Introduction

In Singapore, the increasing prevalence of hyperglycemia, hypertension and hyperlipidemia (3H) is closely linked to a high-calorie diet and a fast-paced lifestyle (Mair et al., 2023; Ministry of Health, Singapore, 2021). Existing 3H management solutions have limitations, such as difficulty in early detection (Grant & Meigs, 2006), lack of individualized interventions (Roach & Marrero, 2005) and ineffective user engagement outside clinical settings (Karekla et al., 2019). The efficacy of interventions depends heavily on patients' self-management beyond hospitals and is influenced by the quality and accessibility of healthcare services provided by primary care settings (Johnson et al., 2014).

To address these challenges, this chapter presents the methodology developed in research and development (R&D) of the AI-assisted 3H care (A3C) system, which integrates AI and service design methodologies to complement primary care and promote healthy lifestyles in Singapore. The research problem is as follows: 'How can service design be effectively incorporated into AI-integrated healthcare to enhance the user experience and inclusivity?' The target population is Singapore residents aged 18 to 74 who either have or are at high risk of developing chronic conditions related to the '3H'. The A3C system framework comprises four layers – user needs, data, intelligent agent and 'HAI touchpoint' – encompassing different AI technology implementations and user empathy strategies. This case study reflects the evolving role of service designers in AI-driven healthcare service design. Through an analysis of this case study, this chapter also reflects on the evolving role of service designers in AI-driven healthcare service design.

Background

In Singapore, several mobile apps have been developed to facilitate 3H care services. For example, the Health Promotion Board (HPB) launched the 'Healthy 365' mobile application, which integrates fitness tracking devices to enable users to log their daily step count and active exercise duration. Another app, 'Health Buddy', developed by SingHealth, assists the public in managing clinic visits and accessing health information and support services. However, these digital service solutions may have limited effectiveness in addressing 3H chronic conditions, mainly because they do not provide relevant personalized interventions for 3H management. Moreover, some fitness and beauty apps commonly used by younger users incorporate functionalities for self-tracking and health behaviour modification. However, they often fail to address more complex chronic health issues (Whelan & Clohessy, 2020).

Researchers and designers have initiated research on how to predict and prevent 3H chronic conditions by leveraging mobile technologies to empower users to self-manage their health status (Logan, 2013; Marshall et al., 2007). These approaches mainly include self-tracking (Baron et al., 2017), conversational agents (Bickmore et al., 2018; Toro-Ramos et al., 2017) and health behavioural change (Ghorai et al., 2014). However, considering the diverse needs, preferences and emotions of end-users to achieve meaningful engagement, the problem of 3H, which is prevalent in large-scale populations, remains challenging (Cresswell et al., 2011).

Researchers have noted that user-centred design is a key factor in achieving user engagement and effectiveness (Birnbaum et al., 2015; McCurdie et al., 2012). Otherwise, it will result in low adherence rates (Karekla et al., 2019).

Empathetic AI in Healthcare

The integration of artificial intelligence (AI) into healthcare can revolutionize patient care, diagnosis and treatment (Poalelungi et al., 2023). However, it is crucial to ensure that AI systems are designed with empathy and compassion because these are fundamental values in patient-centred care (Kerasidou, 2020). Some researchers have emphasized the need for empathetic AI in healthcare, such as Morrow et al. (2022) and Shao (2023), who demonstrate that the integration of empathetic AI in healthcare is crucial for effective patient communication and supporting services. Although AI technologies have the potential to enhance compassion by improving empathetic awareness, response and communication skills in healthcare settings (Mason et al., 2023), there are challenges in how AI expresses empathy because of its nonhuman nature (Girju & Girju, 2022).

Despite the potential benefits of empathetic AI in healthcare, a critical approach is necessary to address the ethical and practical challenges associated with its implementation. One major concern is the potential for AI systems to perpetuate biases and disparities in healthcare (Char et al., 2018). To mitigate these risks, it is essential to ensure that AI algorithms are trained on diverse and representative datasets and that decision-making processes are transparent and explainable (Holtz et al., 2023). Additionally, the development of empathetic AI should be guided by ethical principles, such as respect for patient autonomy, privacy and confidentiality (Gerke et al., 2020).

Empathetic AI and Service Design

Service design has been widely recognized as a valuable approach in healthcare, with numerous studies demonstrating its effectiveness in improving healthcare services, particularly digital services. Freire and Sangiorgi (2012) explore the application of service design in healthcare and found that it can lead to more patient-centred, efficient and innovative services. Similarly, Paul and Glenn (2007) highlight the potential of experience-based codesign as a service design method in transforming healthcare services by involving patients and staff in the design process.

Despite the growing importance of AI in digital healthcare services, research on the human-centred perspective of creating digital services with AI is still lacking (Cruickshank & Trivedi, 2017). However, some researchers have demonstrated the potential of service design as a multidisciplinary approach to support AI development and implementation (Jylkäs, 2020; Rjsé et al., 2023). Jylkäs emphasizes the importance of necessitating the design of new service components within digital service journeys to effectively integrate AI technologies. In the context of healthcare facilities, Tokkuhisa and Morimoto (2023) develop a rapid service design approach based on behavioural change, introducing rapid ethnography and behavioural design techniques. Their work highlights the potential of service design in creating AI-driven healthcare services that are more responsive to user needs and behaviours. Furthermore, Li (2024) has adopted a research-for-service-design approach to investigate methods for aiding designers in designing AI-enabled solutions with high-quality user experience (UX) and responsible, transparent handling of ethical aspects. This research underscores the importance of considering both user-centred design principles and ethical implications when developing AI-powered healthcare services.

The holistic and systemic approach to service design (Sangiorgi, 2011) can help integrate empathetic AI into the broader healthcare service ecosystem by considering the complex interactions between users, technology and organizational contexts. Methods and tools, such as empathy mapping, user journey mapping and codesign workshops (Hwang & Kim, 2023), can gather end-user insights, identify pain points and generate innovative solutions for empathetic AI, facilitating the planning, implementation and evaluation of these solutions in healthcare services (Bailey et al., 2019). This stream of the literature has provided valuable insights into the potential of service design to help build empathetic AI in healthcare. However, there is a need for more empirical studies that validate theoretical frameworks.

A3C (AI-Assisted 3H Care) System

The double-diamond model, which is a staple of innovation and design projects, may require updates to adapt to changing work and thought patterns in the era of generative AI (GenAI) (Damgaard, 2024). The double-diamond model's emphasis on desirability over feasibility and viability can lead to impractical solutions. GenAI is revolutionizing the way we approach complex problem solving by enabling us to consider and prioritize multiple solutions simultaneously rather than relying solely on human thinking.

Researchers have proposed innovation models that align better with AI working modes. For instance, Jylkäs and Borek (2019) observe that the four common phases of 'discover', 'define', 'design' and 'develop' are insufficient for communicating the main activities in designing AI assistants. They propose a seven-phase model for the design of AI assistants consisting of ten process phases: 'discover', 'define', 'ideate', 'design', 'prototype', 'test', 'develop', 'implement', 'operate' and 'scale'. However, the integration of AI technologies, including GenAI, into various domains, such as healthcare, requires careful planning and consideration of potential algorithmic biases (Char et al., 2018; Obermeyer et al., 2019). Furthermore, achieving effectiveness in healthcare service interventions requires a comprehensive understanding of user needs and the integration of empathy into system design (Bazzano et al., 2017). Therefore, new models are needed to construct the collaboration and roles of humans and AI in the healthcare service innovation process.

A3C Built through Multiple-stakeholder Collaboration

During the R&D process of A3C (between 2021 and 2022), the limitations of the traditional double-diamond model were recognized because of pandemic constraints. A3C emphasizes the collaboration of AI framework development involving multiple stakeholders. Small-sample user research was conducted following service design methodologies to develop the framework according to the system architecture. In this context, system architecture refers to the high-level structure and organization of the system, including its components and the principles governing its design (Bass et al., 2012). The system architecture encompasses the integration of AI algorithms, digital platforms and user interfaces to provide effective 3H care.

The AI-assisted 3H care (A3C) system presented in this chapter is the result of collaboration among multiple stakeholders. Experts from diverse fields, including computer science (Joint NTU-UBC Research Centre of Excellence in Active Living for the Elderly, Nanyang Technological University), service design, sociology, medicine (National Healthcare Group Polyclinics, Tan Tock Seng Hospital, Woodlands Health, National Healthcare Group) and community (Woodlands Health, National Healthcare Group and Tan Tock Seng Hospital), collaborated closely to ensure the comprehensive and holistic goal of providing effective 3H care. Computer scientists have contributed technical expertise in developing AI algorithms and integrating them into digital platforms. Designers have developed user-centred approaches and methods. The healthcare professionals provided domain knowledge, clinical insights and an ensuring solution that aligned with evidence-based practices and addressed real-world challenges.

Moreover, the interdisciplinary research team closely collaborated with the focus group of a small-sample user base throughout the research, design and development process, conducting surveys, observations and cocreation sessions. The focus group included eleven participants who represented diverse ethnic backgrounds in Singapore and ranged in age from 20 to 75 years old (mean = 47.5). This diverse representation covered individuals from young, middle-aged and elderly demographics and closely resembled Singapore's population distribution of 76.2 per cent Chinese, 15 per cent Malay, 7.4 per cent Indian and 1.4 per cent from other ethnic backgrounds (Ang et al., 2017).

A3C (AI-Assisted 3H Care) System

The A3C system framework combines intelligent agents and service design to enable dynamic, real-time adjustments to user states by incorporating AI into the information architecture (Rosenfeld et al., 2015). The system comprises four layers: user needs, data, intelligent agents and 'HAI touchpoints' (human-centred AI touchpoints).

The user needs layer forms the foundation, focusing on understanding the user requirements identified from focus group participants. The data management subsystem (DMS) in the data layer functions as the backend database API. The AI subsystem (AIS) within the intelligent agent layer comprises explainable AI and curious AI, featuring assessment and intervention engine subsystems. Intelligent agents are vital for providing high-quality care (Wang et al., 2014) and improving user experiences (Sharma & Ahlawat, 2022).

Intervention plans are sent to the 'HAI touchpoints' layer, seamlessly integrating AI agents and service design touchpoints within the dynamic user journey, encompassing food analysis, dietary recommendations, image recognition and gamification. The A3C system captures comprehensive behavioural data and diet intake history through the 'HAI touchpoints', storing this information in the backend and enriching it with data to train health prediction models within the DMS (Figure 18.1).

Intelligent agents facilitate an eco-service system (frontend and backend integration) in A3C. The eco-service system comprises three main components: the Akeso mobile app and the Health Screening Kiosk for end-users and the hospital dashboard for medical personnel, which places a strong emphasis on usability and inclusivity. By offering solutions through smartphones and self-assessment kiosks, the system ensures that users with diverse preferences and varying levels of technological literacy can effectively engage with the system.

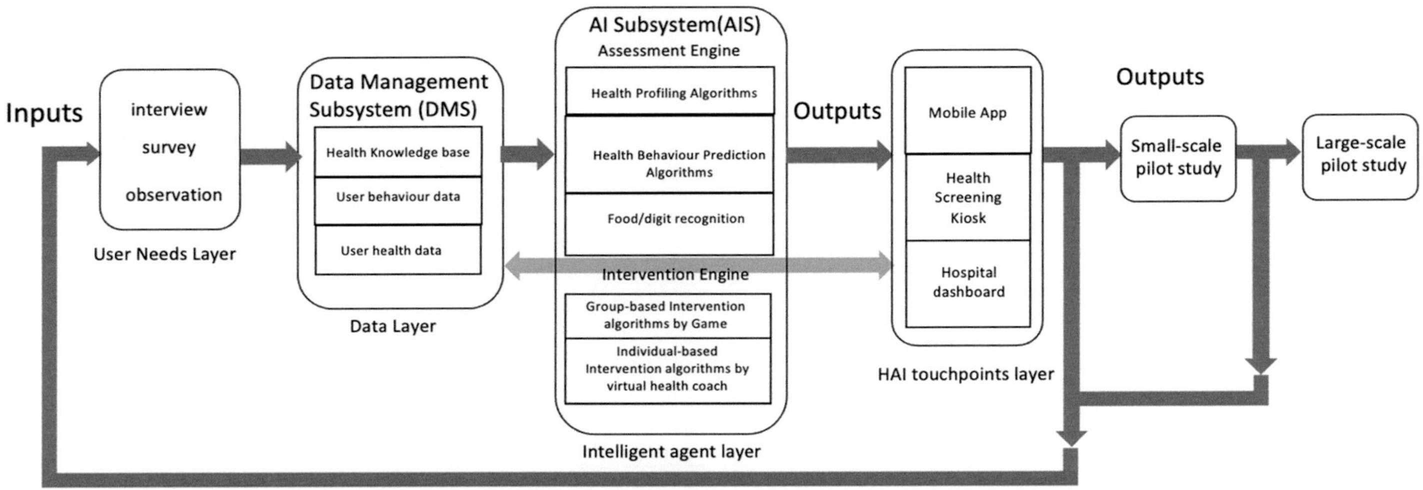

FIGURE 18.1 *Framework of A3C in R&D processes.*

The Akeso mobile app features five tabs: Today (collecting daily data), Akeso (providing personalized recommendations), Game (a farm-themed virtual game), History (visualizing historical performance) and Schedule (displaying tasks and reminders) (Figure 18.2). The Game tab includes tutorials and HELP content to support beginner and elderly users.

The Health Screening Kiosk, which is located in community centres, offers a user-friendly platform for self-assessments, including blood pressure, blood glucose and weight measurements (Figure 18.3). Users can access their personal accounts on the kiosk by scanning a QR code generated by the Akeso mobile app. The kiosk interface is designed with inclusivity, ensuring convenience and safety for users who may require seating, such as individuals in wheelchairs or elderly people. Instructional videos are provided to facilitate user operation, even without staff assistance.

The Akeso app incorporates AI-driven touchpoints, or 'HAI touchpoints', which adapt interventions based on changes in user behaviour while aligning with individual preferences (Jylkäs & Borek, 2019). The app's AI-assisted food recognition function simplifies the assessment of nutritional content and enables the subsequent promotion of healthy eating habits.

FIGURE 18.2 *The Akeso mobile app.*

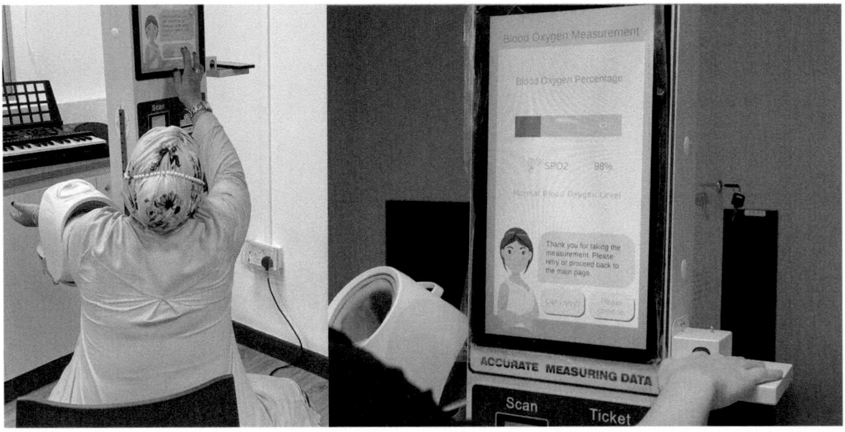

FIGURE 18.3 *The Health Screening Kiosk delivers the service with usability and inclusivity.*

By collecting and labelling over 130,000 images of local Singaporean food, the app achieved high recognition accuracy rates through machine-learning techniques. Users can upload photos of their meals, enabling the AI subsystem (AIS) to provide personalized dietary recommendations based on food categorization, which is in line with the Health Promotion Board's guidelines (HPB, 2014).

The app also employs digit recognition technology, allowing users to upload photos from various 3H-related health measurement devices. By collecting and labelling over 20,000 measurement images, the app achieved high accuracy rates. The app incorporates gamification elements to provide group-based, adaptive, long-term interventions (Johnson et al., 2016). Users' daily performance in achieving recommended lifestyle goals is translated into elements of a farm-themed virtual game, which integrates belongingness theory and gamification elements to create an interactive virtual world with social functions (Figure 18.4).

We designed a virtual health coach, Akeso, to incorporate two primary aspects of empathetic AI. First, the intelligent agent is imbued with a distinct form, personality and backstory, which are integrated into the app and kiosk interfaces. Second, AI technology is employed to facilitate empathic concern and help.

To deliver a positive effect to users, intelligent agents require human likeness, self-representation and personality (Nicenboim et al., 2020). The virtual health coach is presented as an avatar named Akeso. Akeso serves as a convenient and intelligent touch point within a complex service ecosystem (Jylkäs & Borek, 2019). In conversational agents, empathy can be attributed to a range of behaviours, from affective matching to empathic concern and altruistic helping (Coplan & Goldie, 2012; de Waal & Preston, 2017). The

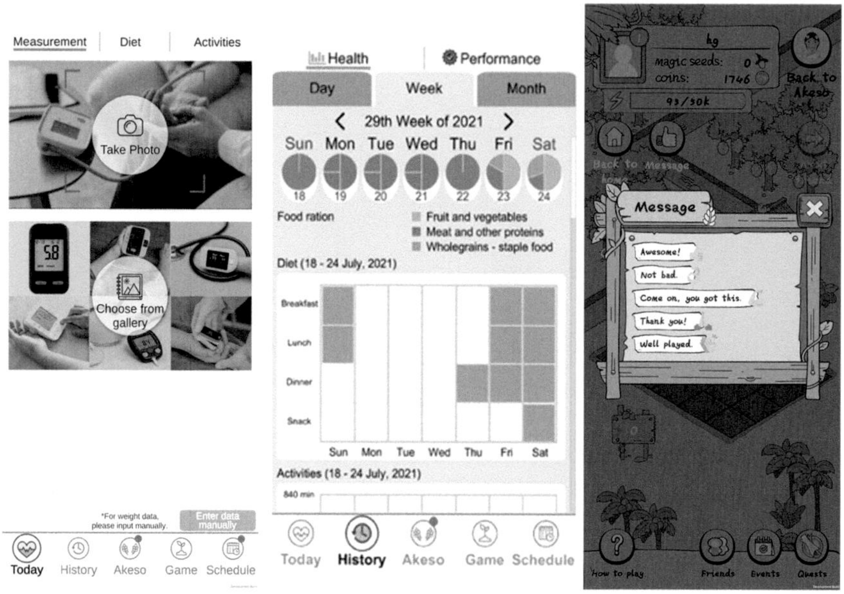

FIGURE 18.4 *The AI-assisted recognition and the farm-themed virtual game in the app.*

virtual health coach (Figure 18.5) demonstrates these behaviours through the following functionalities:

1 Altruistic helping:
   - Helping users maintain health goals: Akeso sends notifications to users, reminding them to upload food data at eight o'clock, twelve o'clock and eighteen o'clock every day.
   - Assisting independent operations in kiosks: As a virtual guide in self-service kiosks, Akeso uses a gentle voice and friendly image to help users complete tasks independently and aid them when needed.

2 Empathic concern:
   - Encouraging user engagement through personalized recommendations: Powered by a curious AI–based recommendation system, Akeso proactively sends customized warnings and suggestions to users. These recommendations are formulated based on personalized insights and reminders provided by individual-based intervention algorithms, which are seamlessly integrated into chatbots with conversational agents.

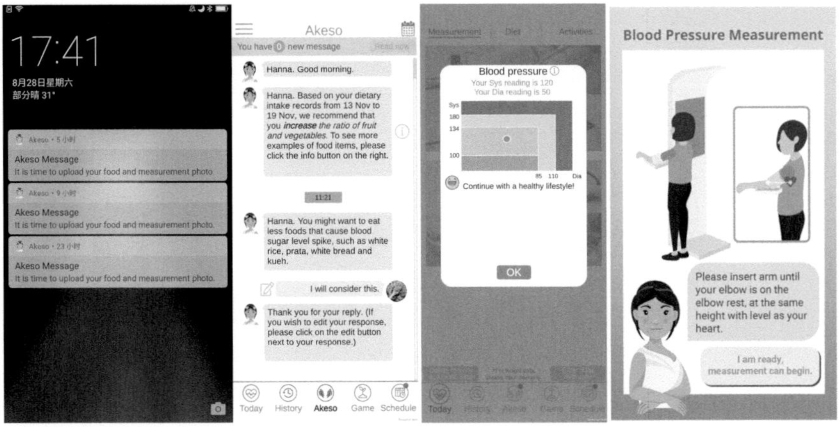

FIGURE 18.5 *Virtual health coach in the app and kiosks.*

- Fast response mechanisms: Akeso not only achieves fast responses in conversational chatbots, but it also employs positive, encouraging and polite language.

Findings

Pilot Study

A pilot study involving thirty participants aged 20 to 35 years old (60 per cent male, 40 per cent female) who used the Akeso app for one month demonstrated above-average usability (SUS score: 70.2). Questionnaires and semi-structured interviews revealed improved health awareness, including dietary changes, increased physical activity and weight loss. The participants appreciated the app's features, such as tips and food recognition, finding the suggestions helpful for learning health knowledge, even among those who were already in good health.

A large-scale pilot study involving 329 participants from multiple healthcare institutions in Singapore evaluated user experience and perceptions of the Akeso app and Health Screening Kiosk (HSK). The study employed a mixed methods approach, combining an online survey with semi-structured interviews. An average system usability scale (SUS) score of 72.5 indicated above-average usability. The diet module received the most positive feedback, with 26.1 per cent of participants finding it helpful for recording and monitoring their diet and increasing awareness of healthy eating habits. The HSK module was also perceived as helpful for understanding and monitoring health indicators. Over 40 per cent of participants indicated that they would be likely to recommend the app to their friends, primarily because of its ability to help monitor health and foster good lifestyle habits. The main areas for improvement include simplifying the food input process, expanding localized food options and enhancing the accuracy of food recognition AI.

Reflections on the Evolving Role of Service Designers through the Case Study

In the era of AI, the role of service designers is undergoing a significant transformation. Traditionally, service designers have been the primary drivers of the design process, relying on user research, feedback and expert walkthroughs to identify problems and improve user experiences. However, with the disruptive evolution of AI, service designers are now transitioning into the facilitators of a cocreative process involving larger-scale diverse stakeholders. This new role requires service designers to possess a unique combination of technical skills, computational thinking and stakeholder collaboration abilities. They act as translators between humans and AI and need to work closely with computer science professionals throughout the R&D process, adapting design methods to align with AI technological requirements.

Service designers collaborate with computer science professionals to align user needs with AI technology implementation. They analyse user behavioural data to gain insights, inform design decisions and develop actionable strategies. Service designers translate user needs into design strategies, bridging the gap between user requirements and technical implementation. This requires analysing user data, identifying patterns and trends and developing data-driven design recommendations. To objectively evaluate and validate design effectiveness, service designers employ methods such as surveys, statistical analysis, A/B testing and quantitative metrics. These methods provide a rigorous, scientific approach to measuring the value and impact of service design outcomes, moving beyond relying on experiential judgement and qualitative data alone.

The evolving role of service designers in the AI era demands an expansion of their work scope. Service designers must advocate for responsible AI practices and ensure the integration of ethical AI solutions to prevent potential AI-related risks, such as privacy concerns, bias, religious sensitivities and political sensitivity considerations, into the design process. This involves assisting AI systems in setting precautions to mitigate and ensuring algorithm transparency, thereby safeguarding users' safety and dignity. Service designers need to strike a balance between leveraging the power of AI technologies and maintaining a human-centred focus, ensuring that the resulting services are not only technologically advanced but also empathetic, inclusive and truly valuable to the users.

Conclusion

This chapter has presented the AI-assisted 3H care (A3C) system, which integrates AI and service design to address 3H management challenges in Singapore. The case study has demonstrated that the four-layered framework of the A3C system enables personalized, predictive and empathetic digital healthcare services. However, limitations include the need for improved food recognition accuracy, localized options and simplified input processes. Future research will consider ethical factors and enhance virtual assistants' empathetic capabilities through service design, affective computing and natural language processing advancements. We will strive to mitigate potential risks, such as misinterpretation of emotions, emotional manipulation and gender stereotypes in virtual assistants.

Takeaway

This case study introduces the AI-assisted 3H care (A3C) system, which integrates AI and service design to provide personalized, predictive and empathetic care for three chronic conditions (hyperglycaemia, hypertension and hyperlipidaemia) in Singapore. The study also reflects on the evolving role of service designers in AI-driven healthcare services, emphasizing the need for collaboration with diverse stakeholders and the adoption of data-driven, user-centred approaches.

Acknowledgements

This research has been supported, in part, by the National Research Foundation (NRF), Singapore, under its AI Singapore Programme (AISG Award No: AISG-GC-2019-003). Any opinions, findings, conclusions or recommendations expressed in this material are those of the authors and do not reflect the views of the National Research Foundation, Singapore.

References

Ang, L. W., Kam, Y. W., Lin, C., Krishnan, P. U., Tay, J., Ng, L. C., James, L., Lee, V. J. M., Goh, K. T., Ng, L. F. P., & Lin, R. T. P. (2017). Seroprevalence of antibodies against chikungunya virus in Singapore resident adult population. *PLOS Neglected Tropical Diseases*, 11(12). https://doi.org/10.1371/journal.pntd.0006163

Bailey, S. G., Bell, K. L., & Hartung, H. (2019). Service design empowering innovative communities within healthcare: Applying service design visualization to facilitate collaborative sense-making and shared decision-making in hospitals. In M. A. Pfannstiel & C. Rasche (Eds.), *Service design and service thinking in healthcare and hospital management* (pp. 127–138). Springer.

Baron, J. S., Hirani, S., & Newman, S. P. (2017). A randomized, controlled trial of the effects of a mobile telehealth intervention on clinical and patient-reported outcomes in people with poorly controlled diabetes. *Journal of Telemedicine and Telecare*, 23(2), 207–216.

Bass, L., Clements, P., & Kazman, R. (2012). *Software architecture in practice* (3rd ed.). Addison-Wesley Professional.

Bazzano, A. N., Martin, J., Hicks, E., Faughnan, M., & Murphy, L. (2017). Human-centred design in global health: A scoping review of applications and contexts. *PloS One*, 12(11). https://doi.org/10.1371/journal.pone.0186744

Bickmore, T. W., Kimani, E., Trinh, H., Pusateri, A., Paasche-Orlow, M. K., & Magnani, J. W. (2018). Managing chronic conditions with a smartphone-based conversational virtual agent. In *Proceedings of the 18th International Conference on Intelligent Virtual Agents* (pp. 119–124). Association for Computing Machinery.

Birnbaum, F., Lewis, D., Rosen, R. K., & Ranney, M. L. (2015). Patient engagement and the design of digital health. *Academic Emergency Medicine*, 22(6), 754–756.

Char, D. S., Shah, N. H., & Magnus, D. (2018). Implementing machine learning in health care—Addressing ethical challenges. *New England Journal of Medicine*, 378(11), 981–983.

Coplan, A., & Goldie, P. (2012). *Empathy: Philosophical and psychological perspectives*. Oxford University Press.

Cresswell, K., Morrison, Z., Crowe, S., Robertson, A., & Sheikh, A. (2011). Anything but engaged: User involvement in the context of a national electronic health record implementation. *Informatics in Primary Care*, 19(4), 191–206.

Cruickshank, L., & Trivedi, N. (2017). When your toaster is a client, how do you design? Going beyond human centred design. *The Design Journal*, 20(sup1), S4158–S4170.

Damgaard, S. (2024). *The AI-powered 'Stingray' model for innovation*. Board of Innovation. https://www.boardofinnovation.com/blog/the-ai-powered-stingray-model-innovation/

de Waal, F. B. M., & Preston, S. D. (2017). Mammalian empathy: Behavioural manifestations and neural basis. *Nature Reviews Neuroscience*, 18(8), 498–509.

Freire, K., & Sangiorgi, D. (2012). Service design and healthcare innovation: From consumption to coproduction and cocreation. Paper presented at *2010 Nordic Conference on Service Design and Service Innovation*, Linkoping, Sweden.

Gerke, S., Minssen, T., & Cohen, G. (2020). Ethical and legal challenges of artificial intelligence-driven healthcare. In *Artificial Intelligence in Healthcare* (pp. 295–336). Academic Press.

Ghorai, K., Jahan, S., Ray, P., & Chylinski, M. (2014). mHealth for behaviour change: Role of a smartphone-based multi-intervention service for hypertension and diabetes in Bangladesh. *International Journal of Biomedical Engineering and Technology*, 16(2), 135–155.

Girju, R., & Girju, M. (2022). Design considerations for an NLP-driven empathy and emotion interface for clinician training via telemedicine. In *Proceedings of the Second Workshop on Bridging Human–Computer Interaction and Natural Language Processing* (pp. 21–27). Association for Computational Linguistics.

Grant, R. W., & Meigs, J. B. (2006). Overcoming barriers to evidence-based diabetes care. *Current Diabetes Reviews*, 2(2), 261–269.

Health Promotion Board. (2014). Nutritious foods for a healthy diet. Retrieved June 18, 2024, from https://www.healthhub.sg/programmes/nutrition-hub/eat-more

Holtz, B., Nelson, V., & Poropatich, R. K. (2023). Artificial intelligence in health: Enhancing a return to patient-centred communication. *Telemedicine Journal and E-Health: The Official Journal of the American Telemedicine Association*, 29(6), 795–797.

Hwang, S., & Kim, D. (2023). JourneyBot: Designing a chatbot-driven interactive visualization tool for design research. *International Journal of Design*, 17(3), 95–110.

Johnson, D., Deterding, S., Kuhn, K.-A., Staneva, A., Stoyanov, S., & Hides, L. (2016). Gamification for health and wellbeing: A systematic review of the literature. *Internet Interventions*, 6, 89–106.

Johnson, J. A., Al Sayah, F., Wozniak, L., Rees, S., Soprovich, A., Qiu, W., Chik, C. L., Chue, P., Florence, P., Jacquier, J., Lysak, P., Opgenorth, A., Katon, W., & Majumdar, S. R. (2014). Collaborative care versus screening and follow-up for patients with diabetes and depressive symptoms: Results of a primary care-based comparative effectiveness trial. *Diabetes Care*, 37(12), 3220–3226.

Jylkäs, T. (2020). SHARED PATH service design and artificial intelligence in designing human-centred digital services [Doctoral dissertation, University of Lapland].

Jylkäs, T., & Borek, A. (2019). Designing with artificial intelligence: AI assistants as a gateway to complex service ecosystems. In S. Miettinen & M. Sarantou (Eds.), *Managing complexity and creating innovation through design* (pp. 79–88). Routledge.

Karekla, M., Kasinopoulos, O., Neto, D. D., Ebert, D. D., Van Daele, T., Nordgreen, T., Höfer, S., Oeverland, S., & Jensen, K. L. (2019). Best practices and recommendations for digital interventions to improve engagement and adherence in chronic illness sufferers. *European Psychologist*, 24(1), 49–67.

Kerasidou, A. (2020). Artificial intelligence and the ongoing need for empathy, compassion and trust in healthcare. *Bulletin of the World Health Organization*, 98(4), 245–250.

Li, F. (2024). *UX & ethics matter! Designing for Human-AI Interaction for individuals lacking AI literacy* [Doctoral dissertation, Eindhoven University of Technology].

Logan, A. G. (2013). Transforming hypertension management using mobile health technology for telemonitoring and self-care support. *Canadian Journal of Cardiology*, 29(5), 579–585.

Mair, J. L., Castro, O., Salamanca-Sanabria, A., Frese, B. F., von Wangenheim, F., Tai, E. S., Kowatsch, T., & Müller-Riemenschneider, F. (2023). Exploring the potential of mobile health interventions to address behavioural risk factors for the prevention of noncommunicable diseases in Asian populations: A qualitative study. *BMC Public Health*, 23(1), 753.

Marshall, A., Medvedev, O., & Markarian, G. (2007). Self management of chronic disease using mobile devices and Bluetooth monitors. In *Proceedings of the Second International Conference on Body Area Networks*. ICST (Institute for Computer Sciences, Social-Informatics and Telecommunications Engineering). Brussels, Article 22, 1–4.

Mason, E., McLoughlin, F., Pagliari, C., Sorell, T., & Wilson, P. (2023). Artificial intelligence technologies and compassion in healthcare: A systematic scoping review. *Frontiers in Psychology*, 13, 971044. https://doi.org/10.3389/fpsyg.2022.971044

McCurdie, T., Taneva, S., Casselman, M., Yeung, M., McDaniel, C., Ho, W., & Cafazzo, J. (2012). mHealth consumer apps: The case for user-centered design. *Biomedical Instrumentation & Technology*, Suppl, 49–56. Biomedical Instrumentation & Technology.

Ministry of Health, Singapore. (2021). *National Population Health Survey, 2021*. Retrieved June 18, 2024 from https://hpb.gov.sg/community/national-population-health-survey.

Morrow, E., Zidaru, T., Ross, F., Mason, C., Patel, K. D., Ream, M., & Stockley, R. (2022). Artificial intelligence technologies and compassion in healthcare: A systematic scoping review. *Frontiers in Psychology*, 13, 971044.

Nicenboim, I., Giaccardi, E., Søndergaard, M. L. J., Reddy, A. V., Strengers, Y., Pierce, J., & Redström, J. (2020). More-than-human design and AI: In conversation with agents. In *Companion Publication of the 2020 ACM Designing Interactive Systems Conference* (pp. 397–400).

Obermeyer, Z., Powers, B., Vogeli, C., & Mullainathan, S. (2019). Dissecting racial bias in an algorithm used to manage the health of populations. *Science*, 366(6464), 447–453.

Paul, B., & Glenn, R. (2007). *Bringing user experience to healthcare improvement: The concepts, methods and practices of experience-based design*. Routledge & CRC Press.

Poalelungi, D. G., Musat, C. L., Fulga, A., Neagu, M., Neagu, A. I., Piraianu, A. I., & Fulga, I. (2023). Advancing patient care: How artificial intelligence is transforming healthcare. *Journal of Personalized Medicine*, 13(8), 1214.

Rjsé, V., Sadeq, T., & Miettinen, S. (2023). AI service model for an airline ecosystem: A systemic design and thematic approach to service-dominant logic: Examining state-of-the-art technologies for service centricity. In U. Z. Abdul Hamid & M. Suoheimo (Eds.), *Service design for emerging technologies product development: Bridging the interdisciplinary knowledge gap* (pp. 117–139). Springer.

Roach, P., & Marrero, D. (2005). A critical dialogue: Communicating with type 2 diabetes patients about cardiovascular risk. *Vascular Health and Risk Management*, 1(4), 301–307.

Rosenfeld, L., Morville, P., & Arango, J. (2015). *Information architecture: For the web and beyond* (4th ed.). O'Reilly Media.

Sangiorgi, D. (2011). Transformative services and transformation design. *International Journal of Design*, 5(2), 29–40.

Shao, R. (2023). An empathetic AI for mental health intervention: Conceptualizing and examining artificial empathy. In *Proceedings of the 2nd Empathy-Centric Design Workshop* (pp. 1–6). Association for Computing Machinery.

Sharma, S., & Ahlawat, A. (2022). Architecture and types of intelligent agent and uses of various technologies. In *2022 3rd International Conference on Issues and Challenges in Intelligent Computing Techniques* (pp. 1–5). IEEE.

Tokkuhisa, S., & Morimoto, T. (2023). Rapid service design method for services with robots in healthcare facilities under Covid-19. In *ServDes.2023 Entanglements & Flows Conference: Service Encounters and Meanings Proceedings* (pp. 1216–1240).

Toro-Ramos, T., Kim, Y., Wood, M., Rajda, J., Niejadlik, K., Honcz, J., Marrero, D., Fawer, A., & Michaelides, A. (2017). Efficacy of a mobile hypertension prevention delivery platform with human coaching. *Journal of Human Hypertension*, 31(12), 795–800.

Wang, D., Subagdja, B., Kang, Y., Tan, A.-H., & Zhang, D. (2014). Towards intelligent caring agents for ageing-in-place: Issues and challenges. In *2014 IEEE Symposium on Computational Intelligence for Human-like Intelligence* (CIHLI).

Whelan, E., & Clohessy, T. (2020). How the social dimension of fitness apps can enhance and undermine wellbeing: A dual model of passion perspective. *Information Technology & People*, 34(1), 68–92.

19

Demonstrating a Two-Method Experiment:

The Use of a Digital Tool and Empathy Map in Designing Empathic Phygital Experiences in a Peripheral Destination

Päivi Hanni-Vaara, Satu Miettinen and Minni Haanpää

Abstract

This qualitative case study demonstrates how a two-method experiment was conducted using a digital tool – a head-mounted mobile eye-tracking device – and an empathy map to explore the visitor's phygital experience of a smartphone-mediated service. The visitor experience is phygital – a blend of physical and digital where the boundaries between physical and digital are eliminated. The experiment was conducted in cocreation with two stakeholders: a nonprofit local heritage museum and the other being a higher tourism education institution. The experiment acted as a testbed using a two-method data collection combination to join visual and audio as visible and 'think aloud' mechanisms. The study provides practical insights and proposals on how to execute such an experiment. An empathy-oriented design thinking approach was chosen for the research. This enabled the study of two dimensions of cognitive empathy. First, simulative empathy was internalized to empathize with the museum's spatiotemporal environment when executing the experiment. This was followed by projective empathy when analysing the audiovisual data on the visitors' phygital experiences. As a result, a two-method data collection approach was introduced as a proof of concept for a technology-aided solution. Although the findings and implications cannot be generalized, the study provides novelty in terms of using a digital tool and empathy map to implement simulative and projective empathy in design thinking.

Introduction

The present empirical case study describes how a two-method data collection experiment was conducted using a combination of a head-mounted mobile eye-tracking device and an empathy map to explore museum visitors' phygital, a blend of physical and digital, experiences. Here, phygital experience reflects a 'local heritage smartphone-mediated service experience continuum' where the boundaries between physical and digital intermesh (Lupo, 2021; Mieli, 2022; Zillinger, 2021).

The methodology was based on research through design to explore a real-life context and form comprehensive insights into implementing an experiment in practice (Redström, 2021). Research through the design process can aim to implement a solution that can be tested and evaluated during the experiment procedure. The current study aimed to experiment with and process a two-method data collection approach and introduce a technology-aided solution as a proof of concept (Zimmerman et al., 2010). In this case study, a higher tourism education institution was willing to organize an experiment as a testbed to collect data to join visual and audio data as visible and 'think aloud' mechanisms. Simultaneously, a nonprofit local heritage museum required insights into its recently developed smartphone-mediated service. This experiment implies a continuum of focusing first on two-method data collection, which is followed by data analysis of the phygital experience (Jobin et al., 2021; Solima & Izzo, 2017). Naturally, this requires stakeholders to empathize with the museum's spatiotemporal environment, its visitors and cocreating stakeholders (Alhonsuo & Soto Hormazábal, 2022; Vink et al., 2021).

This study implemented an empathy-oriented design thinking approach to demonstrate a two-method experiment with five process phases starting with empathizing with the stakeholders and the research space (Beckman & Barry, 2007; Jylkäs et al., 2019; Lee & Park, 2021). This was followed by defining the profound value of cocreating the experiment which was ideated further to include resources, such as framing the roles and responsibilities. The prototyping phase activated the stakeholders to pre-testing. The iteration actions guided the stakeholders to learn the experiment practices when repeating the iteration cycles for the final testing phase (Lee & Park, 2021). Though the phases are presented here as linear, they undergo linear and nonlinear oscillations which is typical in design thinking (Alhonsuo & Soto Hormazábal, 2022; Lee & Park, 2021). Design thinking is well described by Dorst (2011), who suggests that experienced designers have fundamental reasoning patterns based on design and use the core design practices of framing and frame creation. When creating frames, designers emerge in the analytical process. Alongside this, it is essential to understand the broader problem to frame a novel standpoint that can create solutions to the situation. Practically, a proof of concept procures 'an attendant proof of principle', demonstrating the decision-making processes between the stakeholders of devices used in problem-solving practices (Kendig, 2016).

The present study focused on two cognitively elaborated empathies: simulative and projective empathy which provided novelty in the research (Aaltola & Keto, 2018; Ramirez, 2017). In cognitive empathy, one distinguishes and concludes cognitions, such as the feelings and thoughts of the other (Aaltola & Keto, 2018). Regarding simulative empathy, researchers provided integrated insights imitating museums' spatiotemporal environment while the projective empathy was used when analysing museum visitors' phygital experiences (Ramirez, 2017). The significance of empathizing is the ability to react to others' states. Empathy, as the primary sense of morality, helps us form insights into others and maintain social relationships that are valuable in cocreation (Aaltola & Keto, 2018; Hanni-Vaara, 2022). Here, morality requires openness to dissimilarity to recognize value and equality in all those phenomena differing from our needs (Aaltola & Keto, 2018).

Context and Research Space

The present study examined a two-method data collection experiment in a nonprofit local heritage museum at a peripheral destination. Instead of naming the cocreating stakeholders, the current case study concentrates on describing the objectives and design process. This local heritage museum was founded by an association operated by volunteers. The museum engages in cocreation projects with other regional organizations when designing and developing new services to meet the needs of its visitors, for example. The association's mission emphasizes networking to sustain the local heritage and traditions. Its principles and practices value activity, cocreation and openness with community spirit, joint responsibility, trust and equality. Recently, the association has challenged its volunteers to pay attention to the changing visitor base. To respond to this challenge, the volunteers developed a new smartphone-mediated service. Hence, they addressed the needs of the changing visitor base, such as new generations of locals and tourists.

Two-Method Combination: The Eye-Tracking and Empathy Map

The research utilized a mobile head-mounted eye-tracking device, the Tobii™ (Pro Glasses 2 version, Tobii™, Stockholm, Sweden), to collect the research

What do you hear? What affects your decision?	**What do you think and feel?** What is the most important for you?
What do you say and do?	**What do you see?** How does the environment look? What are you offered?
Pain points What frustrates you? Do you avoid anything? Is something hindering you?	**Gain Points** What do you wish to gain? How do you evaluate your success? How do you achieve your objective?

FIGURE 19.1 *The questions of the empathy map prompted the research participants to 'think aloud' their feelings and thoughts during the phygital experience (Ferreira et al., 2015).*

data (Tobii, 2024). The device comprised wearable smart glasses, integrated microphones and used a tablet to follow the research participant to record the data (Tobii, 2024). Based on sensor technology, the eye-tracking device captured the visual information of participants' eye pupil movements. The visual movements of eye fixations formed gaze points, patterns and saccades in the area the participant was looking at (Jung et al., 2020; Rainoldi & Jooss, 2020). The recorded data correlate with a person's cognitive processes, such as attention and perception during the experience (Jung et al., 2020; Rainoldi et al., 2020). These cognitions explain the effect of a person's physical and digital needs constructing the phygital experience (Batat, 2022).

An empathy map was integrated alongside the visual attention data to support the data with audio (Hanni-Vaara et al., 2024). The aim of audio as a 'think aloud' mechanism was to expose aloud what the research participants saw, heard, thought and did during the experiment (Ferreira et al., 2015). As shown in Figure 19.1, the research participants were asked to express their thoughts and feelings aloud by answering the following questions:

- What do you think and feel? What is important to you?
- What do you hear? What affects your decision?
- What do you see? What does the environment look like?
- What do you say and do? (Ferreira et al., 2015)

In addition, the empathy map allowed the participants to express their gains and pain points during the experience. The gains activated participants to express their successes, while the pain points described their failures, such as frustration. The spoken words expressed content and sound when the tone in their voices articulated feelings. As a result, the audiovisual data of the participants' phygital experiences comprised 85 minutes of video material.

Towards Empathy-Oriented Design Thinking

An empathy-oriented design thinking approach (Lee & Park, 2021) was implemented to explain a two-method data collection experiment as a proof of concept. The method's five phases proceed from empathizing to defining, ideating, prototyping and testing the experiment (Lee & Park, 2021). Design thinking first commenced with empathizing to react to the stakeholders' feelings and thoughts concerning the experiment and its spatiotemporal environment. This was followed by a defining phase, in which the stakeholders sought a profound understanding of the experimental context through open discussions and interaction. The third phase involved ideation, in which the stakeholders ideated the data collection practices. The fourth phase included practical prototyping iterations to prepare stakeholders for the testing phase to meet the fifth phase, which was a testing day.

The empathy-oriented design thinking process is typically described as linear though it includes hermeneutic circles which oscillate based on growing knowledge throughout the design process (Aaltola & Keto, 2018; Jobin et al., 2021; Neto et al., 2020). Instead of expressing their ultimate objective for the experiment, the stakeholders agreed to concentrate on each design process phase to learn and redesign it when needed (Aho, 2023).

As shown in Figure 19.2, the study used two dimensions of cognitive empathy in the experiment: simulative and projective empathy.

The variable separating simulative empathy from projective empathy is 'me' or 'I' (Ramirez, 2017). In simulative empathy, one imitates or models the perceptions of empathized objects, while in projective

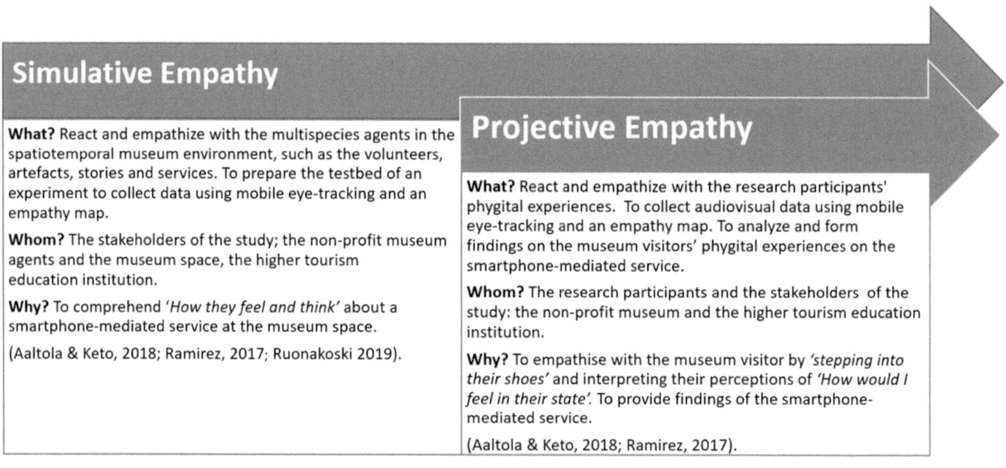

FIGURE 19.2 *The novelty of the study was the implementation of two dimensions of cognitive empathy: simulative and projective (visualization by Hanni-Vaara et al., 2024).*

empathy, one reacts to the other's feelings through one's own experiences (Aaltola & Keto, 2018; Ramirez, 2017). The design process started using simulative empathy by imitating and modelling insights into the museum's spatiotemporal environment, such as perceptions of the museum volunteers and the environment. Later, researchers used projective empathy to analyse the audiovisual data of the research participants' (museum visitors) phygital experiences.

Commencing with Empathy

The first discussions of the experiment were executed six months before the actual testing. The stakeholders had a relationship for several years enabling them to design and implement development in their daily operations. The cocreation model is typical in micro- and small-scale organizations that do not separate their development work from their everyday practices (Garcia-Rosell et al., 2019).

Naturally, former cocreation between the stakeholders and gained knowledge was valuable when forming new insights into the present situation. For example, social relationships were easier to form and strengthen with pre-familiar stakeholders, and previous experiences with cocreation practices generated trust between stakeholders. The values of the cocreation process included equality, human-centredness, others' expertise and a low hierarchy supporting dynamic relationships, mutual respect and openness throughout the experiment (Adner, 2017).

As mentioned, the design process started using simulative empathy which aimed to internalize the museum's spatiotemporal environment in questioning 'how they feel and think' (Aaltola & Keto, 2018; Ramirez, 2017; Ruonakoski, 2019). This created a basis for designing the experiment as a testbed to form insights into perceptions of the smartphone-mediated service. Practically, the simulative empathy phase covered museum stakeholders' stories by replicating and re-experiencing their states through discussions and interactions. Moreover, the museum premises and artefacts affected the reactions in forming insights into the space.

Remarkably, the insight into the museum space changed throughout the experimental phases based on the growth of learning and knowledge. In addition, the temporal moments and subjective interpretations of simulative empathy directed the researchers to redesign the experiment (Aaltola & Keto, 2018; Ramirez, 2017).

Defining the Objectives of the Experiment

To the museum stakeholders, the higher tourism education institution suggested an innovative two-method experiment using a head-mounted mobile eye-tracking device and empathy map to collect data on museum

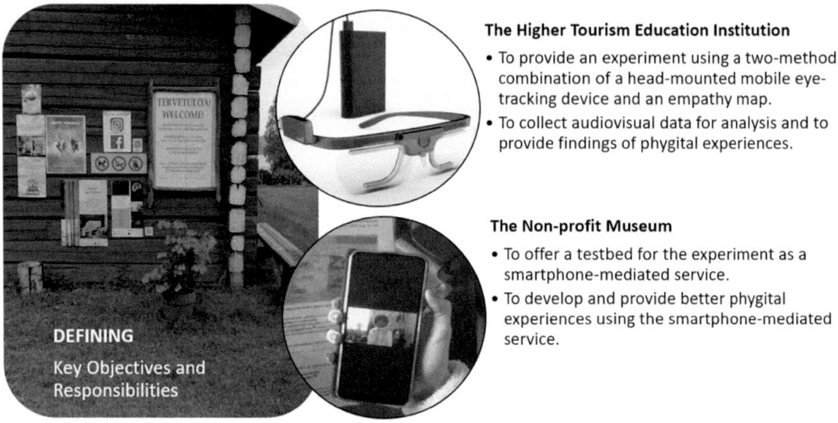

FIGURE 19.3 *The stakeholders defined their key objectives and responsibilities for the experiment (figure by Hanni-Vaara et al., 2024).*

visitors' phygital experiences. The nonprofit museum suggested that its recently launched smartphone-mediated service be tested.

The smartphone-mediated phygital on-site service comprised physical and digital touchpoints using QR codes to immerse the visitors in videos to learn about traditional children's games and duties. Although the smartphone-mediated service included several touchpoints around the museum premises, it was crucial to demarcate the research space for the collected data. The stakeholders agreed on a testbed of three touchpoints which were located near the museum's head building. This demarcation supported the data collection practice by enhancing quick accessibility of the testbed.

The meaningful elements and values, such as the mindset and attitude, worked as drivers, encouraging the stakeholders to participate and lead the experiment without first thinking about the risks of failure (Aho, 2023). The value of an experimental culture is to permit failures. Additionally, other meaningful drivers, such as liberty, creativity, courage and curiosity, guided stakeholders throughout the experiment (Aho, 2023). The experiment inspired the stakeholders to define objectives, share responsibilities and discuss the resources and roles for the ideation phase. It was crucial to empathize with the stakeholders' objectives to understand the value of the cocreation process and to motivate all the participants to proceed with the experiment. Figure 19.3 shows the key objectives of the case study.

Brainstorming and Prototyping: Keeping One's Feet on the Ground

During the ideation and prototyping phases, the stakeholders agreed to use the museum as a research space. The research space was used as a testbed for data collection. As the brainstorming continued, a common understanding of the resources needed was found. The stakeholders took walks around the research space to internalize the museum environment (Ramirez, 2017). The walks are reflected in the prototyping phase and as a pretest of the actual testing. It helped the stakeholders estimate and measure the time, number and length of each research objective. In addition, the practice helped frame the experiment and roles and responsibilities of the stakeholders during the experiment (Alhonsuo & Soto Hormazábal, 2022).

The walks around the museum area created trust and confidence among the stakeholders. While walking, the key was to empathize with each other by generating a common understanding through discussions and interaction. This eases pressure by confirming the plan of action and how to carry it out. Mutual respect increased during the cocreation process, which affected the circle of confidence, enabling the stakeholders to work in a cycle of iterations both together and apart (Aaltonen et al., 2016; Adner, 2017; Aho, 2023). The experiment was experienced as a collective opportunity for a social design that stimulated the stakeholders to interact during their everyday practices (Kuure & Miettinen, 2017).

Celebrating the Testing Day

The closer the test of the two-method data collection experiment became, the more significant it was to assess empathic responses between the stakeholders and spatiotemporal environment. The testing day of the data collection was executed on the museum's premises in September 2022. As shown in Figure 19.4, a marquee in the museum yard invited museum visitors to acquaint themselves with the eye-tracking device. The marquee was the touchpoint used to ask whether the visitors would be interested in participating in the research.

Seven museum visitors agreed to participate. Five were females aged 20 to 55 years, and two were males aged 10 and 70+ years. Four female adults and one male adult participated in the research individually, while one female was accompanied by her 10-year-old son. The pre-discussions with the participants revealed that none of them had previously experienced phygital service (Hanni-Vaara et al., 2024).

The research was explained to the participants, who signed a written consent form (TENK, 2024). To empathize with the research participants, the research assistant gently guided them to the testbed where their smartphone QR code reader was tested, their eyes were calibrated, the sound of their voice was checked, and the use of the empathy map was explained. The adults wore the smart glasses of a head-mounted mobile eye-tracking device. The audio of the 10-year-old child was recorded when accompanied by his mother.

During the testing phase, the experiment could have encountered challenges requiring agility in adapting the plans. Challenges could arise,

FIGURE 19.4 *The marquee in the museum yard invited visitors to familiarize themselves with mobile eye-tracking (Photograph by Paloniemi, 2022).*

especially when implementing a study or digital tool in situ. To guarantee that the experiment works smoothly, preplans were required with ready-provided solutions for use, if needed (Aaltola & Keto, 2018).

Finally, the two-method data collection was executed at the testbed for five hours. The touchpoints of the collection were within easy access to the smartphone-mediated service; thus, the location of the testbed was successful. Moreover, the resources of the testing day were sufficient, and some of the contingency plans were implemented in emergency cases. Approximately 85 minutes of audiovisual video data were provided for further analysis and findings.

When analysing the audiovisual data, the researchers obtained findings using projective empathy. Here, the researchers empathized with the research participants by 'stepping into their shoes' when analysing their feelings and thoughts during the phygital experiences. By positioning themselves in the research participants' states, the researchers explored findings to the question, 'How would I feel in their state?' (Aaltola & Keto, 2018; Ramirez, 2017).

Discussion

The current qualitative case study demonstrates how a two-method data collection experiment was conducted using a digital tool – a head-mounted mobile eye-tracking device – and an empathy map to collect data on phygital experiences using the smartphone-mediated service of a nonprofit museum stakeholder. More specifically, the method was used to collect data on museum visitors' phygital experiences. The present study implemented an empathy-oriented design thinking approach (Lee & Park, 2021; Redström, 2021). As shown in Table 19.1, the research design elaborated on a continuum of two cognitive empathies in the study: simulative and projective empathy. First, the researchers empathized with the spatiotemporal environment to form integrated insights into the volunteers' perceptions of the smartphone-mediated service, the environment of the museum's artefacts and stories of the local heritage (Aaltola & Keto, 2018; Ramirez, 2017). Following the analysis phase of the collected audiovisual experience data, the researchers reacted to the museum visitors' phygital experiences using projective empathy to form findings for further development.

The findings emphasize a two-method data collection practice as a proof of concept. A head-mounted mobile eye-tracking device and an empathy map were used to explore on-site experiences using a smartphone-mediated service. Epistemically, the continuum of simulative and projective empathy created and provided new knowledge in terms of design thinking. The findings substantiate that the empathy-oriented design thinking approach is recommended as a research practice for cognitive empathy, as shown in Table 19.1.

Although the findings and implications of the research cannot be generalized, the study provides novelty in terms of design thinking by combining a two-method data collection practice with a design thinking method using simulative and projective empathy. The current study provides value for researchers, developers and organizations to design and develop empathic phygital services. Moreover, the present study leverages the conceptual understanding of empathy by using simulative and projective empathy in an empathy-oriented design thinking process and practice.

Conclusion

The researchers achieved valuable experience and findings when executing the experiment through the use of two-method data collection with a head-mounted mobile eye-tracking device and an empathy map to collect audiovisual data on museum visitors' phygital experiences. The visual data revealed what the visitors were looking at which was enhanced with sounds and voices of the audio data. The empathy-oriented design thinking approach (Lee & Park, 2021) was implemented in the research design to elaborate on two cognitive dimensions of empathy: simulative and projective empathy (Aaltola & Keto, 2018; Ramirez, 2017).

The cocreation process involving a higher tourism education institution and a nonprofit museum is acknowledged as a contribution. The guiding values of the experiment, such as equality, openness, human-centredness

TABLE 19.1 *The key findings and implications of the experiment*

The Method Phases	Integration of Empathy Dimensions	Actions Done Answering 'What' Questions	Implications Answering 'Why' Questions
Planning two-method data collection approach	Simulative empathy Imitating and modelling with the spatiotemporal environment at the museum, involving volunteers, artefacts, stories and services. Documenting.	Researchers used simulation to form integrated insights into the museum's spatiotemporal environment, such as the perceptions of the museum's volunteers, stories, artefacts, etc.	To comprehend and understand the museum's volunteers and the spatiotemporal environment of *'How they feel and think'*.
Defining two-method data collection approach	Simulative empathy Defining, empathizing and simulating with the spatiotemporal environment to form the testbed. Defining the objectives, resources, roles and responsibilities for the experiment. Documenting this process.	Defining, empathizing and simulating the use of a two-method data collection using a head-mounted mobile eye-tracking device and an empathy map.	Defining the profound value of cocreating the experiment. Motivating all to proceed with the experiment. Instead of thinking about the risks of failure, concentrate on defining the process phases.
Ideating, Brainstorming two-method data collection approach	Simulative empathy Ideating meetings and brainstorming in interactions to enhance open discussions. Empathizing, learning and simulating spatiotemporally. Documenting.	Assessing empathic responses between the stakeholders. Ideating the experiment practices for the data collection, shared roles and responsibilities. Forming the research space as a testbed on the museum premises. Executing introductory walks.	Forming insight and understanding in a circle of confidence enables the stakeholders to work in a cycle of iterations together and apart. Experiencing a collective opportunity for social design during everyday practices.

Table 19.1 *continued*

The Method Phases	Integration of Empathy Dimensions	Actions Done Answering 'What' Questions	Implications Answering 'Why' Questions
Prototyping two-method data collection approach	Simulative empathy Empathizing and simulating with the spatiotemporal testbed to form a deeper insight into '*how the volunteers and environment feel and think*'. Documenting.	Prototyping, iterating and implementing the testbed and the tools of the two-method data collection in practice. Executing introductory walks. Demarcating and measuring lengths and timing of testing phases at the touchpoints. Repeating a cycle of iterations apart and together.	Deepening insights and understanding of the planned and ideated phases through iteration actions in practice. Creating trust and confidence through interaction among the stakeholders.
Testing two-method data collection practice approach	Simulative empathy Empathizing with the spatiotemporal environment throughout the testing phases when implementing a two-method data collection in practice.	Testing the two-method data collection method in practice. Collecting audiovisual data using a head-mounted mobile eye-tracking device and empathy map. Observing and redesigning the spatiotemporal environment ensuring stakeholders' positions. Responding to challenges with agile redesign. Celebrating the testing.	Executing a research practice of a two-method data collection. Implementing empathy-oriented design thinking in the research. Using simulative empathy when planning, defining, ideating, prototyping and testing the testbed. Providing findings of the experiment practice for research.

Table 19.1 *continued*

The Method Phases	**Integration of Empathy Dimensions**	**Actions Done Answering 'What' Questions**	**Implications Answering 'Why' Questions**
Testing collection of the data approach	Projective empathy Empathizing with the research participants (museum visitors) to react to their phygital experience asking, *'How would I feel in their state*'.	Empathizing with the research participants (museum visitors) and data collection practice during the data collection phase.	Collecting audiovisual data to analyse and provide findings of the museum visitors' phygital experiences.
Analysing, Forming Findings audiovisual data	Projective empathy Empathizing with the research participants (museum visitors) when analysing and forming findings of the audiovisual video data on research participants' phygital experiences.	Empathizing and reacting with audiovisual data of the research participants' (museum visitors) phygital experiences. Analysing, forming findings and suggestions for the museum and its further development.	Enhancing projective empathy in testing, analysing and forming findings of the phygital experiences. Providing findings of the audiovisual data analysis for research and development purposes.

and low hierarchy, generated dynamics of mutual trust and respect (Adner, 2017; Aho, 2023). In practice, it enabled the stakeholders to iterate on the experimental process together and apart. 'Together, we have brightened the museum's brand and developed several services by digitalizing them. Cocreation will continue in the new projects!' (Matkailuinstituutti, 2024). Since then, the museum has developed its smartphone-mediated service based on the experimental findings.

Takeaway

The present qualitative research demonstrated how a two-method data collection experiment was conducted in practice using a digital tool, head-mounted mobile eye-tracking device and an empathy map to explore phygital experiences. The current study investigated cocreation between a higher tourism education institution and a nonprofit museum at a peripheral destination. The present study implemented an empathy-oriented design thinking approach to guide the research process, and applied both simulative and projective empathy in practice. The findings substantiate a two-method data collection practice as a proof of concept. Although the findings and implications of the research cannot be generalized, the research provides novelty in design thinking by integrating both simulative and projective empathy in research practice. Moreover, the cocreation process between stakeholders is acknowledged as a contribution. The present study provides value for researchers, developers and organizations to design and develop empathic phygital services.

Acknowledgements

This grant-funded research was supported by the Finnish Cultural Foundation and the Foundation of Economic Education in Finland. The study was part of the EU-funded project (ERDF, Leverage from the EU) *'eHospitality – Empathy and Value Creation in Digital Service Encounters in Tourism'*. The study is part of the Business Finland-funded RRF project *'Empathy Business. How to Digitalize Service Prototyping and Business through Creativity'* with grant agreement No. 7425/31/2022 from 01.11.2022 to 31.12.2024.

References

Aaltola, E., & Keto, S. (2018). *Empatia – Myötäelämisen tiede*. Into Kustannus Oy.

Aaltonen, S., Hytti, U., Lepistö, T. & Mäkitalo-Keinonen, T. (2016). Yhteiskehittäminen: kaikki siitä puhuu, mutta mitä se on ja miten siinä onnistua? [Co-creation: All speak about it, but what it is and how to succeed in it?] (accessed 10 February 2025).

Adner, R. (2017). Ecosystem as structure: An actionable construct for strategy. *Journal of Management*, 43(1), 39–58. https://doi.org/10.1177/0149206316678451

Aho, T. (2023). *Kokeilukulttuuri – Johda kasvua joka päivä*. Alma Talent Oy.

Alhonsuo, M., & Soto Hormazábal, M. (2022). Empathic perspective to understand people's experiences in the service prototyping context in Chile. In M. Sarantou & S. Miettinen (Eds.), *Empathy and business transformation* (pp. 147–163). Routledge.

Batat, W. (2022). What does phygital really mean? A conceptual introduction to the phygital customer experience (PH-CX) framework. *Journal of Strategic Marketing*, 32(8), 1–24. https://doi.org/10.1080/0965254X.2022.2059775

Beckman, S. L., & Barry, M. (2007). Innovation as a learning process: Embedding design thinking. *California Management Review*, 50(1), 25–56. https://doi.org/10.2307/41166415

Dorst, K. (2011). The core of 'design thinking' and its application. *Design Studies*, 32(6), 521–532.

Ferreira, B., Silva, W., Oliveira, E. C., & Conte, T. (2015). *Designing personas with empathy map* [Conference paper]. SEGE 2015. https://www.researchgate.net/publication/276207468_Designing_Personas_with_Empathy_Map

García-Rosell, J.-C., Haanpää, M., & Janhunen, J. (2019). 'Dig where you stand': Values-based co-creation through improvisation. *Tourism Recreation Research*, 44(3), 348–358. https://doi.org/10.1080/02508281.2019.1591780

Hanni-Vaara, P. (2022). Human or nonhuman agent. Experiences of empathy in a digital visitor tourism journey. In M. Sarantou & S. Miettinen (Eds.), *Empathy and business transformation* (pp. 231–245). Routledge.

Hanni-Vaara, P., Haanpää, M., & Miettinen, S. (2024). *Designing new phygital service experiences for hospitality*. In C. Gray, E. Ciliotta Chehade, P. Hekkert, L. Forlano, P. Ciuccarelli & P. Lloyd (Eds.), DRS2024: Boston, 23–28 June, Boston, USA. https://doi.org/10.21606/drs.2024.1345

Jobin, C., Hooge, S., & Le Masson, P. (2021). The logics of double proof in proof of concept: A design theory-based model of experimentation in the unknown. *Proceedings of the Design Society*, 1, 3051–3060. 10.1017/pds.2021.566

Jung, Y. J., Toomey Zimmerman, H., & Pérez-Edgar, K. (2020). Mobile eye-tracking for research in diverse educational settings. In E. Romero-Hall (Ed.), *Research methods in learning design and technology* (1st ed.). Routledge. https://doi.org/10.4324/9780429260919-3

Jylkäs, T., Kuure, E., & Miettinen, S. (2019). Service design creating value for industrial corporates through AI proofs of concept. In E. Bohemia, G. Gemser, N. Fain, C. D. Bont, & R. A. Almendra (Eds.), *Conference proceedings of the*

Academy for Design Innovation Management (pp. 620–628). Academy for Design Innovation Management.

Kendig, C. E. (2016). What is proof of concept research and how does it generate epistemic and ethical categories for future scientific practice? *Science and Engineering Ethics*, 22, 735–753. https://doi.org/10.1007/s11948-015-9654-0

Kuure, E., & Miettinen, S. (2017). Social design for service. Building a framework for designers working in the development context. *The Design Journal*, 20(S1), 3464–3474. https://doi.org/10.1080/14606925.2017.1352850

Lee, H.-K., & Park, J. E. (2021). Designing a new empathy-oriented prototyping toolkit for the design thinking process: Creativity and design sensibility. *International Journal of Art & Design Education*, 40(2), 324–341. http://dx.doi.org/10.1111/jade.12345

Lupo, E. (2021). Design and innovation for the cultural heritage. Phygital connections for a heritage of proximity. *International Journal of Architecture, Art and Design*, 10, 186–199. https://doi.org/10.19229/2464-9309/10172021

Matkailuinstituutti. (2024). *Our partners explain* (Kumppanimme kertovat). Kerttu Oikarinen, Totto volunteer. https://www.matkailuinstituutti.fi (accessed 5 July 2024).

Mieli, M. (2022). Smartphoned tourists in the phygital tourist experience [Doctoral dissertation, Lund University]. Lund University Library. https://lucris.lub.lu.se/ws/portalfiles/portal/124057155/Thesis_Micol_Mieli_e_spik_ex.pdf

Neto, A. J. R., Borges, M. M., & Roque, L. (2020). Flow of knowledge in proof-of-concept activities: Examining the problem of interpretation using hermeneutics. *LIBRES*, 30(2), 45–67. https://doi.org/10.32655/LIBRES.2020.2.1

Rainoldi, M., & Jooss, M. (2020). Introduction to eye-tracking in tourism. In M. Rainoldi & M. Jooss (Eds.), *Eye-tracking in tourism: Tourism on the verge* (pp. 1–9). Springer. https://doi.org/10.1007/978-3-030-49709-5_1

Rainoldi, M., Yu, C., & Neuhofer, B. (2020). The museum learning experience through the visitors' eyes: An eye-tracking exploration of the physical context. In M. Rainoldi & M. Jooss (Eds.), *Eye-tracking in tourism: Tourism on the verge* (pp. 183–199). Springer. https://doi.org/10.1007/978-3-030-49709-5_12

Ramirez, E. (2017). Empathy and the limits of thought experiments. *Metaphilosophy*, 48, 504–526. https://doi.org/10.1111/meta.12249

Redström, J. (2021). Research through and through design. *Artifact: Journal of Design Practice*, 8(1–2), 16.1–16.19. https://doi.org/10.1386/art_00016_1

Ruonakoski, E. (2019). Empatia moraalisen toimijuuden perustana. *Ajatus*, 76(1), 315–330. https://journal.fi/ajatus/article/view/88195 (accessed 5 July 2024).

Solima, L., & Izzo, F. (2017). QR codes in cultural heritage tourism: New communications technologies and future prospects in Naples and Warsaw. *Journal of Heritage Tourism*, 13(2), 115–127. https://doi.org/10.1080/1743873X.2017.1337776

TENK. (2024). *Finnish National Board on Research Integrity*. https://tenk.fi/en (accessed 5 July 2024).

Tobii. (2024). *Tobii. Products. Eye trackers*. https://www.tobii.com/products/eye-trackers (accessed 5 July 2024).

Vink, J., Koskela-Huotari, K., Tronvoll, B., Edvardsson, B., & Wetter-Edman, K. (2021). Service ecosystem design: Propositions, process model, and future research agenda. *Journal of Service Research*, 24(2), 168–186. Sage. https://journals.sagepub.com/doi/10.1177/1094670520952537

Zillinger, M. (2021). *Tourism revisited: The influence of digitalization on tourism concepts* [ETOUR Working Paper 2021:1]. Mid Sweden University, Östersund.

Zimmerman, J., Stolterman, E., & Forlizzi, J. (2010). An analysis and critique of research through design: Towards a formalization of a research approach. In *DIS '10: Proceedings of the 8th ACM Conference on Designing Interactive Systems* (pp. 310–319). Association for Computing Machinery. https://doi.org/10.1145/1858171.1858228.

20

The Role of Emojis in Fostering Empathy in Chatbot Interactions

Emma Kirjavainen, Ashley Colley and Jonna Häkkilä

Abstract

In this chapter, we address the perception of empathy in chatbot conversations in online services, investigating how the chatbot's use of emojis is perceived. We present a user study (n = 20) in which participants had conversations with a simulated dental service chatbot, here comparing cases of no emojis, appropriate emojis and inappropriate emoji use by the chatbot. The salient findings reveal that appropriate emoji use by chatbots had a positive impact on both pragmatic and hedonic user experience.

Introduction

The use of digital services has exploded during the past two decades, and today, the use of online services has progressed from healthcare to entertainment and tourism. In the user interface of online services, conversational agents and chatbots have become common tools with which to interact. Recent developments have shifted chatbots from rule-based to generative systems (Adamopoulou & Moussiades, 2020). With the development of machine learning, language processing capabilities and recent developments in generative AI, these conversational agents can handle complex tasks and conversations, being increasingly trusted to take care of much of the online customer interface.

Previous studies have reported that satisfaction with chatbots is linked to chatbots' ability to efficiently and adequately handle inquiries (Følstad & Skjuve, 2019) and that the most influential factors when using chatbots for customer service are response relevance and problem resolution (Nicolescu & Tudorache, 2022). However, it has been noted that the user experience with chatbots is linked not only to pragmatic factors but also to hedonic factors, especially when younger users are in question (Følstad & Brandtzaeg, 2020).

When emotions and the ability to express them are highly relevant issues, empathy in the service experience is considered. In everyday conversations, nonverbal factors and the ability to express emotions are important factors in communication (Gabbot & Hogg, 2001). With robots, this type of human-like expressive power is limited. In the area of embodied social robots, the question of integrating emotional expressions and behaviour has gained increasing attention (Cauchard et al., 2016; Tielman et al., 2014), so there has been a call for considering emotional expressions in future robot design, for instance, with respect to the uncanniness of robots (Tschöpe et al., 2017). However, the findings in prior studies have suggested that human-like chatbots communicating through text-based User Interfaces (UIs) do not provoke uncanny feelings in users, so people perceive them differently than robots present as embodied (Skjuve et al., 2019). With chatbots, the means of expressing human likeness have been explored with different text typefaces (Candello et al., 2017) and with the use of emojis (Yu & Zhao, 2024). The inclusion of emojis in chatbots' messages has been shown to create perceptions of warmth and increased satisfaction (Yu & Zhao, 2024). On the other hand, Véliz (2023) raises concerns that AI chatbots could utilize emojis to influence human emotions more powerfully than words.

In this chapter, we delve more deeply into the question of user experience with chatbots using emojis in dialogue with digital service users. We address the following research question: 'How is chatbot emoji use perceived for pragmatic, hedonic and overall user experience?' Using the example context of an online service interface for a dentist, we explore

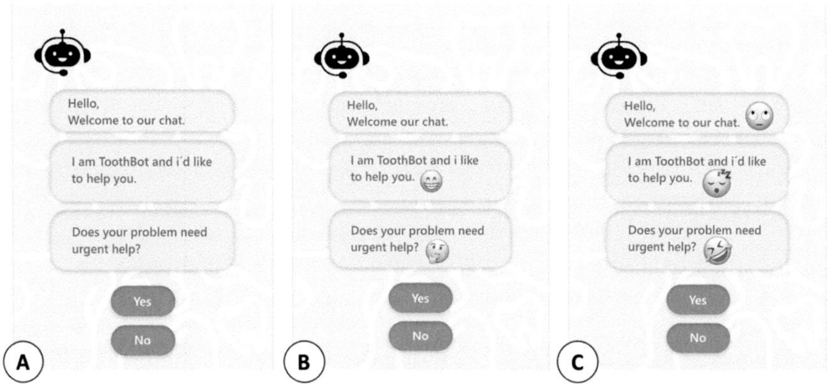

FIGURE 20.1 *Example screens from the three variants of ToothBot evaluated in the study. A) No emojis; B) Suitable emojis; C) Unsuitable emojis.*

user perceptions of interacting with a chatbot in three types of dialogue: a chatbot using no emojis, suitable emojis and unsuitable emojis in conversations with customers, as illustrated in Figure 20.1. Our findings provide insights into a more emphatic chatbot user interface design for online services.

Related Work

To position our research, we review prior work related to communicating with emojis and the perception of emotion in chatbot communication.

Communicating with Emojis

Emojis have become a key part of textual communication in online messaging, acting as an essential tool to provide nonverbal cues that are absent in digital conversations (Walther & D'addario, 2001). When first used in the 1980s in discussion forums, smiley faces indicated humour in comments (Dresner & Herring, 2010) but were later adopted to represent laughter, more closely mirroring face-to-face communication (Provine et al., 2007).

The transition from simple ASCII character strings, such as :) and :(to the visually expressive emojis that originated in Japan in the late 1990s, marked an evolution in their expressive power (Kaye et al., 2017). Today, emojis are used to convey pragmatic information quickly through imagery (Alshenqeeti, 2016), enhance personal expression and reduce message ambiguity (Kaye et al., 2016). They also act as important nonverbal communication cues (Yang, 2020). Emoji use has been adopted not only in personal communication but also in feedback collection, such as in happy-to-sad rating scale push buttons in shops, smiley face Likert scales (Hall et al., 2016) and the emoji-based User Experience Questionnaire (UEQ) (Colley et al., 2023).

There has been much discussion about how the meanings of emojis are understood universally, with some suggesting that they could lead to the creation of a universal symbolic language (Azuma, 2012). Although cultural differences in the understanding of symbolic emojis have been noted (Guntuku et al., 2019), emojis portraying facial expressions (Alshenqeeti, 2016) tend to be more uniformly understood across cultures (Guntuku et al., 2019).

Emotion in Chatbot Interaction

Previous research has explored strategies to enhance emotional expression in chatbots, as illustrated by a literature review of emotional engagement in chatbot interactions (Bilquise et al., 2022). Although many studies have focused on textual emotional enhancements, investigations into the impact of emoji integration are less common. Beattie et al. (2020) evaluate a simulated restaurant recommendation chatbot, finding that incorporating emojis – smiling ones for positive reviews and frowning for negative reviews – enhanced the chatbot's credibility compared with plain text responses. Liu et al. (2023) observe that humorous emojis used during chatbot service failures increased users' willingness to continue interacting with the chatbot and boosted their perceptions of the chatbot's intelligence. However, in some cases, the addition of emojis can create a negative impression. Li and Shin (2023) report that using emojis in chatbots representing traditional luxury brands diminished the perceived prestige of the brand. Interestingly, for a health-related chatbot, Fadhil et al. (2018) found that the inclusion of emojis had a positive impact on mental well-being issues but a negative effect on physical well-being topics.

Previous research has also explored the impact of emojis on the perceived empathy of chatbots (Ravi et al., 2021; Yu & Zhao, 2024). Yu and Zhao (2024) note that emojis in chatbot messages foster a sense of warmth and increase user satisfaction, even though the effect is less pronounced than that of human use of emojis. Ravi et al. (2021) have created a chatbot capable of generating messages that integrate text and emojis and, based on a user study, report that this approach enhanced the empathetic qualities of conversations. Methods for integrating suitable emojis into chatbot messages have been developed by Zuo et al. (2022) and Zhao et al. (2023).

Methods

To investigate our research question, a prototype chatbot for an imaginary dental practice, 'ToothBot', was developed and evaluated in a user study.

Toothbot Concept Design

The user study utilized the ToothBot chatbot concept, which was built using Adobe XD. This programme allowed us to incorporate emojis of the desired size into the prototype. We aimed to evaluate users' reactions to emojis within the chatbot conversation, thus requiring the emojis to be of a noticeable size (Figure 20.1).

During the user testing phase, we conducted three different conversations labelled Case A, Case B and Case C. Each conversation featured identical text content, with the only discrepancies being the emojis used. In Case A, no emojis were included. Case B utilized emojis suitable for conversation, while Case C utilized inappropriate or even provocative emojis. In our ToothBot conversations B and C, we utilized thirteen distinct emojis, as shown in Figure 20.2.

We aimed to use emojis that would be interpreted clearly by the participants and would not contain any offensive elements. Intentionally, we added a significant number of emojis to the conversations, averaging one emoji per message, and increased the size of the emojis to be larger than the font size of the text.

The participants engaged in conversation with ToothBot via mobile phones (Figure 20.3). Prior work has shown that the majority of users

FIGURE 20.2 *The emojis used in the ToothBot concept.*

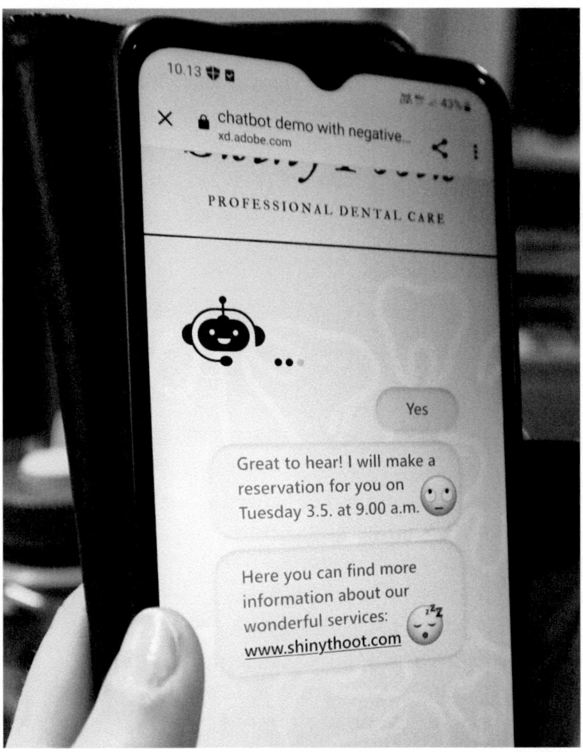

FIGURE 20.3 *The ToothBot used in the user study.*

engage with services via mobile phones (Colley et al., 2023). In this way, we aimed to enhance the conversational experience by fostering a familiar way of interacting and avoiding distractions from anything other than conversations with the ToothBot.

Evaluation

Each study session lasted approximately 15 minutes. The study started by completing a background questionnaire that also included questions about the participants' experiences with emojis and their usage in their communication.

The participants were not informed of the different features of cases A, B or C. The study was conducted within subjects; that is, all participants evaluated all three versions (A, B and C) of the chatbot. To reduce learning effects, the presentation order of the versions was counterbalanced using a Latin square approach. This resulted in seven participants starting with version A, seven with version B and six with version C. The participants went through the discussions individually, and after each session their experiences were assessed using a shortened version of the UEQ. In the final part of the study, an end questionnaire was administered. The final questionnaire included five questions.

In total, twenty participants (9 females, 11 males; aged 18–59) participated in the study. The participants were predominantly university design students and staff but also participants from outside the university. Altogether, 19 out of 20 participants used emojis in messaging applications, such as WhatsApp. Emoji use was most popular in communication with closer acquaintances, and only 2 out of 20 participants claimed to use emojis in email communication with superiors, teachers or other authority figures.

Results

Participant Background

The majority of participants (65 per cent, or 17 out of 20) reported frequently using emojis in their everyday messaging conversations, while only two stated that they used them infrequently, and one participant mentioned using emojis infrequently. According to the findings, all participants used emojis in text-message communication, including messaging apps such as WhatsApp or Snapchat. Half of the participants found using emojis appropriate for social media posts. However, emoji usage in email communication was not common among the participants, with only four participants indicating that they used emojis in email correspondence.

Most participants (65 per cent, or 17 out of 20) reported using emojis to foster a friendlier and more relaxed tone in their messages and highlight the emotions they felt or intended to express to the recipient. One participant highlighted the humorous element that emojis can bring to messaging content. Another participant mentioned using emojis to convey confusion or to end a messaging conversation, indicating a lack of enthusiasm towards emoji usage and stating that they rarely use them.

Quantitative Findings

The UEQ results were analysed, and for each condition, the pragmatic quality, hedonic quality and overall perceived quality levels were calculated (Figure 20.4). Cronbach's alpha exceeded 0.7 for both the pragmatic and hedonic subscales across all conditions, demonstrating a consistent pattern in the responses.

According to the UEQ results, Case B, where appropriate emojis were used, had the highest overall quality and was positively perceived in terms of both pragmatic and hedonic quality. Case A, without emojis, had high pragmatic quality but poor hedonic quality. The condition with inappropriate emojis (Case C) was considered to have the highest hedonic quality in all cases, combined with the lowest pragmatic quality.

When asked to select their preferred version of the chatbot, 65 per cent of participants (13 out of 20) selected Case B, which included appropriate emojis. The participants justified their choice, for example, 'Emojis brought empathy and humanity to the conversation' (Participant 7) and 'It felt more caring than the other conversations' (P5). Case A (no emojis) was selected by 35 per cent of the participants (7 out of 20), who commented, for example, 'Using emojis is annoying, I don't understand it' (P2). One participant noted that the suitability of emojis may depend on the use case: 'Dentist services are a context where emojis do not belong' (P2).

Case C, where inappropriate emojis were used, caused confusion and was not preferred by anyone. However, the use of inappropriate emojis created more raised eyebrows, surprised laughter and participant comments. For example, one participant exclaimed, 'Why is it being mean to me?' (P3). Some participants even stated that they might choose this service provider because of the unexpected emotions it elicited.

Qualitative Findings

The free-text responses from the participants were analysed through an open coding method. One researcher developed the codebook and performed the initial coding of the dataset. Subsequently, another researcher independently coded the dataset using the same codebook. Any disparities in coding were

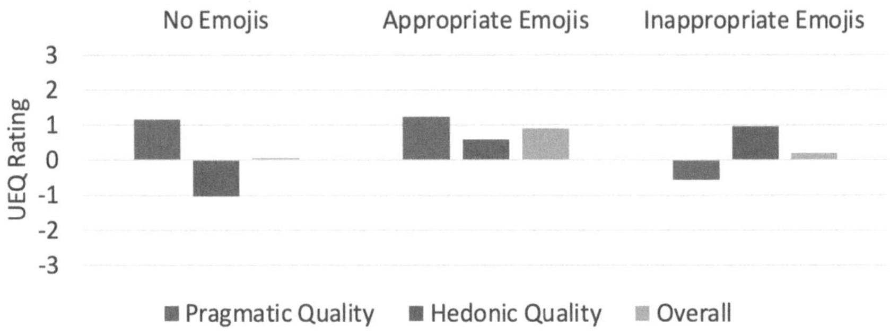

FIGURE 20.4 *UEQ quality rating for the three conditions.*

then discussed between the researchers, leading to the creation of a final coded dataset.

The top four codes identified for benefits and challenges are shown in Table 20.1. The most reported code was emotions, that is, conveying emotions, and the second most mentioned codes were humanity and approachability. One participant mentioned that emojis bring character to the chatbot, and another participant mentioned that 'the chatbot is no longer emotionless' (P13). A fourth benefit that emojis brought to the interaction between participants and the ToothBot was the feeling of presence. One participant described the feeling as follows: 'I had the feeling that someone was listening to me and was present during the ToothBot conversation' (P9); another participant mentioned, 'It felt like ToothBot was truly listening to me' (P10). One of the participants mentioned that 'emojis fit the conversation when used well and made the conversation easier and more comfortable' (P19).

The participants also noted challenges in the use of emojis, even though they were perceived as fewer than the advantages. The challenges that emerged were more related to the application context and how emojis fit into it, as one participant mentioned, for example, 'Emojis are not suitable for formal conversations' (P3). Another participant stated that it is important to consider the context of the conversation when using emojis because they can create an image of an unofficial and unreliable service (P7). Emojis were perceived as challenging if they were generally not liked by the individual participant, for example, a participant who previously stated that they usually do not use emojis in their private messaging: 'Emojis are annoying and complicated' (P2).

We asked whether the use of emojis added value to the ToothBot conversations which 75 per cent (15 out of 20) of the participants agreed with. The emergence of emotions, experiences and encounters were highlighted as significant factors in the positive interaction between machines and humans. The participants perceived the ToothBot as friendlier and more

TABLE 20.1 *The highest mentioned codes for the benefits and challenges of using emojis in the ToothBot. The bracketed number is the number of mentions of the code*

Benefits (Count)	Challenges (Count)
Emotions (10)	Unreliable (1)
Approachability (8)	Unofficial (1)
Humanity (8)	Irritating (1)
Presence (5)	Misunderstanding (1)

caring because of the use of emojis, with the chatbot even appearing to demonstrate an understanding of human emotions during conversations when emojis were used.

Discussion

According to 75 per cent of the participants, emojis added value to the ToothBot conversation. However, only 55 per cent perceived it as suitable for a dental service context. This finding supports similar findings in prior studies, for example, that emojis are not suitable for luxury brand chatbots (Li & Shin, 2023). It should be acknowledged that the overall framework for customer experience with conversational agents and chatbots is larger than just the human–chatbot interaction in the conversation and includes not only chatbot-related but also user- and context-related factors (Nicolescu & Tudorache, 2022). Furthermore, the application use context should be reviewed considering Véliz's (2023) guidance that chatbots should not use emojis because they can be used to manipulate people's emotions (Véliz, 2023). It would be interesting to further investigate this phenomenon with a broader range of service contexts to determine where emoji use is perceived as appropriate and to explore in more detail the reasons why they fit better in one context than in another.

We acknowledge the limitations of our work. The chatbot conversation and added emojis were manually created to ensure that the study settings were consistent among the three cases. The findings from authentic AI-generated conversations and emoji usage may differ. Previous studies have identified differences in the perceptions of emoji use between AI-generated and human-generated chatbots (Yu & Zhao, 2024).

In future work, it would be interesting to create a chatbot conversation in which emoji usage during the conversation is generated by artificial intelligence and to observe whether the conversation is still perceived as warm and human.

Conclusion

In our study, we explored user interactions with a simulated dental service chatbot using three different scenarios: no emojis, appropriately used emojis and inappropriately used emojis. We found that using emojis appropriately in chatbots' responses significantly enhanced both the practical and enjoyable aspects of the user experience. Specifically, the user experience was less satisfactory when no emojis were used and was negatively affected when emojis were used inappropriately in the context of the conversation. These results suggest that when emojis are used effectively, they can make interactions with chatbot services feel more empathetic and engaging.

Takeaway

Appropriate emoji use significantly enhanced both the practical and hedonic aspects of the user experience when interacting with a dental service chatbot. User satisfaction decreased with no emojis and was negatively affected by inappropriate emoji use. Effective emoji use can make chatbot interactions more empathetic and engaging.

Acknowledgements

This work has been partly funded by the European Regional Development Fund, coordinated by the Regional Council of Lapland, as part of the AI.R – Arctic AI & Robotics project.

References

Adamopoulou, E., & Moussiades, L. (2020). Chatbots: History, technology, and applications. *Machine Learning with Applications*, 2, 100006.

Alshenqeeti, H. (2016). Are emojis creating a new or old visual language for new generations? A socio-semiotic study. *Advances in Language and Literary Studies*, 7(6), 56–69. AIAC. https://ssrn.com/abstract=3709343

Azuma, J. (2012). Graphic emoticons as a future universal symbolic language. In *Audiovisual translation and media accessibility at the crossroads* (pp. 61–84). Brill. https://doi.org/10.1163/9789401207812006

Beattie, A., Edwards, A. P., & Edwards, C. (2020). A bot and a smile: Interpersonal impressions of chatbots and humans using emoji in computer mediated communication. In S. Nah, J. E. McNealy, J. H. Kim, & J. Joo (Eds.), *Communicating artificial intelligence (AI): Theory, research, and practice* (pp. 41–59). Routledge.

Bilquise, G., Ibrahim, S., Shaalan, K., et al. (2022). Emotionally intelligent chatbots: A systematic literature review. *Human Behavior and Emerging Technologies*, 2022. https://doi.org/10.1155/2022/9601630

Candello, H., Pinhanez, C., & Figueiredo, F. (2017). Typefaces and the perception of humanness in natural language chatbots. In *Proceedings of the 2017 Chi Conference on Human Factors in Computing Systems* (pp. 3476–3487). ACM.

Cauchard, J. R., Zhai, K. Y., Spadafora, M., & Landay, J. A. (2016). Emotion encoding in human–drone interaction. In *2016 11th ACM/IEEE International Conference on Human–Robot Interaction (HRI)* (pp. 263–270). ACM.

Colley, A., Mayer, S., & Häkkilä, J. (2023). Developing an emoji-based user experience questionnaire: Ueq-emoji. In *Proceedings of the 22nd International Conference on Mobile and Ubiquitous Multimedia* (pp. 65–73). ACM.

Dresner, E., & Herring, S. C. (2010). Functions of the nonverbal in CMC: Emoticons and illocutionary force. *Communication Theory*, 20(3), 249–268. https://doi.org/10.1111/j.1468-2885.2010.01362.x

Fadhil, A., Schiavo, G., Wang, Y., & Yilma, B. A. (2018). The effect of emojis when interacting with conversational interface assisted health coaching system. In *Proceedings of the 12th EAI International Conference on Pervasive Computing Technologies for Healthcare* (pp. 378–383). ACM.

Følstad, A., & Brandtzaeg, P. B. (2020). Users' experiences with chatbots: Findings from a questionnaire study. *Quality and User Experience*, 5(1), 3.

Følstad, A., & Skjuve, M. (2019). Chatbots for customer service: User experience and motivation. In *Proceedings of the 1st International Conference on Conversational User Interfaces* (pp. 1–9). ACM.

Gabbott, M., & Hogg, G. (2001). The role of non-verbal communication in service encounters: A conceptual framework. *Journal of Marketing Management*, 17(1–2), 5–26. Routledge. https://doi.org/10.1362/0267257012571401

Guntuku, S. C., Li, M., Tay, L., & Ungar, L. H. (2019). Studying cultural differences in emoji usage across the East and the West. In *Proceedings of the International AAAI Conference on Web and Social Media* (Vol. 13, pp. 226–235). AAAI.

Hall, L., Hume, C., & Tazzyman, S. (2016). Five degrees of happiness: Effective smiley face Likert scales for evaluating with children. In *Proceedings of the 15th International Conference on Interaction Design and Children* (pp. 311–321). ACM.

Kaye, L. K., Malone, S. A., & Wall, H. J. (2017). Emojis: Insights, affordances, and possibilities for psychological science. *Trends in Cognitive Sciences*, 21(2), 66–68. https://doi.org/10.1016/j.tics.2016.10.007

Kaye, L. K., Wall, H. J., & Malone, S. A. (2016). 'Turn that frown upside-down': A contextual account of emoticon usage on different virtual platforms. *Computers in Human Behavior*, 60, 463–467. https://doi.org/10.1016/j.chb.2016.02.088

Li, Y., & Shin, H. (2023). Should a luxury brand's chatbot use emoticons? Impact on brand status. *Journal of Consumer Behaviour*, 22(3), 569–581.

Liu, D., Lv, Y., & Huang, W. (2023). How do consumers react to chatbots' humorous emojis in service failures. *Technology in Society*, 73, 102244.

Nicolescu, L., & Tudorache, M. T. (2022). Human–computer interaction in customer service: The experience with AI chatbots: A systematic literature review. *Electronics*, 11(10), 1579.

Provine, R. R., Spencer, R. J., & Mandell, D. L. (2007). Emotional expression online: Emoticons punctuate website text messages. *Journal of Language and Social Psychology*, 26(3), 299–307. https://doi.org/10.1177/0261927X06303481

Ravi, A., Yadav, A., Chauhan, J., Dholakia, J., Jain, N., & Singh, M. (2021). SentEmojiBot: Empathising conversations generation with emojis. Preprint arXiv:2105.12399.

Skjuve, M., Haugstveit, I. M., Følstad, A., & Brandtzaeg, P. (2019). Help! Is my chatbot falling into the uncanny valley? An empirical study of user experience in human–chatbot interaction. *Human Technology*, 15(1), 30–54.

Tielman, M., Neerincx, M., Meyer, J.-J., & Looije, R. (2014). Adaptive emotional expression in robot-child interaction. In *Proceedings of the 2014 ACM/IEEE International Conference on Human-Robot Interaction* (pp. 407–414). ACM.

Tschöpe, N., Reiser, J. E., & Oehl, M. (2017). Exploring the uncanny valley effect in social robotics. In *Proceedings of the Companion of the 2017 ACM/IEEE International Conference on Human-Robot Interaction* (pp. 307–308). ACM.

Véliz, C. (2023, March). Chatbots shouldn't use emojis. *Nature*, 375–375. https://doi.org/10.1038/d41586-023-00758-y.

Walther, J. B., & D'addario, K. P. (2001). The impacts of emoticons on message interpretation in computer-mediated communication. *Social Science Computer Review*, 19(3), 324–347.https://doi.org/10.1177/089443930101900307

Yang, Y. (2020). Are you emoji savvy? Exploring nonverbal communication through emojis. *Communication Teacher*, 34(1), 2–7. https://doi.org/10.1080/17404622.2019.1593472

Yu, S., & Zhao, L. (2024). Emojifying chatbot interactions: An exploration of emoji utilization in human-chatbot communications. *Telematics and Informatics*, 86, 102071.

Zhao, S., Jiang, H., Tao, H., Zha, R., Zhang, K., Xu, T., & Chen, E. (2023). Pedm: A multi-task learning model for persona-aware emoji-embedded dialogue generation. *ACM Transactions on Multimedia Computing, Communications and Applications*, 19(3s), 1–21.

Zuo, X., Yu, X., Du, M., & Song, Q. (2022). Generating consistent multimodal dialogue responses with emoji context model. In *2022 5th International Conference on Artificial Intelligence and Big Data (ICAIBD)* (pp. 617–624). IEEE.

INDEX

Bold page numbers indicate tables, *italic* numbers indicate figures.